Current Topics in Microbiology 252 and Immunology

Editors

R.W. Compans, Atlanta/Georgia
M. Cooper, Birmingham/Alabama · Y. Ito, Kyoto
H. Koprowski, Philadelphia/Pennsylvania · F. Melchers, Basel
M. Oldstone, La Jolla/California · S. Olsnes, Oslo
M. Potter, Bethesda/Maryland · P.K. Vogt, La Jolla/California
H. Wagner, Munich

Springer
Berlin
Heidelberg
New York
Barcelona
Hong Kong
London
Milan
Paris
Singapore
Tokyo

B1 Lymphocytes in B Cell Neoplasia

16th Workshop on the
Mechanisms of B Cell Neoplasia, 1999

Edited by M. Potter and F. Melchers

With 97 Figures and 18 Tables

 Springer

Michael Potter, M.D.
Laboratory of Genetics
Building 37, Room 2B04
National Cancer Institute
National Institutes of Health
37, Convent Drive, MSC4255
Bethesda, MD 20892-4255
USA

Fritz Melchers, Ph.D.
Basel Institute for Immunology
Grenzacherstr. 487
CH-4005 Basel
Switzerland

Cover design:
Design & production GmbH, Heidelberg, Germany

ISSN 0070-217X
ISBN 3-540-67567-1 Springer-Verlag Berlin Heidelberg New York

This work is subject to copyright. All rights are reserved, whether the whole or part of the material is concerned, specifically the rights of translation, reprinting, reuse of illustrations, recitation, broadcasting, reproduction on microfilm or in any other ways, and storage in data banks. Duplication of this publication or parts thereof is only permitted under the provisions of the German Law of September 9, 1965, in its current version, and permission for use must always be obtained from Springer-Verlag. Violations are liable for prosecution under the German Copyright Law.

Springer-Verlag Berlin Heidelberg New York
a member of BertelsmannSpringer Science+Business Media GmbH

© Springer-Verlag Berlin Heidelberg 2000
Library of Congress Catalog Number 15-12910
Printed in Germany

The use of general descriptive names, registered names, trademarks, etc. in this publication does not imply, even in the absence of a specific statement, that such names are exempt from the relevant protective laws and regulations and therefore free for general use.

Product liability: The publishers cannot guarantee the accuracy of any information about dosage and application contained in this book. In every individual case the user must chek such information by consulting other relevant literature.

Typesetting: Camera-ready by authors
Printed on acid-free paper   SPIN: 10769478   27/3020 hu - 5 4 3 2 1 0

Preface

This was the 16th Workshop on Mechanisms in B-cell Neoplasia. It was held at the Lister Hill Auditorium, National Library of Medicine, NIH, in Bethesda, Maryland on October 28-29, 1999. As has been the trend in the last few years, we have chosen a single theme as the basis for the Workshop, and this year at the suggestion of Lee Herzenberg we chose the role of B-1 cells in B cell neoplasia. We expanded the context to include natural antibodies as well. The meeting comes at a timely moment, where there is increasing awareness in the immunological community of the role of natural antibodies and innate immunity as a first line of immunological defense of the mammalian vertebrates. For almost twenty years it has been recognized that the subpopulation of B-cells expresses a classic T-cell marker, CD5. This has led to attempts to define subsets of B-cells on the basis of the CD5 marker. One of the products of this effort has been the B-1, B-2 subdivision of the peripheral B-cell population which was largely based on findings in the mouse. Not all immunologists agree on the specifics of this subdivision, but the B-1, B-2 nomenclature has now come into wide usage, and this may be premature. In this meeting, there was considerable discussion of the problem of the origin and nature of B-1 cells. B-1 cells are also known by the characteristics of the immunoglobulins they make, and their B-cell receptors (BCR) or the antibodies they secrete have binding affinities to autoantigens, usually with low affinities, many are also polyreactive. Because of these properties, many of these Igs resemble natural antibodies. There are increasing lines of evidence that the B-cells that produce autoreactive, low affinity binding, polyreactive antibodies can be the precursors of B-cell tumors. The consistent association of CD5 with chronic B-cell lymphocytic leukemia in humans has been one of the long-standing relationships in this field, but there are other associations as well, such as the similarities of the Igs produced in B cell tumors to the Igs produced by B cells involved in autoimmune processes and natural antibody formation.

In the last few years we have published the papers presented at these workshops as copy ready manuscripts in *Current Topics in Microbiology and Immunology* and the *Editions Roche*. We thank the authors of these manuscripts for their careful preparation.

This meeting was sponsored by the Laboratory of Genetics, Division of Basic Sciences of the National Cancer Institute. We thank Ms. Fran Oscar of Palladian Partners, Inc., for her help in arranging the meeting. We are very grateful to Victoria Rogers and Mary Millison of the Laboratory of Genetics for making many of the arrangements for the meeting and for assembling the manuscripts.

Michael Potter,
Fritz Melchers

Table of Contents

List of Contributors xx

I   B-1 Cells: Definitions, Basic Hypotheses

L.A. Herzenberg, N. Baumgarth and J.A. Wilshire
B-1 Cell Origins and V_H Repertoire Determination 3

K. Hayakawa, S.A. Shinton, M. Asano and R.R. Hardy
B-1 Cell Definition 15

II   B-1 Cells: Relationship to Pro B-Pre B Development and Negative Selection

R.R. Hardy, R. Wasserman, Y.-S. Li, S.A. Shinton
and K. Hayakawa
Response by B Cell Precursors to Pre-B Receptor
Assembly: Differences between Fetal Liver
and Bone Marrow 25

M.P. Cancro, A.P. Sah, S.L. Levy, D.M. Allman,
D. Constantinescu, M.R. Schmidt and R.T. Woodland
B Cell Production and Turnover in CBA/Ca, CBA/N
and CBA/N-*bcl*-2 Transgenic Mice: *xid*-mediated Failure
among Pre B Cells is Unaltered by *bcl*-2 Overexpression 31

J.J. Kenny, A. Lustig and D.L. Longo
Positive Selection of Low Affinity Autoreactive B Cells 39

III   B-1 Cells and Positive Selection

R. Berland and H.H. Wortis
A Model for Autoantigen Induction of Natural Antibody
Producing B-1a Cells 49

R.R. Reid, S. Woodcock, A.P. Prodeus, J. Austen,
L. Kobzik, H. Hechtman, F.D. Moore, Jr. and M.C. Carroll
The Role of Complement Receptors CD21/CD35 in Positive
Selection of B-1 Cells 57

F. Agenes, M.M. Rosado and A.A. Freitas
Considerations on B Cell Homeostasis 67

C. Tatu and S.H. Clarke
Selective Maturation of V_H12 B Cells in the Spleen Enriches
for Anti-Phosphatidyl Choline B Cells: Evidence for Receptor
Editing 77

R.G. Mage and R. Pospisil
CD5 and Other Superantigens May Select and Maintain
Rabbit Self-renewing B-lymphocytes and Human
B-CLL Cells 87

F. Martin and J.F. Kearney
Selection in the Mature B Cell Repertoire 97

P.M. Dammers, A. Visser, E.R. Popa, P. Nieuwenhuis,
N.A. Bos and F.G.M. Kroese
Immunoglobulin V_H Gene Analysis in Rat: Most Marginal
Zone B Cells Express Germline Encoded V_H Genes
and Are Ligand Selected 107

IV   CD5

T.L. Rothstein, G.M. Fischer, D.A. Tanguay, S. Pavlovic,
T.P. Colarusso, R.M. Gerstein, S.H. Clarke and T.C. Chiles
STAT3 Activation, Chemokine Receptor Expression,
and Cyclin-Cdk Function in B-1 Cells 121

R. Berland and H.H. Wortis
Role of NFAT in the Regulation of B-1 Cells 131

S. Bondada, G. Bikah, D.A. Robertson and G. Sen
Role of CD5 in Growth Regulation of B-1 Cells 141

D. Donjerkovic, G.B. Carey, C.M. Mueller, S. Liu
and D.W. Scott
Life and Death Decisions in B1 Lymphoma Cells 151

V   B-1 Cells in Inflammation

N. Baumgarth, J. Chen, O.C. Herman, G.C. Jager
and L.A. Herzenberg
The Role of B-1 and B-2 Cells in Immune Protection
from Influenza Virus Infection 163

P.W. Askenase and R.F. Tsuji
B-1 B Cell IgM Antibody Initiates T Cell Elicitation
of Contact Sensitivity 171

T.V. Rajan and N. Paciorkowski
Role of B Lymphocytes in Host Protection Against
the Human Filarial Parasite, *Brugia Malayi* 179

G.J. Silverman, P.X. Shaw, L. Luo, D. Dwyer, M. Chang,
S. Horkko, W. Palinski, A. Stall and J.L. Witztum
Neo-self Antigens and the Expansion of B-1 Cells: Lessons
from Atherosclerosis-prone Mice 189

H.B. Richards, E.A. Reap, M. Shaw, M. Satoh, H. Yoshida
and W.H. Reeves
B Cell Subsets in Pristane-induced Autoimmunity 201

VI   B-1 Cells and Mucosal Immunity

N.A. Bos, J.J. Cebra and F.G.M. Kroese
B-1 Cells and the Intestinal Microflora 211

S. Fagarasan, R. Shinkura, T. Kamata, F. Nogaki, K. Ikuta
and T. Honjo
Mechanism of B1 Cell Differentiation and Migration
in GALT 221

VII   B-1 Cells and Antibodies

M.F. Flajnik and L.L. Rumfelt
Early and Natural Antibodies in Non-mammalian Vertebrates 233

A.L. Notkins
Polyreactive Antibodies and Polyreactive Antigen-Binding B
(PAB) Cells 241

G.J. Silverman, S. Cary, M. Graille, V.E. Curtiss,
R.Wagenknecht, L. Luo, D. Dwyer, C. Goodyear,
A.L. Corper, E.A. Stura and J.-B. Charbonnier
A B-Cell Superantigen that Targets B-1 Lymphocytes 251

M. Potter, G. Jones, W. DuBois, K. Williams
and E. Mushinski
Myeloma Proteins that Bind Hap65 (GroEL) are Polyreactive
and are Found in High Incidence in Pristane Induced
Plasmacytomas 265

VIII B-CLL and B-Cell Tumors

A. Migliazza, E. Cayanis, F. Bosch-Albareda, H. Komatsu,
B. S. Martinotti, E. Toniato, S. Kalachikov, M.F. Bonaldo,
P. Jelenc, X. Ye, A. Rzhetsky, X. Qu, M. Chien,
G. Inghirami, G. Gaidano, U. Vitolo, G. Saglio,
L. Resegotti, P. Zhang, M.B. Soares, J. Russo, S.G. Fischer,
I.S. Edelman, A. Efstratiadis and R. Dalla-Favera
Molecular Pathogenesis of B-Cell Chronic Lymphocytic
Leukemia: Analysis of 13q14 Chromosomal Deletions 275

R.N. Damle, F. Fais, F. Ghiotto, A. Valetto, E. Albesiano,
T. Wasil, F.M. Batliwalla, S.L. Allen, P. Schulman,
V.P. Vinciguerra, K.R. Rai, P.K. Gregersen, M. Ferrarini
nd N. Chiorazzi
Chronic Lymphocytic Leukemia: A Proliferation of B Cells
at Two Distinct Stages of Differentiation 285

R.P. Phipps, S.J. Pollock, K. Kaur, J. Kaufman,
M.A. Borrello, B.A. Graf, D. Nazarenko, L.J. Roberts,
J.D. Morrow, J. Palis, D.J. Ryan and J.M. Bennett
Expression of Cyclooxygenase-2 and Prostaglandins
by B-1 Cells and B-CLL Cells 293

C.-F. Qi, M. Hori, L. Taddasse-Heath, N.A. Jenkins,
N.G. Copeland, H. Shen, T.A. Torrey, J.W. Hartley,
S.K. Chattopadhyay, T.N. Fredrickson and H.C. Morse III
Diffuse Large-Cell and „True" Histiocytic Lymphomas
of Mice 301

M. Potter and F. Melchers
Opinions on the Nature of B-1 Cells and Their Relationship
to B Cell Neoplasia 307

Subject Index 325

List of Contributors

(Their addresses can be found at the beginning of their respective chapters)

Agenes, F. 67
Albesiano, E. 285
Allen, S.L. 285
Allman, D. 31
Asano, M. 15
Askenase, P.W. 171
Austen, J. 57
Batliwalla, F.M. 285
Baumgarth, N. 3, 163
Bennett, J.M. 293
Berland, R. 49, 131
Bikah, G. 141
Bonaldo, M.F. 275
Bondada, S. 141
Borello, M.A. 293
Bos, N.A. 107, 211
Bosch-Albareda, F. 275
Cancro, M.P. 31
Carey, G.B. 151
Carroll, M.C. 57
Cary, S. 251
Cayanis, E. 275
Cebra, J.J. 211
Chang, M. 189
Charbonnier, J.-B. 251
Chattopadhyay, S.K. 301
Chen, J. 163
Chien, M. 275
Chiles, T.C. 121
Chiorazzi, N. 285
Clarke, S.H. 77, 121
Colarusso, T.P. 121

Constantinescu, D. 31
Copeland, N.G. 301
Corper, A.L. 251
Curtiss, V.E. 251
Dalla-Favera, R. 275
Damle, R.N. 285
Dammers, P.M. 107
Donjerkovic, D. 151
DuBois, W. 265
Dwyer, D. 189, 251
Edelman, I.S. 275
Efstratiadis, A. 275
Fagarasan, S. 221
Fais, F. 285
Ferrarini, M. 285
Fischer, G.M. 121
Fischer, S.G. 275
Flajnik. M.F. 233
Fredrickson, T.N. 301
Freitas, A.A. 67
Gaidano, G. 275
Gerstein, R.M. 121
Ghiotto, F. 285
Goodyear, C. 251
Graf, B.A. 293
Graille, M. 251
Gregersen, P.K. 285
Hardy, R.R. 15, 25
Hartley, J.W. 301
Hayakawa, K. 15, 25
Hechtman, H. 57
Herman, O.C. 163

Herzenberg, L.A. 3, 163
Honjo, T. 221
Hori, M. 301
Horkko, S. 189
Ikuta, K. 221
Inghirami, G. 275
Jager, G.C. 163
Jelenc, P. 275
Jenkins, N.A. 301
Jones, G. 265
Kalachikov, S. 275
Kamata, T. 221
Kaufman, J. 293
Kaur, K. 293
Kearney, J.F. 97
Kenny, J.J. 39
Kobzik, L. 57
Komatsu, H. 275
Kroese, F.G.M. 107, 211
Levy, S.L. 31
Li, Y.-S. 25
Liu, S. 151
Longo, D.L. 39
Luo, L. 189, 251
Lustig, A. 39
Mage, R.G. 87
Martin, F. 97
Martinotti, S. 275
Melchers, F. 307
Migliazza, A. 275
Moore, F.D., Jr. 57
Morrow, J.D. 293
Morse, H.C., III 301
Mueller, C.M. 151
Mushinski, E. 265
Nazarenko, D. 293
Nieuwenhuis, P. 107
Nogaki, F. 221
Notkins, A.L. 241
Pacierkowski, N. 179

Palinski, W. 189
Palis, J. 293
Pavlovic, S. 121
Phipps, R.P. 293
Pollock, S.J. 293
Popa, E.R. 107
Pospisil, R. 87
Potter, M. 265, 307
Prodeus, A.P. 57
Qi, C.-F. 301
Qu, X. 275
Rai, K.R. 285
Rajan, T.V. 179
Reap, E.A. 201
Reeves, W.H. 201
Reid, R.R. 57
Resegotti, L. 275
Richards, H.B. 201
Roberts, L.J. 293
Robertson, D.A. 141
Rosado, M.M. 67
Rothstein, T.L. 121
Rumfelt, L.L. 233
Russo, J. 275
Ryan, D.J. 293
Rzhetsky, A. 275
Saglio, G. 275
Sah, A.P. 31
Satoh, M. 201
Schmidt, M.R. 31
Schulman, P. 285
Scott, D.W. 151
Sen, G. 141
Shaw, M. 201
Shaw, P.X. 189
Shen, H. 301
Shinkura, R. 221
Shinton, S.A. 15, 25
Silverman, G.J. 189, 251
Soares, M.B. 275

Stall, A. 189
Stura, E.A. 251
Taddassse-Heath, L. 301
Tanguay, D.A. 121
Tatu, C. 77
Toniato, E. 275
Torrey, T.A. 301
Tsuji, R.F. 171
Valetto, A. 285
Vinciguerra, V.P. 285
Visser, A. 107
Vitolo, U. 275

Wagenknecht, R. 251
Wasil, T. 285
Wasserman, R. 25
Williams, K. 265
Wilshire, J.A. 3
Witztum, J.L. 189
Woodcock, S. 57
Woodland, R.T. 31
Wortis, H.H. 49, 131
Ye, X. 275
Zhang, P. 27

I

B-1 Cells: Definitions, Basic Hypotheses

B-1 cell origins and V_H repertoire determination

L. A. Herzenberg, N. Baumgarth and J. A. Wilshire
Genetics Department, Stanford University Medical School, Stanford, California 94305-5318

We show here that antigen selectively stimulates the progressive increase of B cells expressing a particular V_H gene in the B-1 repertoire. However, the frequencies of cells expressing a series of other V_H genes in antibodies with the same antigen specificity remain constant in the same animals. To establish context for these findings, we first review several key studies that bear on the origins of B-1 cells and the mechanisms that shape the B-1 repertoire.

Distinct Origins of B-1 and B-2 cells

Much has been said and written about the origins of B-1 cells and whether they constitute a distinct lineage. We originally proposed that progenitors for B-1 and B-2 are distinct, and hence that these B cells belong to two distinct lineages[1-6]. Wortis, Haughton, and colleagues[7, 8] later argued that the specificity of the receptor (Ig) expressed by a given B cell determined whether it would differentiate into a B-1 or B-2 cell. Variants of this argument persist today despite consistent evidence, gathered by our laboratory and others, that directly demonstrates that the progenitors for B-1 and B-2 cells are distinct[5, 6, 9-16].

There is little question that under some circumstances, cells whose phenotype classifies them as bone marrow derived follicular B cells can be stimulated to assume phenotype(s) that would classify them as B-1 cells. However, these findings do little more than testify to the plasticity of B cell phenotypes and perhaps the mechanisms that define these phenotypes. Basically, the issue is not whether a B-2 cell can be stimulated to adopt a B-1 phenotype but whether such phenotype shifts occur normally and reflect the origins of substantial numbers of B-1 cells. The collective data from several studies, including our own, argue strongly against this latter hypothesis. These studies, outlined below, show that B-1 and B-2 cells differentiate from distinct progenitors that arise at different points during the ontogeny of the immune system.

Independence of B-1 and B-2 progenitors. Two seminal studies, by John Kearney[12], Miguel Marcos[17] and their colleagues, have shown that progenitors for B-1 cells exist at early fetal sites that do not contain progenitors for B-2 cells. Although these studies did not isolate or phenotypically characterize the B-1 progenitors at these sites, the functional existence of these progenitors independent of progenitors for B-2 cells argues strongly for the existence of separate B cell lineages.

Both the Kearney and the Marcos studies are based on transfer of undisrupted embryonic tissue (fetal omentum in the Kearney paper; splanchnopleura in the Marcos paper) to a protected site under the kidney capsule in SCID mice. Arguments could be made, therefore, that only the B-1 progenitors can function under these conditions. However, a note added in proof to the Kearney paper[12] removes this objection by stating that the same result (B-1 but not B-2 reconstitution) was obtained by transferring a single-cell suspension of cells harvested from omental tissue. Since single-cell suspensions from fetal liver readily reconstitute B-2 cells in similar (SCID) recipients, the demonstration that omental cells transferred in the normal manner reconstitute B-1 but not B-2 cells provides clear evidence that progenitors for B-2 cells are not present in the fetal omentum.

Our studies, which complement these findings (or *vice versa*), show that although functional progenitors for B-2 cells are abundant in adult bone marrow, functional progenitors for B-1 cells are rare at this site[5, 6, 18]. In essence, transfers of adult bone marrow to lethally-irradiated recipients readily reconstitute B-2 cells but reconstitute only a small population of B-1 cells, largely composed of B-1b (CD5$^-$) cells. In contrast, transfers of fetal liver (which usually includes a portion of the fetal omentum) to irradiated or SCID recipients fully reconstitutes both B-1 and B-2 cells. Thus, since functional progenitors of B-1 cells can fully reconstitute the B-1 population in transfer recipients, the minimal reconstitution obtained in typical recipients indicates a paucity of B-1 progenitors in adult bone marrow.

This evidence falls short of conclusively demonstrating that functional progenitors for B-1 cells are rare in adult bone marrow because the transferred bone marrow could either contain an inhibitor of B-1 progenitor differentiation and/or could lack cells necessary to support this differentiation. To test these possibilities, we prepared and mixed single-cell suspensions from bone marrow and allotype-

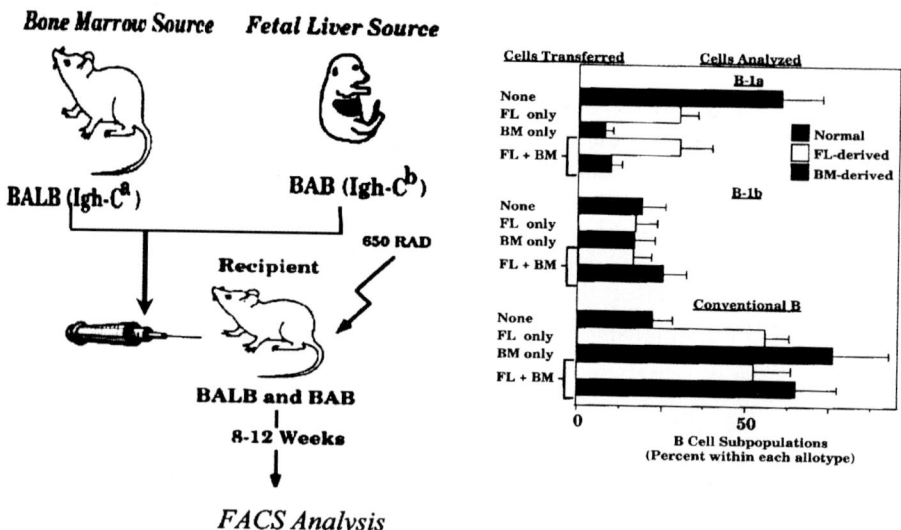

Fig 1. Developmental potential for B cell progenitors for from fetal liver and bone marrow is the same whether transferred alone or together into irradiated recipients[5]. B-cell reconstitution in cotransfer recipients is evaluated by calculating the IgH^a allotype (BM-derived) B-la, B-lb, and B-2 (conventional B) cell populations as percent of total IgH^a B cells and by calculating IgH^b allotype (FL-derived) B-la, B-lb, and B-2 cell populations as percent of total IgH^b allotype B cells. Analyses were done 8-11 weeks after transfer.

congenic fetal liver and transferred the mixture to lethally irradiated recipients (Fig. 1)[5]. As controls, we transferred each cell suspension alone to similar recipients.

Comparison of the B cells that developed in recipients 8 weeks (or more) after transfer demonstrates that B-1 cells derived from fetal liver develop equivalently in the mixture recipients and the recipients of fetal liver alone, thus ruling out the presence of an inhibitor of B-1 development in bone marrow (Fig. 1). Furthermore, the minimal B-1 population derived from bone marrow is equivalent in the mixture recipients and recipients of bone marrow alone, demonstrating that the failure to reconstitute B-1 cells from bone marrow is not due to lack of support for B-1 progenitors. Thus, with the caveat that these mixture experiments were done with fetal liver and adult bone marrow that express different allotypes, we conclude that progenitor activity for B-1 cells is selectively lacking (although not entirely absent) in adult bone marrow.

Importantly, our studies demonstrated progenitors for both B-1 and B-2 cells in fetal liver from embryos as early as day 12. In contrast, Kearney *et al*, recovered progenitors for B-1 but not B-2 cells from day 13 fetal omentum[12]. Therefore, bridging the results from the two studies, progenitors for B-2 cells are already present in fetal liver at a time when they are not detectable in fetal omentum.

In fact, the B-2 progenitors detected in murine fetal liver may be anatomically separated from the B-1 progenitors. Solvason, Kearney and co-workers have pointed out that suspensions of fetal liver cells always contain the endoderm-derived cells that form the bulk of the liver and the mesoderm-derived liver capsule that is contiguous with the omentum. Thus, B-2 progenitors may reside within the liver while B-1 progenitors reside in the liver capsule and related mesoderm tissues.

Together, these studies demonstrate that progenitors for B-1 and B-2 cells are distinct and hence that B-1 and B-2 cells as belong to separate developmental lineages. This conclusion can still accommodate the idea that B-2 progenitors or their progeny contribute to the B-1 compartment during adulthood. Cells that express certain Ig receptors, for example, could differentiate from B-2 progenitors but be triggered to express (or mimic) the B-1 phenotype. This pathway can never be ruled out entirely. However, data from two types of studies put a very close limit on the extent to which it is used:

1) Multiple studies in which adult bone marrow and mature B-1 cells were co-transferred to irradiated mice collectively show that the transferred B-1 cells reconstitute a normal-sized B-1 population that persists indefinitely by self-replenishment. Although a small bone marrow derived B-1 population (mainly B-1b) often develops shortly after transfer, its size, like the size of the majority B-1 population derived from the transferred B-1 cells, remains constant thereafter. Thus, once stabilized, the B-1 population in irradiated recipients is neither replaced nor progressively increased by bone marrow derived cells.

2) Feedback regulation studies show that *de novo* development of B-1 cells is blocked by the presence of mature B-1 cells during the first 3-4 weeks of life[19, 20]. In neonates treated from birth until 4-6 weeks of age with anti-IgM antibodies that remove all endogenous B cells, early introduction of allotype-congenic B-1 cells (that do not react with the treatment antibody) selectively blocks recovery of endogenous B-1 cells when the treatment is terminated. Similarly, in

allotype heterozygotes treated neonatally with anti-allotype antibodies that remove all B cells that express one of the IgM allotypes, recovery of the depleted B-1 population is selectively blocked by the remaining B cells. In both cases, the B-2 population recovers completely and small numbers of B-1 cells (mainly B-1b) may also recover. However, the size of the recovered B-1 population remains constant throughout life, i.e., cells from bone marrow do not add to it.

Collectively, the findings discussed above constitute a solid body of data indicating that B-1 and B-2 cells are derived from different progenitors and have different developmental patterns. Comparative studies of the B cell developmental pathway support this two-lineage model of B cell development. Differential expression of at least two genes distinguish these developmental pathways (PLRLC-myosin-like light chain and MHC I-A [21-23]). Terminal deoxynucleotidyl transferase (TdT) gene expression also distinguishes B cell progenitors in fetal liver and adult bone marrow[24], indicating precommittment to independent development pathways (even if, as has been proposed[7, 8], low TDT predisposes to expression of the B-1 phenotype). Finally, studies by Hardy, Hayakawa and colleagues indicate that B-1 and B-2 progenitors have dramatically different mechanisms for determining which V_H genes ultimately appear in the respective mature B cell repertoires[25].

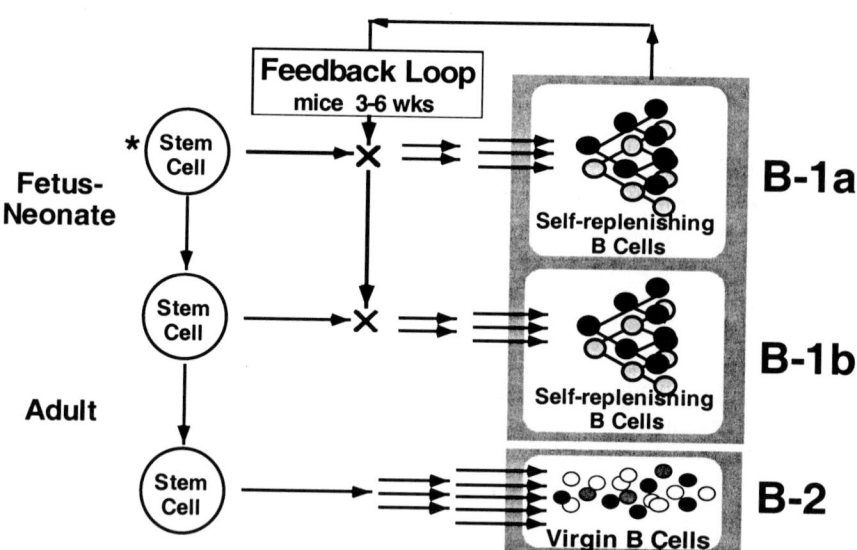

*Stem cells also give rise to T cells, macrophages, etc.

Fig 2. Three B cell lineages: B-1a and B-1b are found in the same locations and are phenotypically similar. However, CD5 is expressed on B-1a but not B-1b. B-2 ("conventional") B cells have a strikingly different phenotype and are distributed differently from B-1 and B-1b.

A third B cell lineage. Current data actually define three murine B cell lineages (Fig. 2), two of which (designated B-1a and B-1b) are similar enough to be treated collectively for most purposes as B-1 cells [26]. B-1a, which normally comprise the majority of B-1 population, express CD5. B-1b do not express detectable surface CD5 but share most of the other properties of B-1a cells, including self replenishment, sensitivity to feedback regulation and localization to the peritoneal and pleural cavities. Nevertheless, studies in which B-1a and B-1b were sorted and transferred to irradiated recipients demonstrate clearly that mature B-1a and B-1b cells are committed to replenish only their respective populations[27].

B-1a also differ from B-1b in that B-1b are somewhat more efficiently reconstituted by bone marrow transfers and tend to recover somewhat better after neonatal depletion in feedback regulation studies[19, 20]. Furthermore, at a functional level, B-1a and B-1b express different antibody repertoires[28], respond to different cytokines and switch to different Ig isotypes[1], i.e., B-1a respond to IL-5[29]and tend to spontaneously produce IgG3, IgG2a and IgG2b[6] while B-1b respond to IL-9 and tend to spontaneously produce IgE and IgG1 [30].

Antigen Selection in the B-1 repertoire

The antibody repertoires produced by B-1 and B-2 cells contain mutually exclusive specificities for antigens[31, 32] and also show differences in antibody variable region (V_H) usage[28]. In perhaps the best studied difference between these repertoires, a significant fraction of B-1 cells (5-15%) produce antibodies that bind phosphatidyl choline (PtC) while B-2 cells that produce antibodies specific for this antigen are not found in normal mice[33].

Antibodies that bind PtC, a ubiquitous membrane phospholipid found in both mammalian and bacterial membranes, are natural antibodies. PtC binding cells, detectable by FACS as cells that bind fluorochrome-labelled liposomes[34], are present at normal numbers in germfree mice[35, 36]. Furthermore, the frequency of PtC-binding cells B cells is not altered by injection of PtC antigen into normal mice[37, 38]. Consistent with this evidence, which suggests that cells producing anti-PtC antibodies play an important role in innate immunity, J. Chen and colleagues[39], using genetically altered mice that cannot secrete IgM antibodies, have shown that injection of anti-PtC antibodies protects against bacterial sepsis induced by cecal ligation and puncture.

The immunoglobulin heavy chain variable (V_H) gene families that encode anti-PtC antibodies are mainly restricted to three V_H gene families (V_H11, V_H12 and V_HQ52)[33, 40]. These three genes collectively encode the bulk of the anti-PtC antibodies produced in all mouse strains tested. The BALB/c and C.B-17 strains, however, differ with respect to which V_H gene family dominates the anti-PtC repertoire (Fig. 3). For example, BALB/c anti-PtC antibodies are predominantly encoded by the V_HQ52 family, while C.B-17 anti-PtC are predominantly encoded by the V_H12

---

[1] Note that although there is a common tendency to think that B-1 cells do not undergo isotype switching, B-1 cells as a whole can switch to produce all of the advanced isotypes.

family. Both strains produce $V_H Q52$ and $V_H 12$ anti-PtC; it is only the representation of these antibodies that differs[40].

To determine the V_H representation in the anti-PtC repertoires in these strains, we used multiparameter FACS analysis and single-cell RT-PCR and sequencing[40]. We co-stained cells with PtC-liposomes and a combination of monoclonal antibody reagents detecting IgH allotypes, B cell surface markers and antibodies encoded by $V_H 11$ or $V_H 12$[41, 42] (The latter reagents were kindly supplied by Geoffrey Haughton and colleagues). Among cells expressing antibodies encoded neither by $V_H 11$ nor $V_H 12$ (collectively designated as "Other"), single-cell RT-PCR and sequencing shows that the $V_H Q52$ family predominates. In fact, among the large proportion of cells that express neither $V_H 11$ nor $V_H 12$ in BALB/c mice, roughly 70% express $V_H Q52$ and many of these express a single member of the $V_H Q52$ family (MMU53526, $V_H Ox-1$)[40].

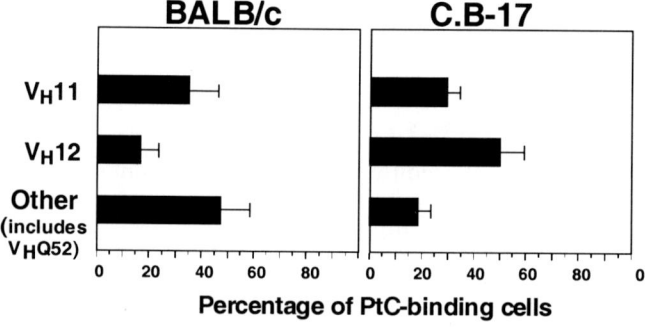

Fig 3. V_H gene family repertoire of PtC-binding B cells. Peritoneal B cells analysed from 5 month old animals. Bars represent the V_H family frequency among PtC-binding cells determined by flow cytometry for individual mice and averaged for 6-12 mice. Standard deviations are shown for each bar. V_H genes other than $V_H 11$ and $V_H 12$, which are individually recognized by monoclonal antibody FACS reagents, are designated "Other".

C.B-17 is a BALB/c IgH allotype congenic strain generated (by Michael Potter) by mating and successively backcrossing a (BALB/c x C57BL/Ka)F_1 hybrid to BALB/c while selecting for maintenance of the C57BL-derived (IgH^b) chromosome region. Thus, this congenic strain carries the IgH^b allotype chromosome region on the BALB/c genetic background. The difference in V_H predominance between the anti-PtC repertoires of BALB/c (IgH^a) and C.B-17 (IgH^b) is linked to the IgH chromosome region (confirmed by backcross analysis, Wilshire et al, in preparation).

The association between the IgH allotype and V_H expression patterns in the BALB/c and C.B-17 anti-PtC repertoires is maintained in the F_1 hybrid between these strains. Staining with monoclonal anti-allotype reagents to distinguish the anti-PtC V_H repertoire encoded by each parental chromosome in the F_1 demonstrates that the IgH^a and IgH^b allotype B cell repertoires in the F_1 are comparable to their corresponding parental repertoires. For example, comparison of the "Other" (mainly $V_H Q52$) anti-PtC data for 5 month old animals in figures 3 and 4 shows that in the F_1, "Other" predominates among IgH^a PtC-binding cells as it does in the

BALB/c parental animals. Furthermore, IgH^b V$_H$12 anti-PtC increases with age in the F$_1$ as it does in C.B-17 (see below).

Analysis of V$_H$ expression during the course of development of the F$_1$ mice from 3 weeks to 8 months of age demonstrates a striking difference between V$_H$12 anti-PtC encoded by the IgH^b chromosome region (derived from C.B-17) and all of the other predominant V$_H$ genes in the anti-PtC repertoire. With the exception of IgH^b V$_H$12, the anti-PtC V$_H$ genes (including IgH^a V$_H$12) represent a fixed percentage of peritoneal B cells at 3 weeks that remains fixed over time (Fig. 4). Only IgH^b V$_H$12 selectively increases with age.

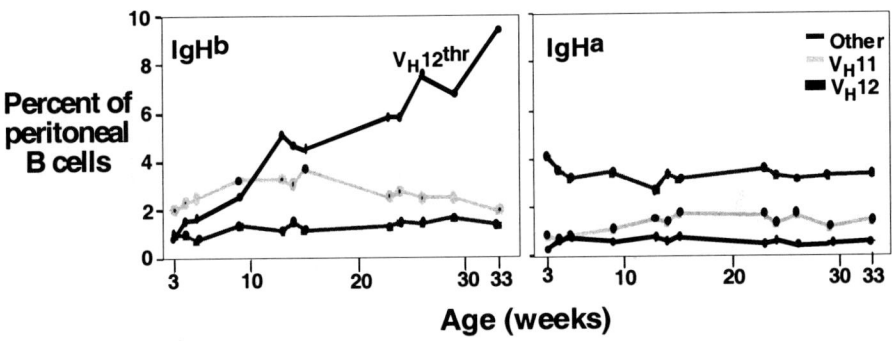

FIG 4. Only cells expressing the V$_H$12 anti-PtC encoded by the Igh^b chromosome (V$_H$12^{thr}) increase with age in (C.B-17 x BALB/c)F$_1$ hybrid mice. **Left panel:** V$_H$ encoded by IgH^b chromosome derived from C.B-17 parent. **Right panel:** V$_H$ encoded by the IgH^a chromosome derived from the BALB/c parent. V$_H$ expression on PtC-binding cells was determined by multiparameter FACS. Each point represents the average of 5-20 mice.

Collectively, studies with F$_1$ mice described above demonstrate that the difference between the parental strain anti PtC repertoires is not due to the presence of different extracellular antigens or cell surface molecules that operate to select the V$_H$ genes. In the F$_1$ mice, selection should operate equally on V$_H$ encoded by both parental chromosomes. However, we find that the V$_H$ expression pattern encoded by each of the chromosomes in the F$_1$ mimics the pattern in the corresponding parental animal. Therefore, selection due to the presence of different antigens cannot account for the difference between the IgH^a and IgH^b anti-PtC repertoires.

In fact, the selective increase in IgH^b V$_H$12 traces to an amino acid difference between IgH^a and IgH^b V$_H$12. Comparison of the V$_H$12 gene sequence from BALB/c and C.B-17 mice demonstrates that there is a single amino acid difference at codon 21 in the framework 1 (FR1) region[40, 43]. There is an alanine residue encoded at this position in BALB/c (V$_H$12^{ala}) and a threonine residue encoded in C.B-17 (V$_H$12^{thr}). Importantly, C57BL/10-related mice, including the "2a4b" mice studied by Haughton and colleagues[44], also carry V$_H$12^{thr}. However, C57BL/6J mice, which have the same IgH^b constant region as C57BL/10 and C.B-17, nonetheless express the V$_H$12^{ala} allele found in BALB/c. Six additional silent nucleotide differences between V$_H$12^{ala} and V$_H$12^{thr} indicate these genes are well separated in evolution.

In contrast to the age-dependent increase in cells expression V_H12^{thr} anti-PtC antibodies, B cells expressing V_H12^{thr} antibodies that do no bind PtC do not increase in frequency over time (Fig. 5). Thus, the specificity of V_H12^{thr} for PtC (and/or related antigens) is required for the selective increase of cells expressing these antibody molecules. In other words, expansion of the V_H12^{thr} anti-PtC population is depend-

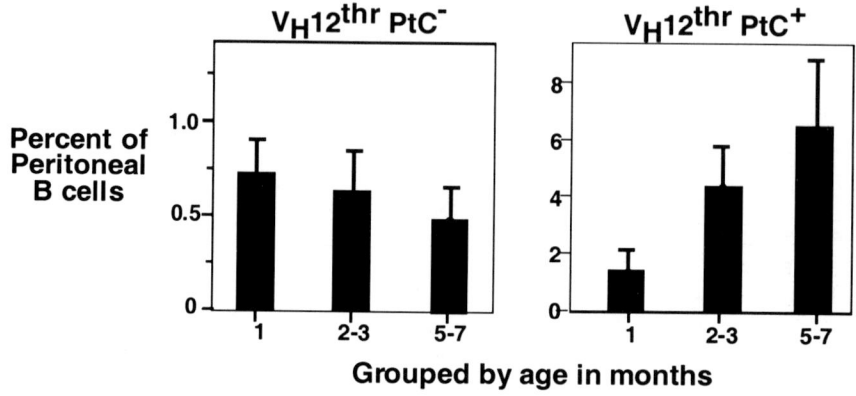

FIG. 5. Cells expressing V_H12^{thr} antibodies that do not bind PtC do not increase in frequency with age. Flow cytometry was used to determine the percent of B cells in C.B-17 mice which express antibodies encoded by V_H12. **Left panel:** Percentage of B cells that *do not* bind PtC. **Right panel:** Percentage of B cells expressing V_H12 encoded antibodies that *do* bind PtC. Points represent the average and standard deviation of the percentage of B cells from 5-20 mice each.

ent on its antigen specificity.

In principle, clonal expansion could explain the V_H12^{thr} anti-PtC findings. However, since 24/25 sequences obtained from V_H12^{thr} PtC-binding cells that were FACS-sorted from 5-month old F_1 mice were unique, and since the frequency of V_H12^{thr} increases with age in all animals tested, these findings rule out both clonal expansion and neoplastic or pre-neoplastic events underlying the increased V_H12^{thr} anti-PtC frequency.

CDR3 sequences associated with V_H12^{thr} anti-PtC could also, in principle, explain the antigen-dependent selective expansion of IgH^b cells expressing this V_H gene. However, sequence data shows that there are no systematic differences between the IgH^a and IgH^b CDR3 sequences, e.g., IgH^a and IgH^b encoded anti-PtC antibodies both have the typical 10/G4 CDR3 region sequences (10 amino acid length with glycine in the fourth position) shown by Mercolino et al[45] to be characteristic of V_H12 anti-PtC antibodies. Thus, although the CDR3 region must contribute to the specificity of these antibodies for PtC, there is no evidence that it is responsible for selective expansion of the IgH^b V_H12. The substitution of threonine for alanine at codon 21 in FR1 therefore emerges as the single defining difference between IgH^a and IgH^b V_H12 anti-PtC.

Since codon 21 occurs in a beta-pleated sheet that is not in proximity to any CDR region (antigen binding site), it seems an unlikely candidate to alter antigen binding avidity. However, several studies indicate that this region may help to increase binding avidity for unusual antigens[46]. For example, human V_H3 encoded antibodies bind to Staphylococcal Protein A using the FR1 region as well as the CDR2 and FR3 regions (reviewed in [47]). Our studies in which B-1 cells were stained graded amounts of PtC-liposomes are also consistent with the idea that the alanine/threonine difference alter avidity. At the same PtC-liposome concentration, V_H12^{thr} B cells bind more PtC-liposomes than V_H12^{ala} cells (Fig. 6). Thus, we propose that the single amino acid difference between the IgH^a and IgH^b allotype encoded V_H12 antibodies alters avidity for PtC and results in the age-dependent increase in V_H12^{thr}.

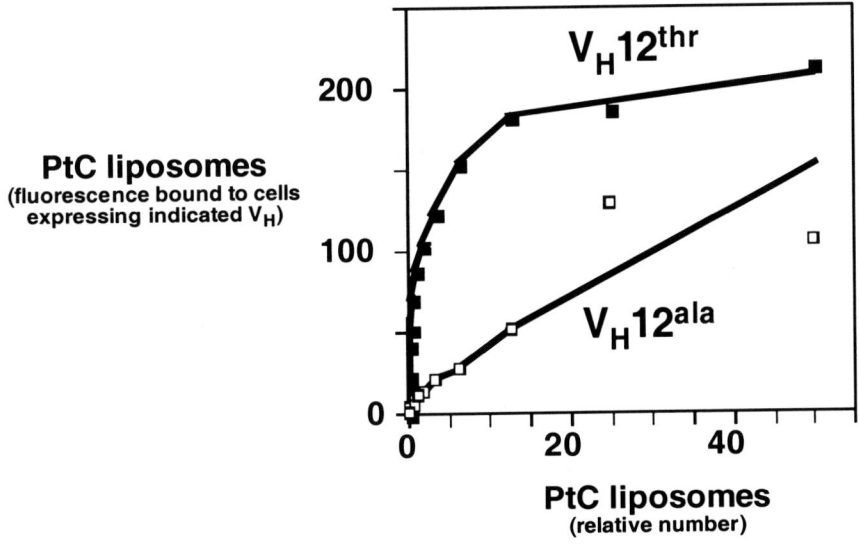

Fig 6. Cells expressing V_H12^{thr} bind PtC-liposomes more efficiently than cells expressing V_H12^{ala}. Peritoneal B cells from BALB/c (V_H12^{ala}) and C.B-17 (V_H12^{thr}) were co-stained with monoclonal antibodies that recognize V_H12 and graded numbers of fluorescent PtC-liposomes. The figure shows liposomes bound (median liposome-derived fluorescence) at each "concentration" of liposomes for the two cell types.

Our findings confirm the age-dependent increase in V_H12^{thr} detected by Clarke and colleagues, who showed that the vast majority of the V_H12 rearrangements isolated from cDNA libraries of adult "2a4b" mice (which express V_H12^{thr}) are productive (rather than non-productive) rearrangements[48]. These investigators reasoned that antigen selection accounts for the increased frequency of the productive rearrangements. In our studies, we show that V_H12^{thr} antibodies that bind PtC increase with age while V_H12^{thr} antibodies that do not bind this antigen remain constant. Thus, we conclusively demonstrate that the increase in V_H12^{thr} is antigen-driven.

Finally, our studies show that except for cells expressing V_H12^{thr}, V_H gene frequencies among PtC-binding B cells within the peritoneal B cell population remain constant from 3 weeks until at least 8 months of age. Thus, although antigen sometimes selectively stimulates continued expansion of components of the B-1 repertoire, the repertoire as a whole does not change dramatically with age.

References

1. Hardy RR, Hayakawa K, Parks DR, Herzenberg LA, Herzenberg LA (1984) Murine B cell differentiation lineages. J Exp Med 159:1169-88
2. Herzenberg LA, Stall AM, Lalor PA, Sidman C, Moore WA, Parks DR, Herzenberg LA (1986) The Ly-1 B cell lineage. Immunological Reviews 93:81-102
3. Herzenberg LA, Kantor AB, Herzenberg LA (1992) Layered evolution in the immune system a model for the ontogeny and development of multiple lymphocyte lineages. Vol 651, p. 1-9: Annals of NYAcad Sci
4. Herzenberg LA, Kantor AB (1993) B-cell lineages exist in the mouse. Immunol Today 14:79-83; discussion 88-90
5. Kantor AB, Stall, AM, Adams S, Herzenberg LA, Herzenberg LA (1992) Differential development of progenitor activity for three B-cell lineages. Proc. Natl. Acad. Sci. USA 89:3320-3324
6. Kantor AB, Herzenberg LA (1993) Origin of murine B cell lineages. Annu Rev Immunol 11:501-38
7. Wortis HH (1992) Surface markers, heavy chain sequences and B cell lineages. Int Rev Immunol 8:235-46
8. Haughton G, Arnold LW, Whitmore AC, Clarke SH (1993) B-1 cells are made, not born. Immunol Today 14:84-7; discussion 87-91
9. Lam K, Stall AM, Kantor AB, Herzenberg LA (1993) Origins of B cell lineages: Aspects of the difference between B-1 and conventional B cells in M54 μ transgenic mice. , p. 39-48. Tokyo, Japan: The Naito Foundation for Medical Research through Academic Press/Harcourt Brace Jovanovich Japan, Inc.
10. Kearney JF (1993) CD5+ B-cell networks. Curr Opin Immunol 5:223-6
11. Kearney JF, Won WJ, Benedict C, Moratz C, Zimmer P, Oliver A, Martin F, Shu F (1997) B cell development in mice. Int Rev Immunol 15:207-41
12. Solvason N, Lehuen A, Kearney JF (1991) An embryonic source of Ly1 but not conventional B cells. Int Immunol 3:543-50
13. Solvason N, Chen X, Shu F, Kearney JF (1992) The fetal omentum in mice and humans. A site enriched for precursors of CD5 B cells early in development. Ann N Y Acad Sci 651:10-20
14. Solvason N, Kearney JF (1992) The human fetal omentum: a site of B cell generation. J Exp Med 175:397-404
15. Godin IE, Garcia-Porrero JA, Coutinho A, Dieterlen-Lievre F, Marcos MA (1993) Para-aortic splanchnopleura from early mouse embryos contains B1a cell progenitors. Nature 364:67-70
16. Marcos MA, Morales-Alcelay S, Godin IE, Dieterlen-Lievre F, Copin SG, Gaspar ML (1997) Antigenic phenotype and gene expression pattern of lymphohemopoietic progenitors during early mouse ontogeny. J Immunol 158:2627-37
17. Marcos MA, Gaspar ML, Malenchere E, Coutinho A (1994) Isolation of peritoneal precursors of B-1 cells in the adult mouse. Eur J Immunol 24:1033-40
18. Kantor AB, Stall AM, Adams S, Watanabe K, Herzenberg LA (1995) De novo development and self-replenishment of B cells. Int Immunol 7:55-68
19. Lalor PA, Stall AM, Adams S, Herzenberg LA (1989) Permanent alteration of the murine Ly-1 B repertoire due to selective depletion of Ly-1 B cells in neonatal animals. Eur J Immunol 19:501-6
20. Lalor P, Herzenberg LA, Adams S, Stall AM (1989) Feedback regulation of murine Ly-1 B cell development. Eur. J. Immunol. 19:507 - 513
21. Oltz EM, Morrow MA, Rolink A, Lee G, Wong F, Kaplan K, Gillis S, Melchers F, Alt FW (1992) A novel regulatory myosin light chain gene distinguishes pre-B cell subsets and is IL-7 inducible. EMBO J 11:2759-2767
22. Hayakawa K, Tarlinton D, Hardy RR (1994) Absence of MHC class II expression distinguishes fetal from adult B lymphopoiesis in mice. J Immunol 152:4801-7
23. Lam KP, Stall AM (1994) Major histocompatibility complex class II expression distinguishes two distinct B cell developmental pathways during ontogeny. J Exp Med 180:507-16
24. Li YS, Hayakawa K, Hardy RR (1993) The regulated expression of B lineage associated genes during B cell differentiation in bone marrow and fetal liver. J Exp Med 178:951-60
25. Wasserman R, Shinton SA, Carmack CE, Manser T, Wiest DL, Hayakawa K, Hardy RR (1998) A novel mechanism for B cell repertoire maturation based on response by B cell precursors to pre-B receptor assembly. J Exp Med (2):259-264

26. Kantor AB, Stall AM, Adams S, Herzenberg LA, Herzenberg LA (1992) Differential development of progenitor activity for three B cell lineages. Proc. Natl. Acad. Sci. U.S.A. 89:3320-24
27. Stall AM, Adams S, Herzenberg LA, Kantor AB (1992) Characteristics and Development of the Murine B-1b (Ly-1 B sister) Cell Population. , p. 33-43. New York: Ann. N.Y. ACAD. SCI.
28. Kantor AB, Merrill CE, Herzenberg LA, Hillson JL (1997) An unbiased analysis of V(H)-D-J(H) sequences from B-1a, B-1b, and conventional B cells. J Immunol 158:1175-86
29. Takatsu K, Takaki S, Hitoshi Y, Mita S, Katoh S, Yamaguchi N, Tominaga A (1992) Cytokine receptors on Ly-1 B cells. IL-5 and its receptor system. Ann N Y Acad Sci 651:241-58
30. Vink A, Warnier G, Brombacher F, Renauld JC (1999) Interleukin 9-induced in vivo expansion of the B-1 lymphocyte population. J Exp Med 189:1413-23
31. Hardy RR, Li YS, Hayakawa K (1996) Distinctive developmental origins and specificities of the CD5+ B-cell subset. Semin Immunol 8:37-44
32. Lalor PA, Morahan G (1990) The peritoneal Ly-1 (CD5) B cell repertoire is unique among murine B cell repertoires. Eur J Immunol 20:485-92
33. Arnold LW, Haughton G (1992) Autoantibodies to phosphatidylcholine. The murine antibromelain RBC response. Ann N Y Acad Sci 651:354-9
34. Mercolino TJ, Arnold LW, Haughton G (1986) Phosphatidyl choline is recognized by a series of Ly-1+ murine B cell lymphomas specific for erythrocyte membranes. J Exp Med 163:155-65
35. Cunningham AJ (1976) Self-tolerance maintained by active suppressor mechanisms. Transplant Rev 31:23-43
36. Poncet P, Huetz F, Marcos M-A, and, Andrade (1990) All VH11 genes expressed in peritoneal lymphocytes encode anti-bromelain-treated mouse red blood cell autoantibodies but othe rVH gene families contribute to this specificity. Eur. J. Immunol. 20:1583-1589
37. Cox KO, Baddams H, Evans A (1977) Studies of the antigenicity and immunogenicity of bromelain-pretreated red blood cells. Aust J Exp Biol Med Sci 55:27-37
38. Pages J, Bussard AE (1975) Precommitment of normal mouse peritoneal cells by erythrocyte antigens in relation to auto-antibody production. Nature 257:316-7
39. Boes M, Prodeus AP, Schmidt T, Carroll MC, Chen J (1998) A critical role of natural immunoglobulin M in immediate defense against systemic bacterial infection. J Exp Med 188:2381-6
40. Seidl KJ, Wilshire JA, MacKenzie JD, Kantor AB, Herzenberg LA, Herzenberg LA. (1999) Prominent V(H) genes expressed in innate antibodies are associated with distinctive antigen-bindings sites (J(H)-CDR3). Proc. Natl. Acad. Sci. 96:2262-2677
41. Arnold LW, Pennell CA, McCray S, Clarke SH (1994) Development of B-1 cells: segregation of phosphatidyl choline-specific B cells to the B-1 population occurs after immunoglobulin gene expression. J. Exp. Med. 179:1585-1595
42. Arnold LW, Spencer DH, Clarke SH, Haughton G (1993) Mechanisms that limit the diversity of antibody: three sequentially acting mechanisms that favor the spontaneous production of germ-line encoded anti-phosphatidyl choline. Int Immunol 5:1365-73
43. Booker JK, Haughton G (1993) Mechanisms that limit the diversity of antibodies. II. Evolutionary conservation of Ig variable region genes which encode naturally occurring autoantibodies. International Immunology 6:1427-1436
44. Haughton G, Arnold LW, Bishop GA, and, Mercolino TJ (1986) The CH Series of Murine B Cell Lymphomas: Neoplastic Analogues of Ly-1+ Normal B Cells. Immunological Reviews 93:35-51
45. Mercolino TJ, Arnold LW, Hawkins LA, Haughton G (1988) Normal mouse peritoneum contains a large population of Ly-1+ (CD5) B cells that recognize phosphatidyl choline. Relationship to cells that secrete hemolytic antibody specific for autologous erythrocytes. J Exp Med 168:687-98
46. Li Y, Spellerberg MB, Stevenson FK, Capra JD, Potter KN (1996) The I binding specificity of human VH 4-34 (VH 4-21) encoded antibodies is determined by both VH framework region 1 and complementarity determining region 3. J Mol Biol 256:577-89
47. Potter KN, Li Y, Pascual V, Capra JD (1997) Staphylococcal protein A binding to VH3 encoded immunoglobulins. Int Rev Immunol 14:291-308
48. Ye J, McCray SK, Clarke SH (1995) The majority of murine VH12-expressing B cells are excluded from the peripheral repertoire in adults. Eur J Immunol 25:2511-21

B-1 Cell Definition

K. Hayakawa, S.A. Shinton, M. Asano, and R.R. Hardy

Institute for Cancer Research, Fox Chase Cancer Center, Philadelphia, PA 19111, USA

Introduction

The nomenclature "B-1", as distinct from "B-2", was proposed at the New York Academy of Science meeting several years ago in an attempt to divorce cell surface phenotype from the development. This was in response to the finding that $CD5^+$ B cell generation was mostly from fetal B cell progenitors, and that some $CD5^-$ B cells derived from the fetal liver appeared closely related to $CD5^+$ B cells, based on other surface markers, tissue localization, and function. Thus, B-1 was viewed as discriminating fetal from adult B cell development in mice, encouraging pursuit of how B cell development alters with ontogeny and the significance of this phenomena in the immune system [1, 2]. However, subsequent reports by many investigators have used the nomenclature "B-1" based on either surface phenotype (as $CD5^+$ B cells or MAC-1/$CD11b^+$ B cells) or location (as peritoneal cavity B cells), criteria associated with, but not exclusive to, the fetal B cell product. This has resulted in significant confusion in interpretation of results [3, 4]. We provide here a scheme (Fig. 1) that we hope will clarify some of this confusion and also introduce data suggesting that positive selection may explain the abundant generation of $CD5^+$ B cells in the peritoneal cavity from fetal/neonatal precursors.

Distinct fetal and adult B cell development, as B-1 (Type I) versus B-2 (Type II)

Fig. 1 presents the B-1 versus B-2 developmental paradigm from the view of distinct fetal and adult B cell development. The B1a, B1b and B-2 nomenclature based on surface phenotypic criteria, as commonly used in the literature, is also marked in the figure. This scheme comes from our pro-B cell transfer experiments where similar reconstitution patterns were seen in spleen, peritoneal cavity, and peripheral blood analyses [5-7]. The point here is that regardless of fetal or adult origin, both transited through an $CD5^-IgM^{hi}IgD^{lo/hi}$ ($\mu^{++}\delta^{+/++}$) stage (i.e. IgD up-regulation occurs for both types of B cell development); however, only adult B cell progenitors eventually yielded mostly $IgM^{lo}IgD^{hi}$ ($\mu^+\delta^{++}$) cells without induction of CD5 ("B-2"). In contrast, the fetal B cell progeny did not enter the "B-2" cell population; instead surviving cells were $IgM^{hi}IgD^{low}$ ($\mu^{++}\delta^{-/+}$), with or without CD5 expression ("B-1a", "B-1b") within the same time frame of reconstitution. Thus, the developmental patterns differed. Note that, in our scheme, the B-1 (or B-2) nomenclature encompasses the entire developmental process involving surface alteration including the $CD5^-$ to $CD5^+$ stages. We hope to replace this nomenclature by "Type I" and "Type II" development, respectively, once distinctions in the mechanism of development are clarified, to avoid overlap with usage based on surface phenotype criteria.

The difficulty of this type of nomenclature is that it is likely that the alteration from B-1 (Type I) to B-2 (Type II) development may be a gradual process, starting from the embryonic/fetal period, neonatal period, weanling period, and finally to the

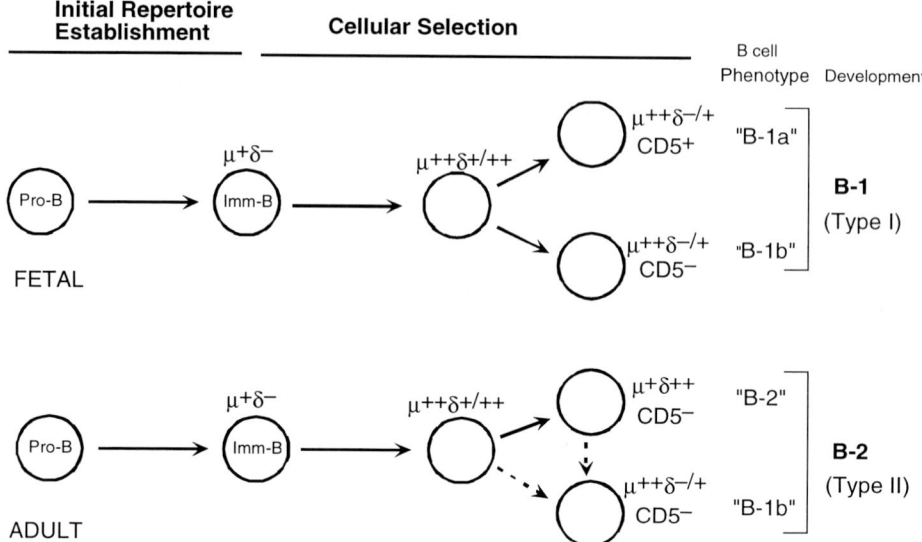

Fig. 1. Fetal versus adult B cell developmental paradigm. IgM^{lo}IgD^{hi} ($\mu^+\delta^{++}$) cell, "B-2", are the predominant B cell phenotype in adult B cell development in contrast with the fetal development. "B-1b" phenotype cells can be generated from adult B cell development as a minor population, which could be either the activation product from B-2 or the result of divergent maturation (dashed lines).

adult period, considering the complexity of developmental checkpoints (see below). Differences in cell dynamics may further complicate investigating B cell populations established at each ontogenic period, as found with the continuous maintenance of B-1 cell progeny in the adult [2]. Nevertheless, mouse B cells generated during the newborn stage are likely to represent the B-1 development product. Fetal pluripotent stem cell transfer reproduced this ontogenic progression from fetal to adult, eventually generating B-2 [8], possibly due to differentiation of self-renewing stem cells which continually provided B cells during the reconstitution period over a couple of months.

B cell development proceeds from an early repertoire establishment stage to a late cellular selection stage. Following describes our recent progress in investigating B-1 versus B-2 distinctions at these different developmental stages.

Early repertoire establishment stage

Since one clear difference between B-1 and B-2 is $V_H11/V_\kappa9^+$ anti-PtC (phosphatidylcholine) autoreactive CD5$^+$ B cell generation from B-1, understanding development of anti-PtC B cells has been an important area of research. Such study revealed a surprising distinction in the mechanism of early repertoire establishment between fetal and adult. Thus, preferential generation of $V_H11\mu^+$ immature B cells occurs in B-1 (fetal) because of a distinct pre-BCR stage selection mechanism, explaining the reason why anti-PtC B cell generation is preferential to B-1 [9]. Obviously, this is a critical research area to understand B-1 versus B-2 cell distinction, as will be described by Hardy et. al. in this volume.

Cellular selection stage

Positive selection and CD5$^+$ phenotype

Autoreactive B cells are enriched in the CD5$^+$ "B-1a" cell population. Involvement of self-antigen in the accumulation of autoreactive CD5$^+$ B cells has been a subject of speculation for a long time in the study of anti-PtC B cells, based on limited CDR3 size and particular amino acids at specific positions in the V_H region [10, 11]. Providing $V_H11\mu$ (or $V_H12\ \mu$) heavy chain as a transgene (Tg) resulted in a striking increase of CD5$^+$ anti-PtC B cells with a usage of Tgμ combined with endogeneous canonical light chain, supporting this idea [9, 12, 13]. However, demonstrating an absolute requirement for self-antigen by eliminating antigen has been difficult or impossible because of their lipid nature or function as key cellular constituents. This has been overcome recently by utilizing anti-thymocyte autoantibody (ATA) reactive to a carbohydrate epitope of the Thy-1/CD90 glycoprotein, encoded by $V_H3609/V_\kappa21C$ germline genes [14, 15]. ATA is another autoantibody enriched in CD5$^+$ B cells [16]. Introducing $V_H3609\mu$ as a transgene resulted in an increase of CD5$^+$ B cells with ATA reactivity generated through pairing of the transgene with endogenous Vk21C, resulting in high ATA levels in serum. The extent of CD5$^+$ATA B cell increase was much lower in all $V_H3609\mu$ transgenic lines compared to $V_H11\mu$ anti-PtC lines; however, it was clearly detectable, particularly in analysis of the peritoneal cavity B cells (Fig. 2, upper right). Importantly, these ATA B cells and increased ATA serum titer did not occur when the self-antigen was eliminated in Thy-1 gene knockout mice[14] (Fig. 2, lower right). Thus, self-antigen (Thy-1 glycoprotein) is indeed positively involved in the generation of natural autoreactive B cells and of serum natural

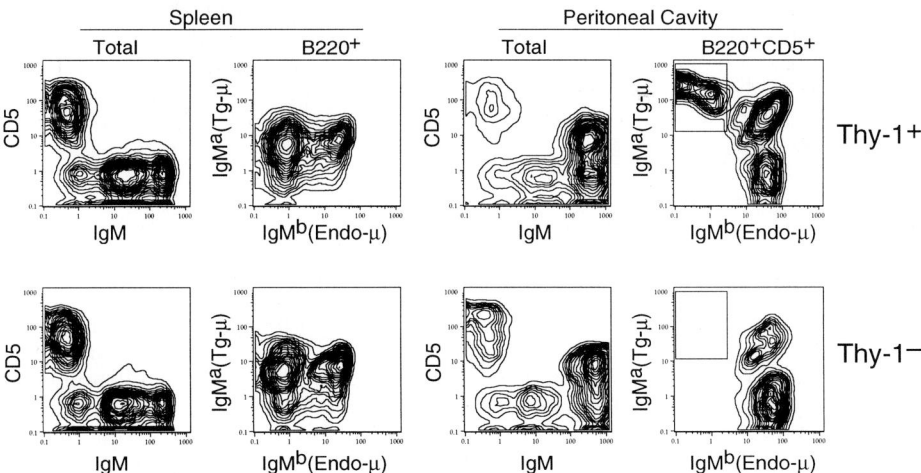

Fig. 2. Specific deficiency of ATA B cell subset in the absence of Thy-1. Flow cytometry analysis of spleen and peritoneal cavity cells in Thy-1$^+$ versus Thy-1$^-$ $V_H3609\mu$ Tg$^+$ littermates (2 mo. old). There was no significant difference in terms of overall lymphocyte populations except for the lack of ATA B cell population, i.e.CD5$^+$ IgM^a (Tgμ)$^+$ IgM^{b-} cells, in Thy-1 deficient mice as marked in boxes on the far right panells (0.65% in Thy-1$^+$ and 0.006% in Thy-1$^-$ mice). Serum ATA titers were also deficient in Thy-1$^-$ background mice and these ATA defects continued throughout life.

autoantibody. Furthermore, the Thy-1 glycoprotein is in fact a major source of physiologic self-antigen ligand for natural autoreactive B cells. Up-regulation of CD5 and eventual accumulation of IgD^{low} cells is likely to be the result of BCR mediated signaling during this process [14, 15].

Central and peripheral positive selection. Whether self-antigen crosslink is preventing apoptosis during maturation, as "positive selection", or only during maintenance of mature B cells remains to be answered. However, we do consider that this CD5 induction is reminiscent of positive selection during T cell development. CD5 is an antigen receptor adapter protein for both T and B cells [17-19]. In T cell development, CD5 up-regulation occurs with maturation in the thymus during the process of self-antigen mediated positive selection [20], but not after maturation, where the highest level of CD5 is expressed. The extent of TCR crosslinking on immature T cells by self-MHC molecules appears to determine the level of CD5 up-regulation, controlling positive and negative selection [20, 21]. Thus, it is not surprising that CD5 up-regulation also occurs during maturation of B-1 cells, similar to T cell development (and possibly controlling self-reactivity after CD5 induction). The level of CD5 up-regulation by B-1 cells is much lower than for T cells. Therefore, it is tempting to speculate that positive selection may be key for immature B-1 cell survival in general, and even CD5$^-$ B cells (as B-1b) may be selected by self-antigen specificity regardless of their undetectable CD5 level.

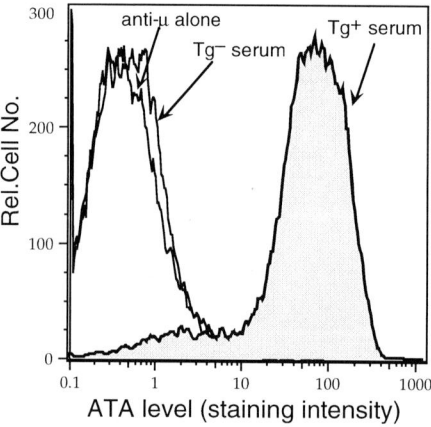

Fig. 3. Detection of serum ATA in 1 wk old V$_H$3609μ Tg mice. Sera were obtained from the heart blood of 1wk old (3609μTg$^{+/-}$ x C.B-17) littermates. The thymocyte staining histograms were generated with undiluted sera, followed by fluorescein-anti μ. All Tg$^+$ neonatal mice sera were ATA$^+$, whereas all Tg$^-$ were ATA$^-$ as represented here.

Development versus surface phenotype. Since CD5 can be induced on mature B-2 cells by in vitro BCR stimulation, the CD5$^+$ phenotype has often been considered to be the result of antigenic stimulated differentiation after the establishment of mature B cells [22, 23]. There is no doubt that such concerns cannot be dismissed, particularly in diseased states. Reactivation of anergic self-reactive B cells may occur, which may induce CD5 under appropriate conditions. Also, loss of surface CD5 can occur with CD5$^+$ B cells. However, such alterations in CD5 phenotype should be discussed separately from the B-1 versus B-2 development issue. Regardless, we would like to recall that, in normal development, the majority of CD5$^+$ B cells established in adults derive from fetal/neonatal precursors and are not constantly replenished from adult B cell precursors. As Fig. 3 shows, serum ATA can be detected in V$_H$3609μ transgenic mice at 1 wk of age, i.e. at an early ontogenic time well-before typical "B-2" mature B cells appear; thus they are not of IgM^{lo}IgD^{hi} cell origin.

Antigen nature

Why does positive selection occur for B cells despite the B cell's well established exquisite sensitivity at the immature stage? One possible reason is the unique antigen nature of these natural antigens. Antigens such as the ATA determinant and PtC may signal differently from high affinity and/or high valency antigens that normally have a negative impact on newly generated B cells, resulting in CD5 up-regulation and allowing continued maturation. Determining the physiologic self/natural antigens that allow positive selection is not straightforward. For example, the precise form of the antigen recognized by ATA B cells or anti-PtC B cells has not been determined as yet. Our data showing early appearance of ATA suggests that positive selection may not require direct contact with Thy-1$^+$ cells, since the only site stained intensely by ATA during the newborn stage is the thymus, well segregated from sites of developing B cells [15]. $V_H3609/V_\kappa21C$ ATA binds to the carbohydrate determinant expressed by affinity purified soluble Thy-1 as well as to thymocytes [15]. Since Thy-1 is a heavily glycosylated GPI (glycophosphatidylinositol)-linked protein, the antigenic determinant may be the carbohydrate moiety cleaved from Thy-1 or the Thy-1 glycoprotein itself in soluble form released into the circulation. This may explain in part the peritoneal cavity preference of natural autoreactive B cells since such soluble antigens may readily effuse into the peritoneal cavity, the site for body fluid accumulation, freely supplying self-antigens required for their maintenance. Anti-PtC appears to be naturally associated with the soluble antigen as previously suggested by sandwich staining, complexed with either surface BCR or secreted antibody [24]. We suggest that continual BCR signaling at some level by antigen or antigen/antibody complex may explain several features associated with peritoneal B-1 cells, possibly promoted by a unique cytokine milieu.

Role of CD5

Another possible reason for positive selection was the expression of CD5, as such CD5 may prevent apoptosis. However, expression of CD5, which occurs after surface IgM expression during development, does not appear to play a key role in the process of positive selection itself, since we found no significant reduction in serum ATA levels when analyzing $V_H3609\mu Tg$ mice on a CD5 knockout background (Fig. 4) similar to the results with anti-PtC transgenic mice(not shown). B-1 cell frequency also appears to be comparable in anti-PtC Tg B cell analyses (unpublished). Thus, it appears that CD5 is not a cause for positive selection, but rather a result. However, as found previously with established CD5$^+$ B cells, it is likely that induction of CD5 may have a role in determining the fate of autoreactive B-1 cells after activation [25], such as prevention of hyperreactivity. Our data in Fig. 4 may suggest at most a modest serum ATA increase in the CD5 deficient mice.

Immature B cell repertoire of B-1

Why are positive selection and induction of CD5 readily obsverved in B-1 development? CD5CD5$^+$ B cells are frequent in newborn mice as a result of B-1 development. Since up-regulation of CD5 appears to be a specific consequence of an antigen-receptor mediated signaling, it is reasonable to speculate that most CD5$^+$ B cells may be self/natural antigen reactive; i.e. an abundance of self-reactive immature B cells are generated from B-1 development. As shown in Fig. 5, when the immature B cell repertoire is constrained by exclusive transgene μ-heavy chain

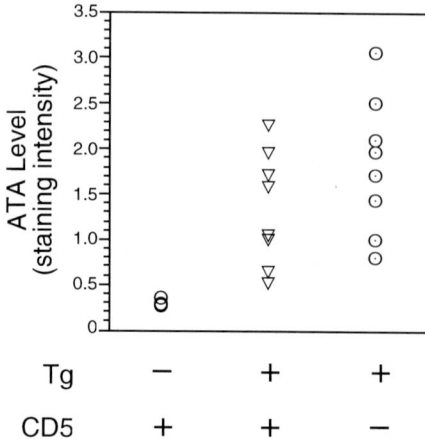

Fig. 4. CD5 deficiency does not affect positive seleciton of B-1. Serum ATA levels of 2-3 mo.old littermates from (3609Tgμ$^{+/-}$ x CD5$^{-/-}$) x CD5$^{-/-}$. Staining intensity by 1/10 diluted sera. n=8 for each group

expression, but with little chance to express ATA specificity due to diverse light chain association, then most B cells are CD5$^-$, even in the peritoneal cavity (6d, Tg+). This contrasts with non-transgenic littermate controls which allow expression of a much more diverse self-reactive repertoire by diverse H/L combinations, a large fraction of which express CD5 (6d, Tg–). Within a week, endogenous μ$^+$ B cells (with or without Tgμ co-expression) appear in transgenic mice which are mostly CD5$^+$, followed by CD5$^+$ ATA B cell (Tgμ$^+$ Endoμ$^-$) accumulation (14d, Tg+). These data may indicate that the neonatal immature B cell repertoire normally favors positively selectable ones, and that there is a strong preference to generate B cells with self-reactivity when the repertoire is genetically altered by transgenesis.

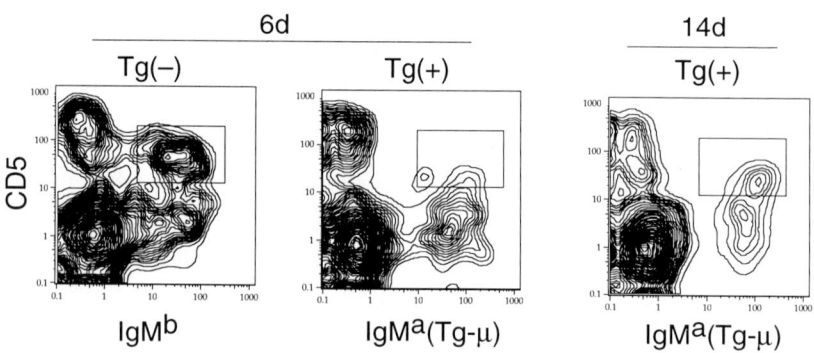

Fig. 5. Delay of CD5$^+$ B cell increase in VH3609μ Tg neonates compared with Tg$^-$ littermates. Peritoneal cavity washout cells from 6 and 14 day old Tg$^+$ or Tg$^-$ (3609μTg$^{+/-}$ x C.B-17) littermates were analysed for IgM allotype and CD5 expression. Most B cells in 6 day Tg$^+$ mice were CD5$^-$ Tgμ$^+$Endμ$^-$, in contrast to the high CD5$^+$ B cell frequency in Tg$^-$ control. Accumulation of ATA B cells became apparent by the increase of CD5$^+$ Tgμ$^+$ ATA B cells as shown in analysis of 14 day neonate (gated for Endo μ$^-$).

B-1 versus B-2, repertoire and selection: Concluding remarks

Lastly, it is possible that B cell tolerance susceptibility may differ between B-1 and B-2 even if the same specificity is expressed. BCR signaling at the immature stage of B-2 development leads to tolerance induction, anergy or deletion; otherwise the receptor is edited. Thus, B-2 cell development appears to proceed mostly by an "ignorance" mechanism, generating "B-2" cells as final products (Fig. 1). Whether the apparent difference at cellular selection stage between B-1 and B-2 is solely pre-determined by the difference in immature B cell repertoire remains unclear. Tolerance susceptibility and B cell receptor signaling thresholds can be modulated by co-receptors whose expression may change with ontogeny. Differences in the levels of several surface molecules on pre-B/immature B cells between the neonate and adult have been noted previously. These include lower MHC class II, Qa-2, complement receptor, and CD40 in early ontogeny, all of which have the potential to influence selection at the immature B cell stage [7, and unpublished data]. Whether B-1 and B-2 development involves differences in the mechanism of cellular selection, i.e. whether positive selection is a unique feature of B-1, is a critical question that needs to be answered to complete our understanding of the scheme presented in Fig. 1.

Acknowledgements

We thank Dr. M. Gui for ATA study and Ms. J. Dashoff for technical help. This work was supported in part by NIH grants, AI31412, AI26782, and AI40946.

References

1. Herzenberg LA (1992) CD5 B cells in development and disease. Ann. N.Y. Acad. Sci. 651:Preface
2. Hardy RR, Hayakawa K (1994) CD5 B cells, a fetal B cell lineage. Adv. Immunol. 55:297-339
3. Haughton G, Arnold LW, Whitmore AC, Clarke SH (1993) B-1 cells are made, not born. Immunol Today 14:84-7; discussion 87-91
4. Hayakawa K, Hardy RR Development and function of B-1 cells. Curr. Opi. Immunol. (in press):
5. Hardy RR, Hayakawa K (1991) A developmental switch in B lymphopoiesis. Proc. Natl. Acad. Sci. USA 88:11550-11554
6. Hayakawa K, Li Y-S, Wasserman R, Sauder S, Shinton S, Hardy R (1997) B lymphocyte developmental lineages. Ann. N.Y. Acad. Sci. 815:15-29
7. Hayakawa K, Tarlinton D, Hardy RR (1994) Absence of MHC class II expression distinguishes fetal from adult B lymphopoiesis in mice. J. Immunol. 152:4801-4807
8. Hayakawa K, Hardy RR, Herzenberg LA, Herzenberg LA (1985) Progenitors for Ly-1 B cells are distinct from progenitors for other B cells. J Exp Med 161:1554-68
9. Wasserman R, Li YS, Shinton SA, Carmack CE, Manser T, Wiest DL, Hayakawa K, Hardy RR (1998) A novel mechanism for B cell repertoire maturation based on response by B cell precursors to pre-B receptor assembly. J Exp Med 187:259-64
10. Pennell CA, Mercolino TJ, Grdina TA, Arnold LW, Haughton G, Clarke SH (1989) Biased immunoglobulin variable region gene expression by Ly-1 B cells due to clonal selection. Eur J Immunol 19:1289-95
11. Hardy RR, Carmack CE, Li YS, Hayakawa K (1994) Distinctive developmental origins and specificities of murine CD5$^+$ B cells. Immunol. Rev. 137:91-118

12. Arnold LW, Pennell CA, McCray SK, Clarke SH (1994) Development of B-1 cells: segregation of phosphatidyl choline-specific B cells to the B-1 population occurs after immunoglobulin gene expression. J Exp Med 179:1585-95
13. Lam KP, Rajewsky K (1999) B cell antigen receptor specificity and surface density together determine B-1 versus B-2 cell development. J Exp Med 190:471-8
14. Hayakawa K, Asano M, Shinton SA, Gui M, Allman D, Stewart CL, Silver J, Hardy RR (1999) Positive selection of natural autoreactive B cells. Science 285:113-116
15. Gui M, Wiest DL, Li J, Kappes D, Hardy RR, Hayakawa K (1999) Peripheral $CD4^+$ T cell maturation recognized by increased expression of thy-1/CD90 bearing the 6C10 carbohydrate epitope. J Immunol 163:4796-804
16. Hayakawa K, Carmack CE, Hyman R, Hardy RR (1990) Natural autoantibodies to thymocytes: origin, VH genes, fine specificities, and the role of Thy-1' glycoprotein. J. Exp. Med. 172:869-878
17. Gringhuis SI, de Leij LF, Coffer PJ, Vellenga E (1998) Signaling through CD5 activates a pathway involving phosphatidylinositol 3-kinase, Vav, and Rac1 in human mature T lymphocytes. Mol Cell Biol 18:1725-35
18. Perez-Villar JJ, Whitney GS, Bowen MA, Hewgill DH, Aruffo AA, Kanner SB (1999) CD5 negatively regulates the T-cell antigen receptor signal transduction pathway: involvement of SH2-containing phosphotyrosine phosphatase SHP-1. Mol Cell Biol 19:2903-12
19. Sen G, Bikah G, Venkataraman C, Bondada S (1999) Negative regulation of antigen receptor-mediated signaling by constitutive asociation of CD5 with the SHP-1 protein tyrosine phosphatase in B-1 B cells. Eur J Immunol 29:3319-28
20. Azzam HS, Grinberg A, Lui K, Shen H, Shores EW, Love PE (1998) CD5 expression is developmentally regulated by T cell receptor (TCR) signals and TCR avidity. J Exp Med 188:2301-11
21. Tarakhovsky A, Kanner SB, Hombach J, Ledbetter JA, Muller W, Killeen N, Rajewsky K (1995) A role for CD5 in TCR-mediated signal transduction and thymocyte selection. Science 269:535-7
22. Ying-zi C, Rabin E, Wortis HH (1991) Treatment of murine CD5-B cells with anti-Ig, but not LPS, induces surface CD5: two B cell activation pathways. Int. Immunol. 3:467
23. Tatu C, Ye J, Arnold LW, Clarke SH (1999) Selection at multiple checkpoints focuses V(H)12 B cell differentiation toward a single B-1 cell specificity. J Exp Med 190:903-14
24. Carmack CE, Shinton SA, Hayakawa K, Hardy RR (1990) Rearrangement and selection of $V_H 11$ in the Ly-1 B cell lineage. J. Exp. Med. 172:371-374
25. Bikah G, Carey J, Ciallella JR, Tarakhovsky A, Bondada S (1996) CD5-mediated negative regulation of antigen receptor-induced growth signals in B-1 B cells. Science 274:1906-9

II

B-1 Cells: Relationship to Pro B-Pre B Development and Negative Selection

Response by B cell precursors to pre-B receptor assembly: differences between fetal liver and bone marrow

R. R. Hardy, R. Wasserman[1], Y.-S. Li, S. A. Shinton, and K. Hayakawa

Institute for Cancer Research, Fox Chase Cancer Center, Philadelphia, PA 19111
[1] present address: Merck Research Laboratories, Merck & Co., Inc., West Point, PA

Summary The expression of different sets of immunoglobulin specificities by fetal and adult B lymphocytes is a longstanding puzzle in immunology. In the past few years it has become clear that production of μ heavy chain and subsequent assembly with surrogate light chain to form the pre-B cell receptor complex is critical to promote development of adult B cell precursors in mouse bone marrow. Recently we found that instead of promoting pre-B cell expansion as in adult bone marrow, this complex inhibits pre-B cell growth in fetal liver, providing a previously unrecognized mechanism for alteration of the B cell repertoire with age. The consequence is very distinct primary repertoires for development of fetal B1 cells and adult bone marrow B2 cells.

Immune responses during the neonatal period show significant differences from those in the adult, with striking deficiencies in ability to respond to certain antigens [1, 2], although the mechanism for this shift is as yet incompletely understood. Consistent with their restricted ability to respond to antigenic challenge, fetal and neonatal B cell precursors utilize a restricted set of heavy chain V_H genes, preferentially from segments proximal to J_H such as $V_H 81X$ in mice [3, 4]. The expansion from a restricted set of genes employed by neonatal B cells to the wide variety employed in the adult [5, 6] is referred to as "repertoire maturation" and while the phenomenon has been appreciated for many years, no adequate explanation has been provided. Since fetal B cells preferentially express V_H segments proximal to J_H, ordered accessibility to recombination has been suggested as a possibility. Yet this cannot be the full explanation, since preferential rearrangement of J-proximal V_H segments is also observed in B precursor cells in adult bone marrow, whereas productive rearrangements of such genes predominate only in fetal cells [7-10]. Furthermore, some V_H genes biased to B cells generated fetally/neonatally in mice are not J_H proximal, e.g., $V_H 11$ [11, 12]. These observations imply that an additional mechanism beyond rearrangement accessibility shapes the distinctive repertoires of fetal and adult times, possibly the differential control of B cell development by particular V_H genes.

Early B-lineage development is critically dependent on expression of Ig μ heavy chain at the pre-B stage, where it associates with surrogate light chain, composed of two molecules, λ5 and VpreB [13, 14], that collectively form a complex known as the pre-B cell receptor (pre-BCR). The importance of the pre-BCR in this process is illustrated by the developmental arrest (see Fig 1a for an example) induced upon elimination of any of these components in gene targeted mice

[15-17]. Successful pre-B receptor assembly induces several hallmark events associated with progression from the pro-B to large pre-B stage in the bone marrow [18], including down-regulated expression of genes involved in Ig rearrangement, such as TdT shown in Fig. 1b and the recombinase activating genes, Rag-1 and Rag-2 [19-21]. These changes coincide with a sharp proliferative expansion in bone marrow, at precisely the stage where representation of the fetally-biased V_H81X shows a profound decrease [8]. Thus it seems possible that the V region of the µ heavy chain itself is able to differentially regulate growth of B cell precursors in fetal and adult development. We have tested this idea by investigating differences in the efficiency of pre-BCRs containing differnt V_H genes to influence B cell progression and clonal expansion in fetal and adult life.

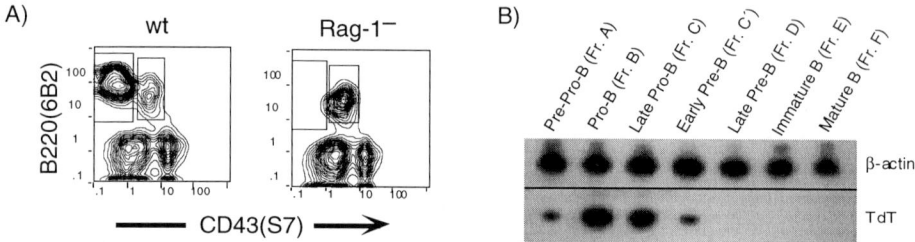

Fig. 1. a) Block in B cell development resulting from inability to rearrange µ heavy chain in Rag-1 deficient mice. Cells do not proceed past the B220+CD43+ pro-B stage. b) TdT is sharply down-regulated at the large pre-B stage where pre-B cell receptor is expressed.

Transfection of µ heavy chain into pro-B cell lines results in the down-regulation of TdT expression [20], providing a model system to investigate this issue. As Fig. 2 presents, expression of several µ heavy chains with different V_H (V_HDJ_H) regions, such as Sp6 and 3H9, in the pro-B cell line ret02/1 results in diminished message levels of TdT as well as Rag-1. These changes mimic those normally seen in the differentiation of pro-B cells to the early pre-B cell stage in vivo [19, 21]. However, not all µ heavy chains are equally efficient in this assay: transfectants with V_H81X or V_H11 µ chains do not show significant TdT or Rag-1 downregulation when compared to the parental line. As Previous work has shown that V_H81x µ chains frequently do not associate well with λ5 [22]. Our immunoprecipitation analysis with µ heavy chain and surrogate light chain components revealed that V_H11 was poorly assembled with VpreB in the transfectant [23].

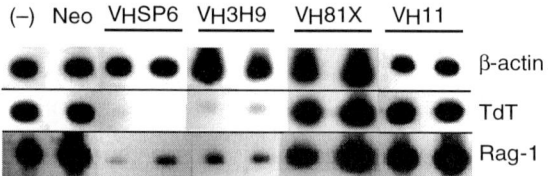

Fig. 2. Effect of µ heavy chains with different V regions on downregulation of TdT and Rag-1 message levels as revealed by RT-PCR. RNA prepared from the ret02/1 cell line parent (lane "–"), from a clone transfected (29) with neo alone ("neo"), and from two clones each generated by transfection with the four different V_H-µ constructs indicated. β-actin levels were used to compare efficiency of RNA/cDNA preparation.

Non-transformed pre-B cells developing in vivo also show a comparable dependence on particular V_H μ as shown in analysis of different lines of μ transgenic mice. To eliminate any effect by endogenous μ expression and to restrict analysis to the pre-B cell stage, we used Rag-1 deficient mice bearing Ig transgenes (Tg). As Fig. 3a shows, a differential ability to down-regulate Rag-2 in adult bone marrow was clearly evident when comparing low copy number BR5 μ with 3H9 μ Tg mice, in agreement with data from the cell line transfection experiments. In addition, a human μ Tg (hereafter referred to as "Hu μ"), previously shown to promote B cell development in mice [16], also induced down-regulated Rag-2 expression. Thus the extent of bone marrow pre-B cell progression in Ig transgenic mice, as monitored by changes in gene expression that we measured, appears to be dependent on V_H mediated pre-B receptor assembly, with $V_H 11$ being particularly ineffective.

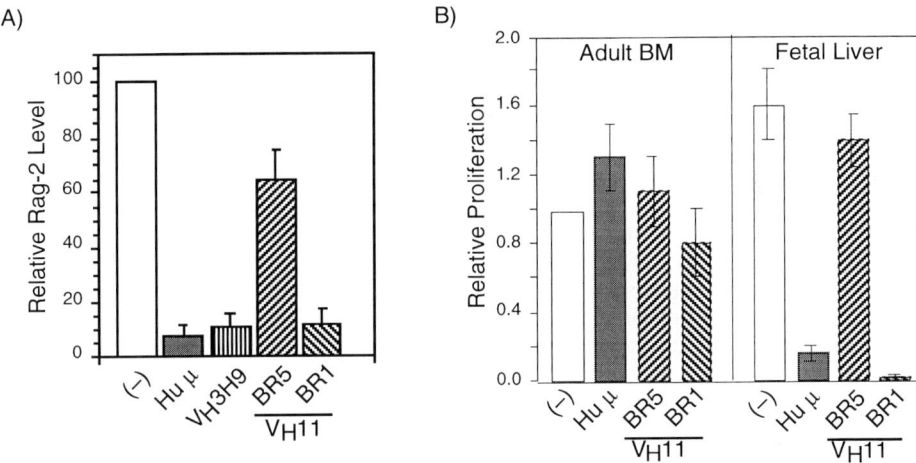

Fig. 3. a) Analysis of Rag-2 gene expression in pre-B cells from transgenic mouse lines. RNA prepared from B220+CD43+HSA+ cell fraction sorted from bone marrow of Rag-1– μ Tg mice. All values expressed relative to the maximum level seen with RNA from comparable fraction in Rag-1⁻ mice (lane "–"). b) Growth effect of various μ transgenes (Rag-1⁻ background) after four day stromal cell dependent cultures of B220⁺CD43⁺HSA⁺ cells sorted from bone marrow and fetal liver. Live cell number recovered after culture of Rag-1⁻ bone marrow (average 6-fold increase) is set to 1.0.

The high copy number $V_H 11$ μ Tg mouse line BR1 showed good Rag-2 down-regulation, different from the transfection data (Fig. 3a). Anti-μ immunoprecipitation analysis of pre-B cells isolated from bone marrow of the BR1 and BR5 $V_H 11$ Tg mouse pre-B cells showed that significant amounts of assembled pre-BCR was generated in the BR1 line, different from the BR5 line where inefficient assembly resembled the $V_H 11$ transfectants. Total μ protein in BR1 pre-B cells is much higher compared with BR5 and also significantly higher than that in normal pre-B cells from non-Tg BALB/c mice. Semi-quantitative RT-PCR analysis showed approximately three-fold greater message in BR1 pre-B cells compared with BR5 or non-transgenic mice (not shown). Thus the difference between BR5 and BR1may result from super-physiological over-expression of the "non-competent" $V_H 11$ μ, beyond the level achieved in our transfection experiments, such that some pre-B receptor assembly occurs.

We next tested the growth response of fetal and adult pre-B cells to μ expression in stromal cell culture (Fig. 3b), since a proliferative burst is another characteristic associated with early pre-B cell progression in the bone marrow. As Fig. 3a shows, analysis of short term cultures of pre-B cells sorted from bone marrow of competent (Hu μ or BR1 μ) and incompetent (BR5 μ) adult Tg mice (on a Rag-1⁻ background) revealed that transgene expression had relatively little effect on cell growth, with any enhancement in proliferation largely balanced by differentiation and exit from cell cycle. Striking, however, analysis of the comparable pre-B fraction isolated from fetal liver of the same transgenic mouse lines revealed that efficient generation of pre-BCR arrested cell growth. In contrast, the BR5 V_H11 line showed little inhibition of fetal liver B-lineage proliferation, instead allowing continued pre-B cell growth. This variation in growth response with ontogeny by pre-B cells to heavy chain - surrogate light chain assembly would be predicted to alter the representation of B cells in transgenic mice depending on the animal's age.

In fact, analysis of transgene expression in the BR5 line using mice at different ages is consistent with this prediction. As shown in Fig 4, more than half of the B cells in spleen of one week old animals are transgene positive (the rest expressing exclusively endogenous μ heavy chain), but by two months of age the transgene surface positive cells decrease to a small portion of the B cell pool, with the vast majority expressing only endogenous heavy chain on their cell surface. We assume that the bulk of the transgene positive cells in the adult animals, most of which co-express CD5 and bind to PtC self-antigen, are persisting cells selected from the fetal-generated population. Our results and data from other laboratories clearly show that a bias in V_H representation can occur after successful V_H-DJ_H rearrangement due to interaction (or lack thereof) with surrogate light chain [8, 24]. The additional point that we make here is that this assembly mediated response can differ between fetal and adult pre-B cells. Taken together, our data provide a reasonable explanation for why B cells expressing both V_H81X and V_H11 are preferentially produced during the fetal/neonatal period and predict a significant (and different) change in VDJ representation at the late pre-B cell stage during both fetal and adult B cell development.

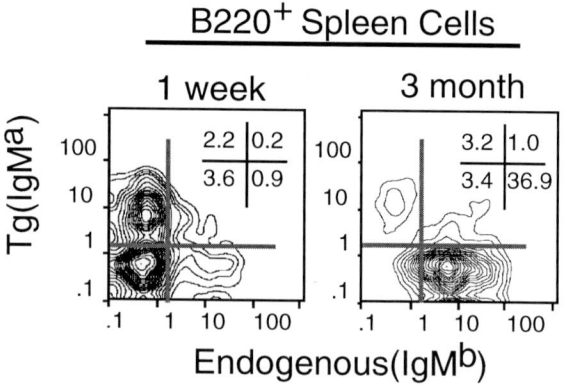

Fig. 4. Difference in VH11 transgene expressing B cell frequency with age. Flow cytometry analysis of B220+ cell in spleen of BR5 mice at 1 week and 3 months. Numbers show percent live cells in a lymphocyte size gate for indicated quadrants.

As animals mature from the fetal to the adult stage, the potential to generate a more diverse $V_H \mu$ repertoire increases due to TdT expression [25, 26]. This increasing heterogeneity of Ig heavy chain, important for generating more diversity in the adult, may require a more intricate mechanism for selection of "appropriate" $V_H \mu$, such as screening V_H structures for ability to pair with light chains [24, 27]. If this is the case, then surrogate light chain could provide a template for an "average" light chain structure. This process would be less important during fetal B lymphopoiesis where TdT is absent and homology mediated recombination is favored [28, 29], resulting in expression of a more restricted germline-encoded set of V_H regions. Thus a dependence of the pro-B to pre-B transition on efficient pre-B cell receptor assembly in adult mouse bone marrow could be viewed as an elaboration on a simpler mode of B cell development represented in fetal liver. We propose that over evolution, certain useful V_H regions have been selected into the germline repertoire and that the distinctive response of fetal pre-B cells to surrogate light chain association provides a mechanism for their preferential expression. Determining why fetal B cells should express a primordial repertoire distinct from the adult remains an interesting subject for future studies.

Acknowledgements

We thank Drs. M. Weigert, J. Kearney, and R. Perry for providing heavy chain constructs. This work was supported by a grant (AI-26782) from the NIH. Correspondence and requests for materials should be sent to R. R. Hardy, E-Mail rr_hardy@fccc.edu

References

1. Sigal NH, Pickard AR, Metcalf ES, Gearhart PJ, Klinman NR (1977) Expression of phosphorylcholine-specific B cells during murine development. J. Exp. Med. 146:933-48
2. Klinman N, Linton P (1988) The clonotype repertoire of B cell subpopulations. Adv. Immunol. 42:1
3. Yancopoulos GD, Desiderio SV, Paskind M, Kearney JF, Baltimore D, Alt FW (1984) Preferential utilization of the most J_H-proximal V_H gene segments in pre-B-cell lines. Nature 311:727-33
4. Perlmutter RM, Kearney JF, Chang SP, Hood LE (1985) Developmentally controlled expression of immunoglobulin V_H genes. Science 227:1597-601
5. Perlmutter RM (1987) Programmed development of the antibody repertoire. Curr. Top. Microbiol. Immunol. 135:95-109
6. Yancopoulos GD, Malynn BA, Alt FW (1988) Developmentally regulated and strain-specific expression of murine V_H gene families. J. Exp. Med. 168:417-35
7. Decker D, Boyle N, Klinman N (1991) Predominance of nonproductive rearrangements of $V_H 81X$ gene segments evidences a dependence of B cell clonal maturation on the structure of nascent H chains. J. Immunol. 147:1406
8. Decker DJ, Kline GH, Hayden TA, Zaharevitz SN, Klinman NR (1995) Heavy chain V gene-specific elimination of B cells during the pre-B cell to B cell transition. J. Immunol. 154:4924-35
9. Marshall AJ, Paige CJ, Wu GE (1997) V_H repertoire maturation during B cell development in vitro: Differential selection of Ig heavy chains by fetal and adult B cell progenitors. J. Immunol. 158:4282-4291
10. Marshall AJ, Wu GE, Paige CJ (1996) Frequency of $V_H 81X$ usage during B cell development: initial decline in usage is independent of Ig heavy chain surface expression. J. Immunol. 156:2077

11. Reininger L, Ollier P, Poncet P, Kaushik A, Jaton JC (1987) Novel V genes encode virtually identical variable regions of six murine monoclonal anti-bromelain-treated red blood cell autoantibodies. J. Immunol. 138:316-23
12. Hardy RR, Carmack CE, Shinton SA, Riblet RJ, Hayakawa K (1989) A single V_H gene is utilized predominantly in anti-BrMRBC hybridomas derived from purified Ly-1 B cells. Definition of the V_H11 family. J. Immunol. 142:3643-51
13. Melchers F, Strasser A, Bauer SR, Kudo A, Thalmann P, Rolink A (1989) Cellular stages and molecular steps of murine B-cell development. Cold Spring Harb. Symp. Quant. Biol. 1:183-9
14. Karasuyama H, Kudo A, Melchers F (1990) The proteins encoded by the VpreB and lambda 5 pre-B cell-specific genes can associate with each other and with mu heavy chain. J. Exp. Med. 172:969-72
15. Reichman-Fried M, Hardy RR, Bosma MJ (1990) Development of B-lineage cells in the bone marrow of scid mice following the introduction of functionally rearranged immunoglobulin transgenes. Proc. Natl. Acad. Sci., USA 87:2730-2739
16. Spanopoulou E, Roman CA, Corcoran LM, Schlissel MS, Silver DP, Nemazee D, Nussenzweig MC, Shinton SA, Hardy RR, Baltimore D (1994) Functional immunoglobulin transgenes guide ordered B-cell differentiation in Rag-1-deficient mice. Genes & Dev. 8:1030-42
17. Kitamara D, Kudo A, Schaal S, Muller W, Melchers F, Rajewsky K (1992) A critical role of λ5 protein in B cell development. Cell 69:823-831
18. Hardy RR, Carmack CE, Shinton SA, Kemp JD, Hayakawa K (1991) Resolution and characterization of pro-B and pre-pro-B cell stages in normal mouse bone marrow. J. Exp. Med. 173:1213-25
19. Li YS, Hayakawa K, Hardy RR (1993) The regulated expression of B lineage associated genes during B cell differentiation in bone marrow and fetal liver. J. Exp. Med. 178:951-60
20. Wasserman R, Li YS, Hardy RR (1997) Down-regulation of terminal deoxynucleotidyl transferase by Ig heavy chain in B lineage cells. J. Immunol. 158:1133-8
21. Chang Y, Bosma GC, Bosma MJ (1995) Development of B cells in scid mice with immunoglobulin transgenes: implications for the control of V(D)J recombination. Immunity 2:607-16
22. Keyna U, Beck-Engeser GB, Jongstra J, Applequist SE, Jack HM (1995) Surrogate light chain-dependent selection of Ig heavy chain V regions. J. Immunol. 155:5536-42
23. Wasserman R, Li YS, Shinton SA, Carmack CE, Manser T, Wiest DL, Hayakawa K, Hardy RR (1998) A novel mechanism for B cell repertoire maturation based on response by B cell precursors to pre-B receptor assembly. J Exp Med 187:259-64
24. ten Boekel E, Melchers F, Rolink AG (1997) Changes in the V(H) gene repertoire of developing precursor B lymphocytes in mouse bone marrow mediated by the pre-B cell receptor. Immunity 7:357-68
25. Landau NR, Schatz DG, Rosa M, Baltimore D (1987) Increased frequency of N-region insertion in a murine pre-B-cell line infected with a terminal deoxynucleotidyl transferase retroviral expression vector. Mol. Cell. Biol. 7:3237-43
26. Gilfillan S, Dierich A, Lemeur M, Benoist C, Mathis D (1993) Mice lacking TdT: mature animals with an immature lymphocyte repertoire. Science 261:1175-8
27. Kline GH, Hartwell L, Beck-Engeser GB, Keyna U, Zaharevitz S, Klinman NR, Jack HM (1998) Pre-B cell receptor-mediated selection of pre-B cells synthesizing functional mu heavy chains. J Immunol 161:1608-18
28. Chukwuocha RU, Feeney AJ (1993) Role of homology-directed recombination: predominantly productive rearrangements of V_H81X in newborns but not in adults. Molec. Immunol. 30:1473-9
29. Gerstein RM, Lieber MR (1993) Extent to which homology can constrain coding exon junctional diversity in V(D)J recombination. Nature 363:625-7

B cell production and turnover in CBA/Ca, CBA/N and CBA/N-*bcl-2* transgenic mice: *xid*-mediated failure among pre B cells is unaltered by *bcl-2* overexpression.

M. P. Cancro[1], A. P. Sah[1], S. L. Levy[1], D. M. Allman[1], D. Constantinescu[1], M. R. Schmidt[2], and R.T. Woodland[2]

[1]Univ. of Pennsylvania School of Medicine, Philadelphia, PA 19104
[2]Univ. of Massachusetts Medical Center, Worcester, MA 01655

Abstract

The CBA/N strain carries *xid*, a murine *btk* missense mutation that reduces peripheral B cell numbers. Using *in vivo* BrdU labeling and cytofluorimetry, we have compared the magnitude, production rates, and turnover rates of each B lineage subset in the marrow and periphery of CBA/Ca and CBA/N mice. Our results show the pro- B compartment is largely unaffected by *xid*. In contrast, the pre-B cell pool is markedly reduced, reflecting a diminished production rate and unaltered turnover time. Despite diminished pre-B cell formation, the size of the immature B cell pool is relatively normal in CBA/N mice, due to increased proportional survival of pre-B cells. In addition, we have assessed the marrow and peripheral B cell subsets of CBA/N mice transgenic for *bcl-2*. These results indicate that while the *bcl-2* transgene promotes lengthened survival in most B cell subsets, the pro/pre-B cell losses mediated by *xid* are not abrogated by b^1-2 overexpression. Taken together, these findings suggest that the initial point of action for *xid* is during transit from the pro- to pre-B cell pools, and that anomalies in subsequent compartments likely reflects the action of homeostatic mechanisms compensating for compromised pre-B cell production.

Introduction

Bruton's tyrosine kinase (Btk) is a signaling molecule critical for human B cell development. Male patients with *btk* mutations have a severe B cell immunodeficiency, X-linked agammaglobulinemia (XLA), characterized by reduced numbers of mature circulating B cells, diminished serum immunoglobulin levels, and disrupted secondary lymphoid architecture (Reviewed in Conley *et al.*, 1994). While the commitment of hematopoietic stem cells to the B cell lineage seems normal in XLA, B cell development is arrested at the cytoplasmic μ^+ ($c\mu^+$) pre-B cell stage. Mice also have a B lineage specific, X-linked immune defect (*xid*) originally described in the CBA/N strain. Although *btk* is affected in both diseases, B cell depletion in murine XID is less severe than that seen in human XLA; *xid* mice have a half to a third the conventional peripheral follicular B cells (B2 cells) of normal mice and lack peritoneal B1 subpopulations (Hardy *et al.*, 1984; Scher, 1982). A similar phenotype is found when *btk* is deleted by gene targeting (Khan *et al.*, 1995). Because Btk is expressed at all stages of B cell development (Weers *et al.*, 1993), this defect may act independently at multiple points in B cell differentiation and activation.

In XLA, the *btk* mutation acts at the pro- to pre-B transition. Determining the critical site(s) of action for the murine Btk, however, has proven more difficult. Btk is involved in multiple signaling pathways (Go *et al.*, 1990; Hitoshi *et al.*, 1993; Kurosaki, 1997; Miyake *et al.*, 1994; Santos-Argumedo *et al.*, 1995) and XID inhibits activation of peripheral B cells. Btk signaling defects may affect the late stages of B cell development. For instance, *xid* mice exhibit a higher than normal proportion of immature B cells (sIgM^{hi}, sIgD^{lo}) in the spleen and reduced numbers of mature recirculating lymph node B cells, supporting the contention that XID restricts entry of immature peripheral B cells into the long-lived mature B cell pool (Klaus *et al.*, 1997). In contrast to XLA, the importance of XID-mediated developmental effects in the BM have been minimized by the lack of striking numerical differences when *xid* or *btk* knockout mice are compared to wild type (Forrester *et al.*, 1987; Khan *et al.*, 1995; Reid and Osmond, 1985).

To probe the mechanism of *xid*-mediated B cell deficiency, we have examined the size, turnover, and production rates of marrow and splenic B-lineage populations among *xid* and wild type mice. In addition, we also examined mice in which a human*bcl-2* transgene (tg) was introduced into the *xid* background (Woodland *et al.*, 1996).

Methods

Mice and cell suspensions: CBA/Ca and CBA/N (*xid*) mice were obtained from the Jackson Laboratory Bar Harbor, ME or from the small animal production unit of the National Cancer Institute, Frederick MD. Normal mice expressing a human *bcl-2* tg were originally obtained from Dr. Stanley Korsmeyer (Wash. U. Medical School, St. Louis) and were maintained at UMMC by passage on a CBA/Ca background. Mating CBA/N females with transgenic males from CBA/N x CBA/Ca *bcl-2* crosses produced mice coexpressing the CBA/N defect and *hubcl-2* (*xid/bcl-2*). Splenic and bone marrow lymphocyte suspensions were prepared as described previously (Allman *et al.*, 1992; Lentz *et al.*, 1996).

Antibodies to cell surface antigens and immunofluorescent analyses: The following reagents were purchased from PharMingen (San Diego, CA): phycoerythrin (PE) and FITC-labeled anti-CD24 (heat stable antigen) (M1/69); PE-labeled anti-CD43 (leukosialin) (S7); APC and PE-labeled anti-CD45R (B220) (RA3-6B2). Biotin-labeled goat anti-mouse IgM, PE-labeled anti-IgD (SBA-1), and streptavidin-alkaline phosphatase were purchased from Southern Biotechnology (Birmingham, AL). FITC-labeled anti-BrdU (B44) was purchased from Becton Dickinson (San Jose, CA). Cell surface staining was done as described (Allman *et al.*, 1992; Lentz *et al.*, 1996)

BrdU labeling and cell cycle analyses: Cells from BrdU-treated mice were stained for surface markers and BrdU incorporation as previously described (Allman, 1993, Lentz *et al.*, 1996). Cytometric analyses were performed by gating on all nucleated cells. The percentage of BrdU-labeled cells in each subset was measured by cytofluorimetry and multiplied by the total cells in the subset to give the number of labeled cells. The means for these percentages and numbers were plotted versus time, and least squares regression done to obtain the turnover and production rates respectively. Cell cycle progression in bone marrow subpopulations from CBA/Ca, CBA/N, normal/*bcl-2* and *xid/bcl-2* donors was determined by propidium iodide staining and FACS analysis.

Determination of huBcl-2 expression in bone marrow populations: Extracts from FACS separated bone marrow cell fractions, or transgene positive splenic B cells were assessed for HuBcl-2 expression by Western blotting as previously described (Hockenbery et al., 1990; Woodland et al., 1996).

Spontaneous apoptosis in bone marrow cell cultures: Survival kinetics of FACS isolated pro, pre, immature and mature BM B cells from *xid* and *xid/bcl-2* donors were determined for cultured cells. Thirty thousand cells from each subpopulation were cultured in 96 well round bottomed tissue culture dishes in 100µl of complete medium containing: RPMI-1640, 10% fetal calf serum, 2mM glutamine, 50µM 2-mercaptoethanol, 100µg/ml streptomycin, 100u/ml penicillin, and 1x MEM nonessential amino acids. Cell viability was determined daily for 5 days by trypan dye exclusion.

Results

Magnitude of marrow B cell subsets in CBA/Ca, CBA/N, and xid/bcl-2 mice: The numbers of each B lineage subset in the marrow and spleen were determined. The number of pro B cells ($B220^{lo}$, $CD43^+$, $sIgM^-$) is similar in all three strains (Table 1). Thus neither XID nor introduction of the *bcl-2* transgene alters B lineage commitment. In contrast, pre B cell ($B220^{lo}$, $CD43^-$, $sIgM^-$) numbers differed. Marrow from CBA/N mice yielded a mean of 3.2 million pre-B cells, significantly fewer than the 6.5 million found in CBA/Ca. Introduction of the *bcl-2* tg restored pre-B cell numbers in *xid/bcl-2* donors to a mean of 6.3 million cells. The number of immature ($B220^{lo}$, $CD43^-$, $sIgM^+$) marrow B cells was similar in all strains, ranging from 1.1 to 1.5 million. Thus, despite the reduction in pre-B cells induced by *xid*, compensatory homeostatic mechanisms appear to replenish the immature B cell compartment in CBA/N mice. Finally, CBA/N marrow contained only half as many recirculating mature B cells as CBA/Ca (~0.6 vs. ~1.2 million per 2 hind-limbs), whereas *xid/bcl-2*tg mice had 17 million mature recirculating B cells.

Table 1. Magnitude of bone marrow subsets in CBA and transgenic mice

		Bone marrow subsets		
	Strain	Pro-B	Pre-B	Immature B
	CBA/Ca	1.2±0.3	6.5±1.8	1.5±0.6
Number of Cells[1] ($\times 10^{-6}$)	CBA/N	1.3±0.6	3.2±2.1*[2]	1.1±0.8
	CBA/N *bcl-2* tg+	1.5±0.5	6.3±1.8	1.3±0.5

[1] Bone marrow was harvested and stained for sIgM CD43, and B220 as described in *Methods*. Numbers given are mean total cells per 2 hind limbs, and represent the means of at least 15 for each strain.
[2] Means were compared using Student's t-test; *, $P< 0.01$

Table 2. Magnitude of splenic subsets in CBA and transgenic mice

		Splenic B cell subsets	
	Strain	Immature	Mature
Number of Cells[1] ($\times 10^{-6}$)	CBA/Ca	8.4	45
	CBA/N	6.9	16*[2]
	CBA/N bcl-2 tg+	28.4*	110*

[1] Splenocytes were harvested and stained as described in *Methods*. Values given are total number of each subset per spleen, and represent means of at least 10 mice.
[2] Means were compared using Student's t-test; *, $P < 0.01$

Magnitude of splenic B cell subsets in CBA/Ca, CBA/N, and *xid/bcl-2* mice: Newly emerging B cells exit the bone marrow and spend several days in the periphery before final maturation and entry to the long-lived mature B cell pool (Allman et al., 1992; Allman et al., 1993). Numbers of immature ($B220^{lo}HSA^{hi}sIgM^{hi}$) splenic *xid* B cells were slightly reduced relative to wild type, but this difference was not statistically significant. The number of peripheral immature B cells in *xid/bcl-2* mice was higher than that found in either CBA/N or CBA/Ca. Despite the similarity of immature peripheral B cells pools in CBA/N and CBA/Ca, the number of mature splenic B cells differed substantially among all strains. CBA/N mice had only one-third the number of mature splenic B cells found in normal CBA/Ca mice. In contrast, *xid/bcl-2* mice had more than twice the number of mature B cells found in normal mice, and more than six fold that in *xid*s.

Transgene expression and function in BM subsets: Because ectopic expression of the *hubcl-2* transgene seemed to affect some subpopulations but not others, we wanted to exclude biased transgene expression as a basis for our results. Thus we determined transgene protein production in FACS selected pro, pre, immature and mature marrow B cells by Western blot. The results (data not shown) indicated the HuBcl-2 protein is expressed in all BM subpopulations. With the exception of pro-B cells, the amount of transgene was similar for marrow and splenic B cells. Functionally, the HuBcl-2 produced was sufficient to significantly retard spontaneous apoptosis in cultures of all marrow B cell subpopulations (Table 3).

Table 3. Spontaneous apoptosis of cultured BM subpopulations[1]

Day of Culture	Pro-B	Pre-B	Immature B	Mature B
2	42/78%[2]	46/87	12/86	27/80
3	28/68	6/83	4/90	4/72
5	6/49	1/46	1/62	2/60

[1] Survival of FACS isolated BM B cell subsets were determined as described in *Methods*. Viability of input cells ranged from 92-99%.
[2] Each entry gives the mean percent viable cells from *xid* vs. *xid/hubcl-2* mice.

Table 4. Kinetics of bone marrow subsets in CBA strain and transgenic mice

	Strain	Bone marrow subsets		
		Pro-B	Pre-B	Immature B
Renewal Rate (% of pool/day)[1]	CBA/Ca	35	36	26
	CBA/N	33	35	27
	CBA/N bcl-2 tg+	28	9*[2]	10*
Production rate (cells/day x 10^{-6})[1]	CBA/Ca	0.40	2.1	0.28
	CBA/N	0.25	0.5*	0.22
	CBA/N bcl-2 tg+	0.30	0.7*	0.13*

[1] BrdU labeling and analyses were performed as described in *Methods*. Rates were determined by least squares regression within the linear portion of each labeling curve
[3] Labeling rates were compared using Student's t-test; *, P< 0.01

Turnover and production rates in bone marrow B lineage subsets of CBA/Ca, CBA/N, and *bcl-2* transgenic CBA/N mice: The influence of *xid* on marrow B cell production and turnover was determined by comparing the *in vivo* BrdU labeling rates among marrow B lineage subsets in CBA/Ca and CBA/N mice. The linear portions of proportional and absolute labeling curves were use to estimate renewal and production rates respectively. Because of the differences in marrow cell numbers across the three strains studied, the absolute labeling rates are expressed in terms of the cells recovered per two hind limbs.

Both the renewal and production rates of pro-B cells are similar in CBA/Ca and CBA/N mice (Table 4). In both strains, the pro-B compartment has a renewal rate of ~35% per day, corresponding to a production rate in the range of ~0.3 million cells per day per two hind limbs. This finding is in accord with the similar numbers of pro-B cells observed in the CBA/Ca and CBA/N strains, and further strengthens the conclusion that *xid* has little influence on B lineage commitment.

In contrast, pre-B cell labeling kinetics in CBA/Ca *vs.* CBA/N mice differ significantly. CBA/Ca adults generate ~ 2 million pre-B cells daily; whereas CBA/N individuals produce only ~0.5 million each day (Table 4). Nonetheless, the pre-B cell renewal rates are identical, showing that the loss rate within the pre-B cell pool is unaffected by *xid*, since increased attrition would yield a more rapid renewal rate. This indicates XID mediates excessive losses among cells during transit from the pro-B to the pre-B cell stages. The basis for this is currently under investigation. In spite of this four-fold reduction in pre-B cell generation, neither the generation rate nor turnover rates of immature marrow B cells differ significantly between CBA/Ca and CBA/N. The relatively normal size and kinetics of immature marrow B cells in CBA/N mice indicates that a larger proportion of the pre-B cells generated successfully transit to the immature compartment, likely due to yet undefined homeostatic mechanisms.

Ectopic expression of the *bcl-2* transgene restored the pre-B cell compartment of *xid/bcl-2* mice to that of wild type mice. This could reflect a reversal of the XID-mediated attrition at the pro- to pre- B cell transition, or instead indicate changes in the turnover rates of pre-B cells. These possibilities were distinguished by determining the turnover and production rates of B lineage subsets in *xid/bcl-2* mice. The renewal and generation rates of pro-B cells in *xid/bcl-2* mice are similar to CBA/N and CBA/Ca (Table 4). Since Bcl-2 is normally expressed within pro-B cells (Li et al., 1993), it is not surprising that ectopic Bcl-2 expression had little effect on cellular turnover within this population. In contrast, the absolute labeling rate indicated that the production of *xid/bcl-2* pre-B cells was similar to that in CBA/N, but is only 25 to 40% that observed in CBA/Ca (Table 4). Thus, HuBcl-2 overexpression does not ameliorate XID-mediated cell loss at the pro- to pre-B transition, otherwise production rates equivalent to CBA/Ca would have been observed. However, the cellular dynamics within the compartment were profoundly affected by Bcl-2 expression: pre-B cells in *xid/bcl-2* transgenic mice were renewed at a rate of only 9% per day, which differs markedly from the 35% per day observed in both CBA/Ca and CBA/N mice. Indeed, at this renewal rate it takes ~ 6 days to achieve 50% labeling compared with ~36 hours for either CBA/N or CBA/Ca. These results suggest that *bcl-2* restores the pre-B cell compartment by lengthening the survival and/or residency time within the pre-B cell pool, rather than by reversing the effects of *xid* per se. Furthermore, while differences in cellular turnover of pre-B cells from CBA/N versus CBA/Ca were readily detectable, renewal and production rates of immature BM B cells were identical in both strains. Together, these data support the view that Btk functions specifically in a pathway required for differentiation at the late pro and early pre-B cell stages.

The BM immature B cell pool shows a similar pattern (Table 4), in that the renewal rate is significantly slowed in the bcl-2 tg, although the entrance rate is marginally changed. Thus, the lengthened survival mediated by Hubcl-2 overexpression extends across all marrow differentiation stages after pro-B cells.

Kinetics of splenic B cell subsets in normal, *xid*, and transgenic mice:. While our data show an early point of Btk action at the pro- to pre-B stages, our finding that compensatory mechanisms yield an immature marrow pool of normal magnitude makes it difficult to accommodate the reduced mature peripheral B cell numbers unless a second, extramedullary, point of action also exists. We therefore determined renewal and production rates of splenic B cell compartments (Table 5).

Table 5. Kinetics of splenic subsets in CBA and transgenic mice

		Splenic B cell subsets	
	Strain	Immature	Mature
Renewal Rate (% of pool/day)[1]	CBA/Ca	3.2	1.2
	CBA/N	3.3	1.7
	CBA/N *bcl-2* tg+	1.9*[2]	0.8*
Production rate (cells/day x 10^{-6})[1]	CBA/Ca	0.3	0.2
	CBA/N	0.2	0.1*
	CBA/N *bcl-2* tg+	0.6*	0.8*

[1] BrdU labeling and analyses were performed as described in *Methods*. Rates were determined by least squares regression within the linear portion of each labeling curve
[3] Labeling rates were compared using Student's t-test; *, P< 0.01

Neither the number, renewal rate, nor production rate of the immature peripheral B cells differ between CBA/N and CBA/Ca strains. This indicates that the compensatory mechanisms that restore the magnitude of the immature marrow pool are sufficient to provide a normal rate of marrow egress, suggesting Btk has minimal impact upon this process. In contrast, the production rate, and consequently the steady-state number, of mature peripheral B cells are significantly lower in CBA/N mice than in CBA/Ca. Since the renewal rates between these strains are similar, *xid* likely alters the threshold of an event requisite for recruitment into the long-lived mature peripheral pool, such that fewer cells transit this stage successfully, but those that do exhibit a relatively normal average lifespan. The substantially increased size of both the immature and mature peripheral B cell compartments in the CBA/N-*bcl-2* transgenic mice appears due to lengthened lifespan, as evidenced by the reduced renewal rate; and a correspondingly increased production rates in both pools. These results suggest that, unlike the XID-mediated failure at the pro- to pre B cell transit that resists ectopic bcl-2 expression; the XID-mediated deficit in survival at the transition from immature to mature peripheral pools is largely ameliorated by *bcl-2* expression.

Discussion

Our results show that *xid* produces inordinate cell losses at the early pre-B cell stage, thus placing the earliest manifestation of the murine *btk* mutation at the same developmental stage as that in human XLA. Surprisingly, we find nearly normal numbers of immature marrow and immature peripheral B cells, likely reflecting the action of compensatory homeostatic mechanisms controlling the magnitude of the immature B cell pool. Consistent with the findings of others (Klaus *et al.*, 1997; Loder *et al.*, 1999), we also observe an XID-mediated deficit in the transit from immature peripheral to mature peripheral pools, which reduces the mature B cell production rate and decreases mature peripheral B cell numbers. Finally, ectopic Bcl-2 expression restores both the pre-B cell numbers in the marrow and mature B cell numbers in the periphery. However, for pre-B cells this reflects increased survival rather than reversal of the *xid* lesion *per se*; whereas the expanded mature B cell population comes about by extending the lifespan of both immature and mature peripheral B cells, presumably facilitating their entry into and maintenance within the long lived B cell pool.

A striking feature of our results is the remarkably consistent number of immature B cells in the marrow and periphery, suggesting that strong homeostatic mechanisms operate to insure an immature compartment of consistent size. In the CBA/N, the *xid* lesion severely reduces the number of pre-B cells available, yet the immature bone marrow population is restored to near normal magnitude by enhanced proportional survival at the pre-B to immature B transit. Conversely, while the *hubcl-2* transgene affords lengthened lifespan within the pre- and immature pools, the proportional survival of pre-B cells in this transit is reduced, again yielding steady state numbers of immature marrow B cells. Thus, the numbers of cells admitted to the immature B cell pool seem highly plastic, and largely dictated by mechanisms aimed at maintaining relatively constant steady state numbers regardless of influx from the pre B pool or efflux to the immature peripheral pool. This finding requires that either a large proportion of cell death within this pool reflects neglect rather than negative selection, or that the threshold for negative selection be variable based on counteracting positive signals.

References

Allman, D. A., Ferguson, S. E., and Cancro, M. P. (1992). Peripheral B cell maturation 1.Immature peripheral B cells in adults are heat-stable antigen^{hi} and exhibit unique signaling characteristics. J. Immunol. 149, 2533-2540.

Allman, D. M., Ferguson, S. E., Lentz, V. M., and Cancro, M. P. (1993). Peripheral B cell maturation II. Heat-stable antigen^{hi} splenic B cells are an immature developmental intermediate in the production of long-lived marrow-derived B cells. J. Immunol. 151, 4431-4444.

Conley, M. E., Parolini, O., Rohrer, J., and Campana, D. (1994). X-linked agammaglobulinemia: New approaches to old questions based on the identification of the defective gene. Immunol. Rev. 138, 5-22.

Forrester, L. M., Ansell, J. D., and Micklem, H. S. (1987). Development of B lymphocytes in mice heterozygous for the X-linked immunodeficiency (xid) mutation: xid inhibits development of all splenic and lymph node B cells at a stage subsequent to their initial formation in bone marrow J. Exp. Med. 165, 949-58.

Go, N. F., Castle, B. E., Barrett, R., Kastelein, R., Cang, W., Mossmann, T. R., Moore, K. W., and Howard, M. (1990). Interleukin 10, a novel B cell stimulatory factor: Unresponsiveness of X chromosome-linked immunodeficient B cells J. Exp. Med. 172, 1625-1632.

Hardy, R. R., Hayakawa, K., Parks, D. R., Herzenberg, L. A., and Herzenberg, L. A. (1984). Murine B cell differentiation lineages. J. Exp. Med. 159, 1169-1178.

Hitoshi, Y., Sonoda, E., Kikuchi, E., Yonehara, S., Nakauchi, H., and Takatsu, K. (1993). Il-5 receptor positive B cells, but not eosinophils, are functionally and numerically influenced in mice carrying the X-linked immune defect. Int. Immunol. 5, 1183-1190.

Hockenbery, D., Nunez, G., Milliman, C., Schreiber, R. D., and Korsmeyer, S. J. (1990). Bcl-2 is an inner mitochondrial membrane protein that blocks programmed cell death. Nature 348, 334-6.

Khan, W., Alt, F., Gerstein, R. M., Malynn, B. A., Larsson, I., Rathbun, G., Davidson, L., Muller, S., Kantor, A T. (1997). Curr. Opin. Immunol. 9, 309-18.

Lentz, V. M., Cancro, M. P., Nashold, F. E., and Hayes, C. (1996). Bcmd governs recruitment of new B cells into the stable peripheral B cell pool in the A/WySnJ mouse J. Immunol. 157, 598-606.

Li, Y. S., Hayakawa, K., and Hardy, R. R. (1993). The regulation of B lineage associated genes during B cell differentiation in bone marrow and fetal liver J. Exp. Med. 178, 951-60.

Loder, F., Mutschler, B., Ray, R. J., Paige, C. J., Sideras, P. Torres, R., Lamers, C. and Carsetti R. B Cell Development in the Spleen Takes Place in Discrete Steps and Is Determined by the Quality of B Cell Receptor–derived Signals (1999) J. Exp. Med. 1999 190: 75-90.

Miyake, K., Yamashita, Y., Hitoshi, Y., Takatsu, K., and Kimoto, M. (1994). Murine B cell proliferation and protection from apoptosis with an antibody against a 105-kD molecule: unresponsiveness of X-linked immunodeficient B cells J. Exp. Med. 180, 1217-24.

Reid, G. K., and Osmond, D. G. (1985). B lymphocyte production in the bone marrow of mice with X-linked immunodeficiency (xid) J Immunol 135, 2299-302.

Santos-Argumedo, L., Lund, F. E., Heath, A. W., Solvason, N., Wu, W. W., Grimaldi, J. C., Parkhouse, R. M., and Howard, M. CD38 unresponsiveness of xid B cells implicates Bruton's tyrosine kinase (btk) as a regular of CD38 induced signal transduction (1995). Int. Immunol. 7, 163-70.

Scher, I. (1982). The CBA/N mouse strain: An experimental model illustrating the influence of the X-chromosome on immunity Adv. Immunol. 33, 1-53.

Weers, M. d., Verschuren, M. C. M., Kraakman, M. E. M., Mensink, R. G. J., Schuurman, R. K. B., Dongen, J. J. M. v., and Hendriks, R. W. (1993). The Bruton's tyrosine kinase gene is expressed throughout B cell differentiation, from early precursor B cell stages preceding immunoglobulin gene rearrangement up to mature B cell stages. Eur. J. Immunol. 23, 3109-3114.

Woodland, R. T., Schmidt, M. R., Korsmeyer, S., and Gravel, K. A. (1996). Regulation of B cell survival in xid mice by the protooncogene bcl-2 J. Immunol. 156, 2143-2154.

Positive Selection of Low Affinity Autoreactive B Cells

J. J. Kenny, A. Lustig, and D. L. Longo
B Cell Development Section, Laboratory of Immunology, Gerontology Research Center, National Institute on Aging, National Institutes of Health, 5600 Nathan Shock Drive, Baltimore, MD 21224

Introduction

B cells that produce autoreactive Ig-receptors are generally tolerized. When the autoreactive Ig-receptor is extensively crosslinked by a membrane displayed self-antigen, the B cells undergo clonal deletion in the bone marrow. If the autoantigen is a soluble antigen or the receptors are not extensively crosslinked, the B cells down regulate their IgM receptors but still migrate to the peripheral lymphoid organs. These autoreactive B cells are generally short-lived in the periphery. They fail to compete with normal B cells in homing to the follicle and appear to be trapped by binding to antigen in the T-cell areas of the lymphoid tissue (B cell tolerance is reviewed in ref. [1]. Autoreactive B cells can escape tolerance by editing their light (L) chain. Binding to self-antigen can reactivate the recombinase activating genes (Rag) and the L-chain gene can be replaced [2,3]. This editing process replaces the original L-chain with one that can pair with the heavy (H) chain to form a receptor that binds an antigen other than a self antigen [4,5].

We have previously presented data from x-linked immune deficient (Xid) mice indicating that phosphocholine- (PC)-specific M167-idiotype positive (id[+]) B cells behave like autoreactive B cells in that they are clonally deleted in the absence of a functional Bruton's tyrosine kinase (btk) signaling pathway while normal mice positively select these B cells into peripheral lymphoid tissues [6,7]. Positive selection of these autoreactive M167-id[+] B cell was demonstrated in mice expressing only the M167H chain [8]. In these mice 5 to 10% of the total splenic B cells expressed an endogenous κ24 L-chain resulting in a 50 to 100 fold higher production of PC-specific, M167-id[+] B cells than would be expected from random rearrangement and expression of the κ24 L-chain.

In Rag knockout (KO) mice, B cells expressing rearranged H and L chain transgenes encoding autoreactive Ig-receptors cannot edit their transgene encoded Ig-receptors because they cannot rearrange their endogenous Ig-genes. We have recently demonstrated [9] that PC-specific B cells that utilize M167Hκ24L chains fail to develop when these genes are expressed in Rag mice. The developmental arrest of these M167-id[+] B cells in Rag mice confirmed that they were autoreactive and suggested that in normal mice these B cells must escape developmental arrest by rearranging and expressing more than one H or L chain. Rather than replacing its L-chain, the B cell escapes tolerance by a process of receptor dilution; thus, the number of anti-PC specific receptors on an individual B cell is lowered below the threshold for tolerance induction. We tested this hypothesis by showing that M167-id$^+$ PC-specific B cells were rescued in Rag mice in which we expressed a second L-chain (λ L-chain) in addition to the κ24L chain. The association of the M167 H-chain with the λ L-chain produces an antigen-binding-site that is not specific for PC; thus, the number of PC-specific binding sites displayed on the B cell membrane is reduced below that required to induce deletion of the B cell. To demonstrate that this process was also working in normal mice, we crossed the M167Hκ24L chain genes into both μMT and κ KO mice. In M167 μMT mice, 50% of the PC-specific B cells coexpressed λ L-chains in addition to the M167Hκ24L chains, and in M167 κ KO mice 100% of these B cells coexpressed endogenously encoded λ L-chains. When the B cell is unable to express an endogenous H-chain or an endogenous κ L-chain, it rearranges and expresses a λ L-chain. The coexpression of this endogenous L-chain dilutes the autoreactive receptor and permits the B cell to excape tolerance. Thus, all these mice appear to compromise allelic exclusion and the paradigm of one antigen-specific Ig-receptor per lymphocyte to retain B cells that are autoreactive but potentially beneficial. Anti-PC antibodies are highly protective against pathogens such as *Streptococcus pneumoniae* and filariae and the inability to produce anti-PC antibodies results in a dramatic increase in susceptibility to infection by *S. pneumoniae* [10-12]. By coexpressing multiple sIg-receptors with different antigen specificities, one autoreactive and one not autoreactive, B cells retain the autoreactive yet potentially beneficial sIg-receptor and still escape negative selection by clonal deletion.

It was of interest to know whether positive selection and receptor dilution were developmental mechanisms restricted to mice expressing M167Hκ24L chain receptors or whether these processes might be operating in other low affinity autoreactive B cells. To address this question, we have analyzed other transgenic mice expressing H and L chains which associate to produce anti-PC receptors. In this manuscript, we show data on the analysis of mice expressing anti-PC antibodies of the M603-id.

Results and Discussion

To produce the M603-like H-chain transgenic mice, a T15 H-chain gene was mutated at amino acid 95H from Asp to Asn using site-directed-PCR [13]. M603 H-chain mice were produced by injecting this gene into FVB/N oocytes. These H-chain transgenic mice were analyzed by flow cytometry to determine whether PC-specific B cells were being positively selected in normal mice.

Positive Selection of PC-Specific B Cells in M603H Transgenic Mice

FACS analysis of spleens from M603H mice is shown in Fig. 1. Staining with anti-B220 and anti-IgM (panel A) shows that 13% of the spleen cells express sIgM and that 25% of the B220$^+$ cells appear to have little or no sIgM. Panel B shows that most of the IgM$^+$ cells express the H-chain transgene; however, there are clearly two populations that express different amounts of the transgene encoded H-chain but similar amounts of total sIgM. This suggests that some of the B cells are expressing endogenous μ-chains. Since the FVB/N mouse expresses the same μ^a allotype as the transgene-encoded H-chain, we were not able to determine the precise number of double μ-chain expressing B cells. However, more than 25% of these cells expressed endogenously endoded δ-H chains (Fig. 1, panel C) suggesting that these B cells have extensively rearranged their endogenous Ig-genes. When we analyzed the bone marrow (BM) of these mice, virtually no δ^+ cells were detected (data not shown); thus, B-cells expressing endogenous Ig were preferentially surviving in the spleen. If the transgene-encoded H-chain randomly associates with any one of the 100 to 200 possible endogenous L-chains, one would expect 0.5 to 1 % of the B cells to express a given L-chain. The M603H chain produces a PC-specific binding site when it associates with a κ8 L-chain. Fig. 1, panel D shows that 7.5% of the splenic B cells in M603H transgenic mice bind PC indicating that PC-specific B cells are also being positively selected as was previously seen in M167H transgenic mice (8).

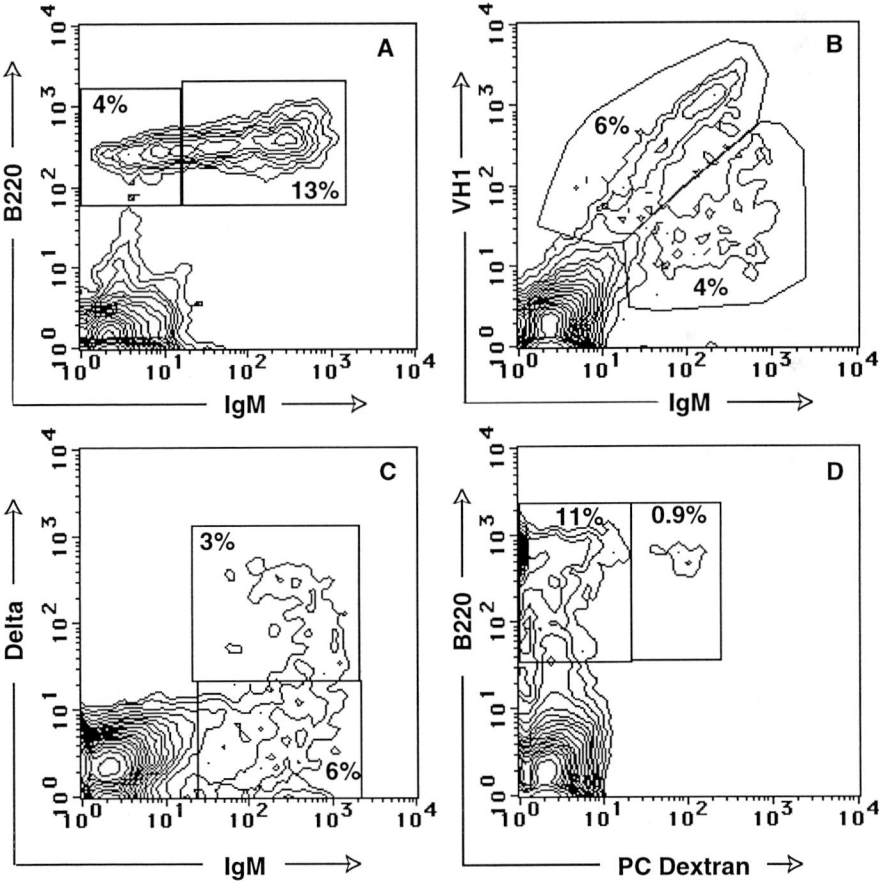

Fig. 1. FACS analysis of spleen cells from mice expressing an M603-like H chain. Spleen cells were prepared and stained as previously described [7]

Elimination of Competition from B Cells Expressing Endogenous H-chains Increases Positive Selection of PC-Specific B Cells

Goodnow and his colleagues [1,14,15] have shown that competition for survival in the spleen plays a major role in the elimination of autoreactive B cells. If the PC-specific B cells are being excluded from follicular recirculation by the presence of B cells expressing endogenous sIg, eliminating B cells expressing endogenous H chain should result in an increase in survival of the autoreactive PC-specific B cells. To test this hypothesis, we crossed the M603H chain into the mu membrane KO mouse (μMT) [16]. These mice can rearrange their endogenous H-chain genes but they cannot insert the endogenous H-chain into the B cell membrane. As seen in Fig 2., elimination of endogenous H-chain expression results in a 3 fold increase in the number of PC-specific cells in the spleens of M603H μMT mice. Up to 20% of the spleen cells in these mice bind PC-Dex (panel B) which is 100 to 200 times higher than expected based on random rearrangement of the κ8 L-chain. We are currently determining whether the PC-specific cells seen in these mice express two L-chains, as would be predicted from our hypothesis of receptor dilution.

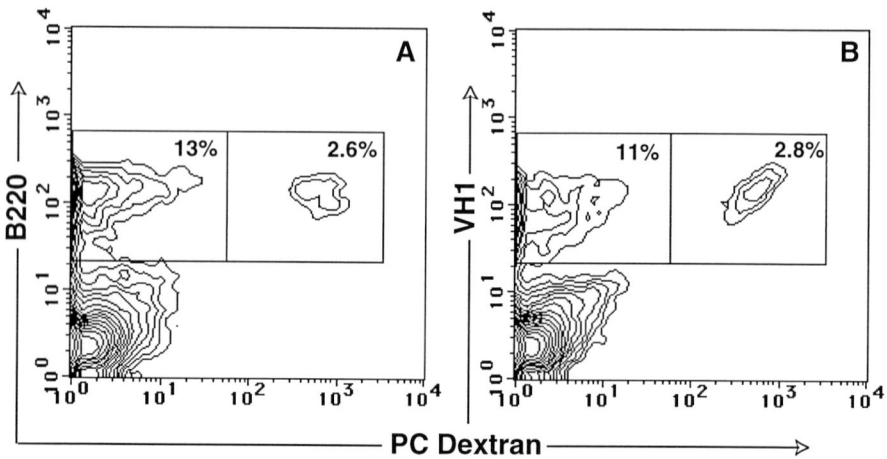

Fig. 2. Increased positive selection of PC-specific B cell occurs in the spleens of M603H μMT mice. M603H mice were crossed to μMT mice and the F1 progeny backcrossed to μMT mice. BC1 mice lacking expression of endogenous δ H-chains were selected for both experimentation and production of BC2 mice.

M603Hκ8L Double Transgenic Mice Rearrange and Express Endogenous Ig

In transgenic mice expressing both a H and L transgene, virtually all the B cells express both chains and the endogenous Ig genes have not been rearranged and expressed. However, if the Ig-receptor is autoreactive, the recombinase system will be activated in an attempt to edit the receptor [2,3]. M603H and κ8L transgenic mice were crossed to produce double transgenic mice and their spleen and BM cells analyzed to elucidate whether endogenous Ig was being coexpressed with the transgene-encoded Ig or expressed to the exclusion of the transgene product. The data in Fig. 3, panels A and D, show that almost all the spleen and BM cells express the transgene endoded V1 H-chain; however, in the spleen a population of B cells expressing reduced levels of the V_H1-id is present (panel A) suggesting co-expression of endogenous H-chains, whereas this population is absent in the bone marrow (panel D). When the spleen and BM were analyzed for the expression of endogenous sIgD (panels B and E) almost 40% of the splenic B cells expressed δ on their surface while less than 1% of the bone marrow cells were $δ^+$. These data strongly suggest that the BM cells are being induced to rearrange endogenous Ig genes following the expression of the transgene-encoded receptor, and that these double H-chain expressing B cells are preferentially selected to survive in the spleen.

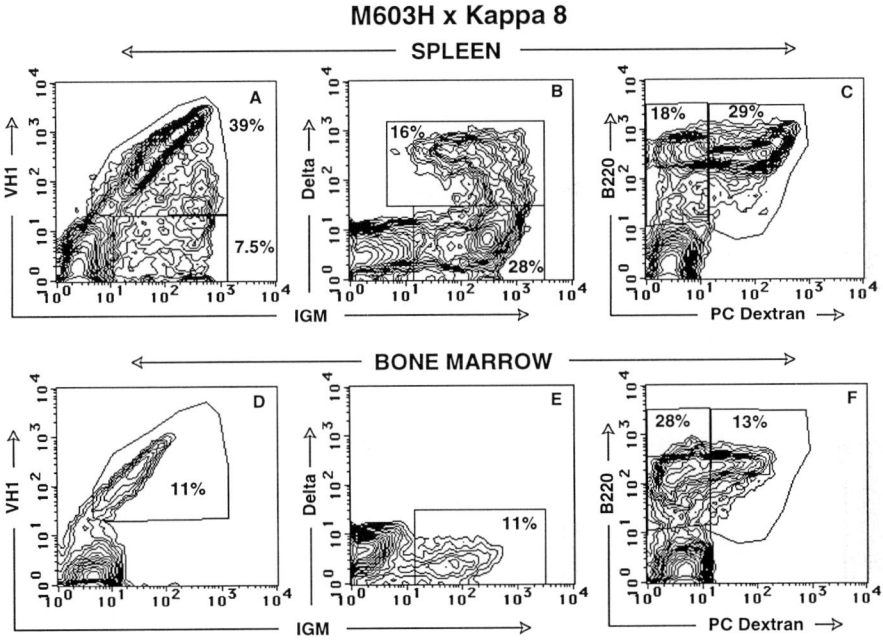

Fig. 3. Expression and selection of B cell expressing endogenous Ig occurs following expression of transgene encoded H and L chains. M603H mice were crossed to mice expressing a kappa 8 L-chain and double transgenics were selected based on the expression of large numbers of PC-Dex binding cell in the peripheral blood.

However, as seen in Fig. 3 panel C, 62% of the spleen cells retain the ability to bind PC. In the bone marrow, only 32% of the $B220^+$ B cells bind PC and the level of PC-binding appears to be reduced. This suggests that the B cell may be arrested in its development until it rearranges and expresses either a second H-chain or a second L-chain. In the M167Hκ24L anti-PC transgenic mice [9], we were able to demonstrate co-expression of λ L-chain on 10% of the PC-specific B cells. However, the λ L-chain when combined with the M603 H-chain produces an autoreactive receptor (Kenny et al., unpublished data); thus, λ L-chain co-expression would not rescue the M603Hκ8 B cells via receptor dilution.

Analysis of B Cell Development in M603Hκ8 Rag Mice

The above data suggest that the PC-specific B cells produced by expression of M603Hκ8L are autoreactive. However, to elucidate whether these B cells would develop in the absence of endogenous Ig co-expression, both the M603H and κ8 transgenes were crossed onto the Rag KO background and then these Rag$^{-/-}$ mice were crossed to produce H + L double transgenic Rag KO mice. As seen in Fig. 4, (panel A) very few sIgM$^+$ B cells develop in the spleens of these mice and the level of receptor expression is very low suggesting that the receptors are being down regulated. Panel B shows that these B cells are capable of binding only low levels of PC-Dex. Thus, B cell development in the M603Hκ8L anti-PC transgenics resembles that previously seen in the M167Hκ24L anti-PC transgenic mice [9], and one can conclude that the M603Hκ8L anti-PC B cells are also autoreactive B cells.

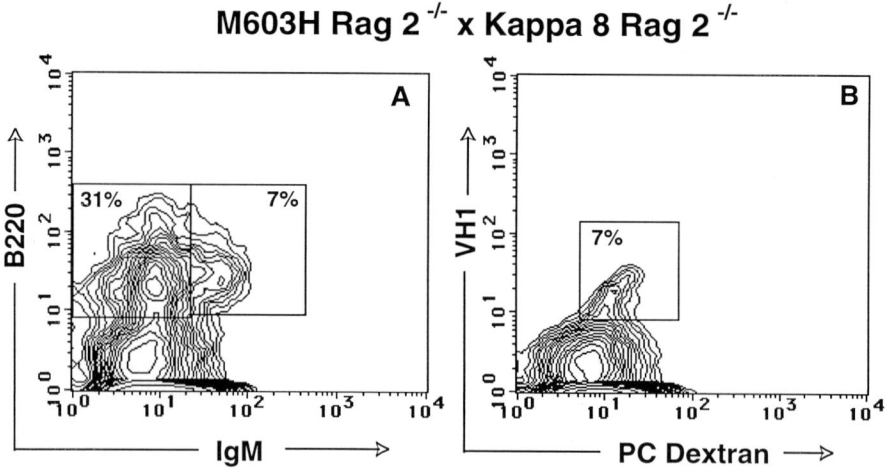

Fig. 4. M603Hκ8 PC-specific B cells are autoreactive. M603H and κ8 transgenic mice were crossed onto the Rag2 KO background and then crossed to each other. Double H + L transgenic were selected by PCR analysis of tail DNA and their spleen cell analyzed for the presence of PC-specific B cells.

The M167 and M603 H-chain sequences are not encoded in the germ-line but are generated via somatic mutation [17]. It is possible that this mutational process generates a receptor that crossreacts with an autoantigen that may or may not contain PC itself. The affinity for this unknown autoantigen may be higher in M167 and M603 receptors than it is in the germ-line T15 receptor. It will therefore be of interest to determine whether the germ-line encoded T15-id is also autoreactive. Preliminary studies on development and survival of T15-id$^+$ B cells in both Rag and normal mice indicate that T15Hκ22L B cells develop in the BM and spleen of Rag mice, but these B cells do not compete well against normal B cells. Thus, T15-id$^+$ B cells are lost as the animal ages (Lustig et al., unpublished data). These data suggest that T15, which has the highest affinity for PC of the idiotypes studied [17], may have the least affinity for the crossreactive autoantigen that appears to be driving positive selection of PC-specific B cells in normal mice. B cell competition studies are in progress to determine which PC-specific B cells will compete the best for survival when they are competing with each other.

References

1. Goodnow, C. C., J. G. Cyster, S. B. Hartley, S. E. Bell, M. P. Cooke, J. I. Healy, S. Akkaraju, J. C. Rathmell, S. L. Pogue, and K. P. Shokat. 1995. Self-tolerance checkpoints in B lymphocyte development. *Adv. Immunol.* 59:279.
2. Gay, D., T. Saunders, S. Camper, and M. Weigert. 1993. Receptor editing: An approach by autoreactive B cells to escape tolerance. *J. Exp. Med.* 177:999.
3. Radic, M. Z., J. Erikson, S. Litwin, and M. Weigert. 1993. B lymphocytes may escape tolerance by revising their antigen receptors. *J. Exp. Med.* 177:1165.
4. Tiegs, S. L., D. M. Russell, and D. Nemazee. 1993. Receptor editing in self-reactive bone marrow B cells. *J. Exp. Med.* 177:1009.
5. Palanda, R., S. Schwers, R. M. Torres, D. Nemazee, and K. Rajewsky. 1997. Receptor editing in a transgenic mouse model:site,efficiency, and role in B cell tolerance and antibody diversity. *Immunity* 7:765.
6. Kenny, J. J., A. M. Stall, D. G. Sieckmann, M. C. Lamers, F. Finkleman, L. Finch, and D. L. Longo. 1991. Receptor-mediated elimination of phosphocholine-specific B cells in X-linked immune deficient mice. *J. Immunol.* 146:2568.
7. Kenny, J. J., R. T. Fischer, A. Lustig, H. M. Dintzis, M. Katsumata, J. C. Reed, and D. L. Longo. 1996. bcl-2 alters the antigen-driven selection of B cells in uk but not in u-only xid transgenic mice. *J. Immunol.* 157:1054.
8. Kenny, J. J., C. O'Connell, D. G. Sieckmann, R. T. Fischer, and D. L. Longo. 1991. Selection of antigen-specific, idiotype positive B cells in transgenic mice expressing a rearranged M167-μ heavy chain gene. *J. Exp. Med.* 174:1189.
9. Kenny, J. J., L. J. Rezanka, A. Lustig, R. T. Fischer, J. Yoder, S. Marshall, and D. L. Longo. 1999. Autoreactive B cells escape clonal deletion by expressing multiple antigen receptors. *J. Immunol.*
10. Briles, D. E., C. Forman, S. Hudak, and J. L. Claflin. 1982. Anti-phosphorylcholine antibodies of the T15 idiotype are optimally protective against *Streptococcus pneumoniae*. *J. Exp. Med* 156:1177.
11. Kenny, J. J., G. Guelde, R. T. Fischer, and D. L. Longo. 1994. Induction of phosphocholine-specific antibodies in X-linked immune deficient mice: In vivo protection against a *Streptococcus pneumoniae* challenge. *Int. Immunol* 6:561.
12. Fischer, R. T., D. L. Longo, and J. J. Kenny. 1995. A novel phosphocholine antigen protects both normal and X-linked immune deficient mice against Streptococcus pneumoniae: Comparison of the 6-O-phosphocholine hydroxyhexanoate-conjugate with other phosphocholine-containing vaccines. *J. Immunol* 154:3373.
13. Ho, S. N., H. D. Hunt, R. M. Horton, J. K. Pullen, and L. R. Pease. 1989. Site-directed mutagenesis by overlap extension using the polymerase chain reaction. *Gene* 77:51.
14. Cyster, J. G., S. B. Hartley, and C. C. Goodnow. 1994. Competition for follicular niches excludes self-reactive cells from the recirculating B-cell repertoire. *Nature* 371:389.
15. Townsend, S. E., B. C. Weintraub, and C. C. Goodnow. 1999. Growing up on the streets: why B-cell development differs from T-cell development. *Immunol. Today* 20:217.
16. Kitamura, D., J. Roes, R. Kuhn, and K. Rajewsky. 1991. A B-cell-deficient mouse by targeted disruption of the membrane exon of the immunoglobulin μ chain gene. *Nature* 350:423.
17. Rudikoff, S. 1983. Immunoglobulin structure-function correlates: Antigen binding and idiotypes. *Contemp. Top. Mol. Biol.* 9:169.

III

B-1 Cells and Positive Selection

A Model for Autoantigen Induction of Natural Antibody Producing B-1a Cells

Robert Berland and Henry H. Wortis
Department of Pathology and Graduate Program in Immunology, Tufts University School of Medicine, Boston, MA 02111

We propose a model of murine B-1a cells that accounts for their function, repertoire, expression of CD5 and origins. We believe the model is simple, internally consistent and compatible with the available data. It is based on the idea that B-1a cells arise from mature non B-1 (B-2 or B-0) cells subsequent to ligation with an appropriate antigen.

B-1a Cells Produce Protective Natural Antibodies

Despite their relatively small numbers B-1a cells are responsible for most serum IgM (1). Much of this antibody appears to react with conserved antigenic epitopes. In particular, these IgM non-hypermutated antibodies react with polysaccharides, phospholipids and with conserved protein epitopes (2-5). The result is that they frequently bind to antigens found on bacteria or other infectious organisms and/or autoantigens (6-9). These antibodies, which arise in the absence of experimental manipulation or obvious infection, have come to be known as natural antibodies. B-1a cells are constitutive secretors of antibody (10, 11). Presumably, it is autoantigen that drives B-1a cells to secrete antibody. (Unless the progenitors of B-1a cells are programmed to mature into secreting cells in the absence of antigenic stimulation.) By this means, even in the absence of exogenous stimulation, B-1a cells function to produce natural antibodies.

Natural antibodies produced by B-1a cells play a crucial role in the defense of the host against pathogens. A mutation that prevents the secretion of IgM, yet permits isotype switching and secretion of non-IgM antibody, severely compromises the response to bacterial pathogens (12-14) and Carroll in this volume. In work reported from the Herzenberg laboratory at this meeting (see also(15)) optimal antibody responses to a pathogen required both natural and acquired antibody. Furthermore, the X-linked immune deficiency (xid) caused by mutations of Bruton's tyrosine kinase (Btk) prevents the development of B-1a cells (16), decreases the levels of IgM and IgG3 antibodies (17), prevents natural antibody formation and increases susceptibility to bacterial infection (18-20).

B-1a development is dependent on BCR signaling

Many mutations of genes encoding proteins critical for BCR signaling reduce the number of B-1a cells. This has been demonstrated for Vav (21, 22), Btk (16), CD19 (23-25), CD21 (26), CD81 (27), PKCβ (28), p85a subunit of PI-3 kinase (29) and BLNK (SLP-65) (30, 31). Conversely, mutation of genes encoding proteins that are negative regulators of BCR signaling increases the number of B-1a cells. The identified mutations involve SHP-1 (motheaten and motheaten^v) (32), CD22(25, 33, 34), Lyn (35-37) and CD72 (38). Not surprisingly, mutations that prevent the function of negative regulators of BCR activation also result in increased production of autoantibodies. CD5 is itself a negative regulator of BCR signaling and its deletion increases the response of B-1a cells to BCR signaling (see below).

As stimulation through the BCR induces CD5 expression (39) and loss of BCR signaling prevents B-1a development, it is reasonable to conclude that B-1a cells result from BCR signaling. A study supporting this is presented in this volume by Clarke and colleagues. Mice transgenic for the light and heavy chains of an antibody that binds phosphatidyl choline produce large numbers of transgene expressing B-1a cells (40) (see below). Clarke's group showed that maturation of the complete B-1a phenotype is cyclosporin A sensitive, implying that it is calcineurin and BCR dependent (see Clarke, this volume).

B-1a cell development is dependent on BCR specificity

Expression of transgenes encoding antibodies that are normally produced by B-2 cells results in B cells with a B-2 phenotype (e.g. (41)). Expression of transgenes encoding an antibody usually produced by B-1a cells results in expanded production of B-1a cells. That is, the specificity of the antibody determines the phenotype of the cell expressing it. Similar observations were made in mice with a transgene encoding only the heavy chain of a potential autoantibody. In these situations, the phenotype of the B cell reflects the specificity of the antibody generated by the endogenous light chain that has combined with the transgenic heavy chain (40). B cells expressing heavy/light chain combinations that result in a B-1a specificity have the B-1a phenotype.

Two recent studies indicate that in addition to BCR specificity, BCR surface density is critical for B-1a development. In heterozygous Ig-transgenic mice expressing an IgM with a B-1a specificity there is a decrease in the fraction of transgene positive B-1a cells and an increase in the fraction of transgene positive B-2 cells compared to homozygous mice carrying the same transgene and expressing about two fold higher levels of the transgene encoded IgM (42). A similar result is seen when comparing Ig-knockin mice expressing two IgMs with different specificities, one B-1a and the other B-2, to knockin mice expressing a single IgM with a B-1a specificity. Thus the strength of signal through the BCR determines whether a cell will exhibit the B-1a phenotype (43).

The importance of BCR specificity and signaling capacity in B-1a development suggest that B-2 cells are driven to the B-1a phenotype by engagement of the BCR by autoantigen. If this is so, then in the absence of a ligating antigen cells would not switch from B-2 to B-1. Work presented at this meeting by Hardy and Hayakawa is consistent with this. In otherwise normal mice transgenic for the H chain of an antibody to the T cell antigen thy-1, the transgene-encoded autoantibody producing cells are CD5+. This phenotype is antigen-dependent as anti thy-1 producing CD5 + cells do not appear on a thy-1 negative background (44) (and this volume).

B-1a Cells Share Some Properties With Anergic B Cells

In mice doubly transgenic for anti Hen's Egg Lysozyme (HEL) and soluble HEL, B cells become anergic. At this meeting Behrens presented evidence that these cells, like B-1a cells, are CD5+. In addition, we have shown (Berland and Wortis, this volume) that B-1a cells have constitutively elevated nuclear NF-AT as do anergic B cells (45). Both anergic B cells and B-1a cells have encountered antigen in the absence of T cell help and both are refractory to activation through the BCR. They do differ in some ways, particularly with respect to sIgM levels which are high in B-1a cells and low in anergic B cells. CD5 levels are also lower in anergic B cells than in B-1a cells. Nevertheless, it is striking that in both cases, encounter with autoantigen has resulted in CD5 expression.

Our own unpublished observations with mice transgenic for anti H-2^K provide another example of this. In the presence of an autoantigen there is an increase in CD5 positive cells. (Dong, Berland and Wortis unpublished observations).

CD5 Expression Prevents Dangerous Autoreactivity

The expression of CD5 on both B-1a cells and anergic B cells may have been selected as a mechanism to prevent pathogenic autoantibody formation. In CD5 knockout mice, B-1a cells, normally refractory to activation through the BCR, can be activated by BCR crosslinking (46). In the same study, it was shown that even in CD5+ mice, B-1a cells can be activated through the BCR if CD5 is sequestered from the BCR. Behrens (this volume) showed that when anti HEL/sHEL double transgenic mice are bred onto a CD5 negative background, anergy is lost in a significant number of mice. In these mice, B cells can be activated in vitro with HEL and in vivo high anti-HEL titres are observed.

It has recently been shown that CD5 contains an ITIM which can recruit SHP-1 and thereby act as a negative regulator of B cell activation through the BCR (46, 47). (Also see Bondada in this volume.) Whether CD5 is the sole mechanism to explain the inability of B-1a cells to be fully activated via the B cell receptor (48) is not known.

A Model

The simplest formulation of these observations is that an individual B cell producing antibody that binds to (auto) antigen with undefined but moderate affinity will be driven into the B-1a phenotype. Cells with rearranged immunoglobulin genes that encode antibodies that bind autoantigens with moderate affinity are more likely than other cells to be triggered in this way.

A portion of the germline repertoire of immunoglobulin V genes has been selected to encode for conserved antigens that are shared by pathogens and host tissue. This assures that autoantigens will activate some B cells prior to contact with pathogen.

Cells that rearrange their immunoglobulin genes in the absence of TdT are more likely to express these reactivities than are TdT expressing cells. This explains why fetal tissue is more likely than adult bone marrow to produce B-1a cells.

The induction of CD5 provides these typically self-reactive B cells with a negative regulator of activation through the BCR, thus providing assurance against increased antibody production in the absence of cognate T helper cell interactions.

An alternative model based on the postulate that there exists in the fetal liver and bone marrow subsets of B cell precursors that express genes, other than TdT, favoring the development of either a B-1a or B-2 phenotype can not be ruled out. Evidence for the existence of these postulated genes may yet be generated. Until then, there is no evidence that counters the idea that the ligation of mIgM by autoantigen initiates the induction of the B-1a phenotype.

References

1. Forster, I., and K. Rajewsky. 1987. Expansion and functional activity of Ly-1+ B cells upon transfer of peritoneal cells into allotype-congenic, newborn mice. European Journal of Immunology 17:521-528.
2. Dighiero, G., A. Lim, P. Poncet, A. Kaushik, X.R. Ge, and J.C. Mazie. 1987. Age-related natural antibody specificities among hybridoma clones originating from NZB spleen. Immunology 62:341-347.
3. Hayakawa, K., C.E. Carmack, R. Hyman, and R.R. Hardy. 1990. Natural autoantibodies to thymocytes: origin, VH genes, fine specificities, and the role of Thy-1 glycoprotein. Journal of Experimental Medicine 172:869-878.
4. Kaushik, A., R. Mayer, V. Fidanza, H. Zaghouani, A. Lim, C. Bona, and G. Dighiero. 1990. Ly1 and V-gene expression among hybridomas secreting natural autoantibody. Journal of Autoimmunity 3:687-700.
5. Mayer, R., H. Zaghouani, O. Usuba, and C. Bona. 1990. The LY-1 gene expression in murine hybridomas producing autoantibodies. Autoimmunity 6:293-305.
6. Reid, R.R., A.P. Prodeus, W. Khan, T. Hsu, F.S. Rosen, and M.C. Carroll. 1997. Endotoxin shock in antibody-deficient mice: unraveling the role of natural antibody and complement in the clearance of lipopolysaccharide. Journal of Immunology 159:970-975.
7. Lalor, P.A., and G. Morahan. 1990. The peritoneal Ly-1 (CD5) B cell repertoire is unique among murine B cell repertoires. European Journal of Immunology 20:485-492.
8. Tornberg, U.C., and D. Holmberg. 1995. B-1a, B-1b and B-2 B cells display unique VHDJH repertoires formed at different stages of ontogeny and under different selection pressures. Embo Journal 14:1680-1689.
9. Hayakawa, K., R.R. Hardy, M. Honda, L.A. Herzenberg, and A.D. Steinberg. 1984. Ly-1 B cells: functionally distinct lymphocytes that secrete IgM autoantibodies. Proc Natl Acad Sci U S A 81:2494-2498.
10. Kaushik, A., A. Lim, P. Poncet, X.R. Ge, and G. Dighiero. 1988. Comparative analysis of natural antibody specificities among hybridomas originating from spleen and peritoneal cavity of adult NZB and BALB/c mice. Scand J Immunol 27:461-471.
11. Nakajima, P.B., S.K. Datta, R.S. Schwartz, and B.T. Huber. 1979. Localization of spontaneously hyperactive B cells of NZB mice to a specific B cell subset. Proceedings of the National Academy of Sciences of the United States of America 76:4613-4616.
12. Carroll, M.C., and A.P. Prodeus. 1998. Linkages of innate and adaptive immunity. Curr Opin Immunol 10:36-40.
13. Boes, M., C. Esau, M.B. Fischer, T. Schmidt, M. Carroll, and J. Chen. 1998. Enhanced B-1 cell development, but impaired IgG antibody responses in mice deficient in secreted IgM. Journal of Immunology 160:4776-4787.
14. Ehrenstein, M.R., T.L. O'Keefe, S.L. Davies, and M.S. Neuberger. 1998. Targeted gene disruption reveals a role for natural secretory IgM in the maturation of the primary immune response. Proceedings of the National Academy of Sciences of the United States of America 95:10089-10093.
15. Baumgarth, N., O.C. Herman, G.C. Jager, L. Brown, and L.A. Herzenberg. 1999. Innate and acquired humoral immunities to influenza virus are mediated by distinct arms of the immune system. Proc Natl Acad Sci U S A 96:2250-2255.
16. Hayakawa, K., R.R. Hardy, and L.A. Herzenberg. 1986. Peritoneal Ly-1 B cells: genetic control, autoantibody production, increased lambda light chain expression. European Journal of Immunology 16:450-456.
17. Scribner, C.L., C.T. Hansen, D.M. Klinman, and A.D. Steinberg. 1987. The interaction of the xid and me genes. Journal of Immunology 138:3611-3617.

18. O'Brien, A.D., D.L. Rosenstreich, and B.A. Taylor. 1980. Control of natural resistance to Salmonella typhimurium and Leishmania donovani in mice by closely linked but distinct genetic loci. Nature 287:440-442.
19. Yother, J., C. Forman, B.M. Gray, and D.E. Briles. 1982. Protection of mice from infection with Streptococcus pneumoniae by anti-phosphocholine antibody. Infection & Immunity 36:184-188.
20. Zaldivar, N.M., and I. Scher. 1979. Endotoxin lethality and tolerance in mice: analysis with the B-lymphocyte-defective CBA/N strain. Infection & Immunity 24:127-131.
21. Gulbranson-Judge, A., V.L. Tybulewicz, A.E. Walters, K.M. Toellner, I.C. MacLennan, and M. Turner. 1999. Defective immunoglobulin class switching in Vav-deficient mice is attributable to compromised T cell help. Eur J Immunol 29:477-487.
22. Zhang, R., F.W. Alt, L. Davidson, S.H. Orkin, and W. Swat. 1995. Defective signalling through the T- and B-cell antigen receptors in lymphoid cells lacking the vav proto-oncogene. Nature 374:470-473.
23. Rickert, R.C., K. Rajewsky, and J. Roes. 1995. Impairment of T-cell-dependent B-cell responses and B-1 cell development in CD19-deficient mice. Nature 376:352-355.
24. Krop, I., A.R. de Fougerolles, R.R. Hardy, M. Allison, M.S. Schlissel, and D.T. Fearon. 1996. Self-renewal of B-1 lymphocytes is dependent on CD19. European Journal of Immunology 26:238-242.
25. Sato, S., N. Ono, D.A. Steeber, D.S. Pisetsky, and T.F. Tedder. 1996. CD19 regulates B lymphocyte signaling thresholds critical for the development of B-1 lineage cells and autoimmunity. Journal of Immunology 157:4371-4378.
26. Ahearn, J.M., M.B. Fischer, D. Croix, S. Goerg, M. Ma, J. Xia, X. Zhou, R.G. Howard, T.L. Rothstein, and M.C. Carroll. 1996. Disruption of the Cr2 locus results in a reduction in B-1a cells and in an impaired B cell response to T-dependent antigen. Immunity 4:251-262.
27. Tsitsikov, E.N., J.C. Gutierrez-Ramos, and R.S. Geha. 1997. Impaired CD19 expression and signaling, enhanced antibody response to type II T independent antigen and reduction of B-1 cells in CD81-deficient mice. Proceedings of the National Academy of Sciences of the United States of America 94:10844-10849.
28. Leitges, M., C. Schmedt, R. Guinamard, J. Davoust, S. Schaal, S. Stabel, and A. Tarakhovsky. 1996. Immunodeficiency in protein kinase cbeta-deficient mice. Science 273:788-791.
29. Suzuki, H., Y. Terauchi, M. Fujiwara, S. Aizawa, Y. Yazaki, T. Kadowaki, and S. Koyasu. 1999. Xid-like immunodeficiency in mice with disruption of the p85alpha subunit of phosphoinositide 3-kinase. Science 283:390-392.
30. Jumaa, H., B. Wollscheid, M. Mitterer, J. Wienands, M. Reth, and P.J. Nielsen. 1999. Abnormal development and function of B lymphocytes in mice deficient for the signaling adaptor protein SLP-65. Immunity 11:547-554.
31. Minegishi, Y., J. Rohrer, E. Coustan-Smith, H.M. Lederman, R. Pappu, D. Campana, A.C. Chan, and M.E. Conley. 1999. An Essential Role for BLNK in Human B Cell Development. Science 286:1954-1957.
32. Sherr, D.H., M.E. Dorf, M. Gibson, and C.L. Sidman. 1987. Ly-1 B helper cells in autoimmune "viable motheaten" mice. Journal of Immunology 139:1811-1817.
33. Otipoby, K.L., K.B. Andersson, K.E. Draves, S.J. Klaus, A.G. Farr, J.D. Kerner, R.M. Perlmutter, C.L. Law, and E.A. Clark. 1996. CD22 regulates thymus-independent responses and the lifespan of B cells. Nature 384:634-637.
34. Nitschke, L., R. Carsetti, B. Ocker, G. Kohler, and M.C. Lamers. 1997. CD22 is a negative regulator of B-cell receptor signalling. Current Biology 7:133-143.
35. Chan, V.W., F. Meng, P. Soriano, A.L. DeFranco, and C.A. Lowell. 1997. Characterization of the B lymphocyte populations in Lyn-deficient mice and the role of Lyn in signal initiation and down-regulation. Immunity 7:69-81.
36. Hibbs, M.L., and A.R. Dunn. 1997. Lyn, a src-like tyrosine kinase. International Journal of Biochemistry & Cell Biology 29:397-400.
37. Nishizumi, H., I. Taniuchi, Y. Yamanashi, D. Kitamura, D. Ilic, S. Mori, T. Watanabe, and T. Yamamoto. 1995. Impaired proliferation of peripheral B cells and indication of autoimmune disease in lyn-deficient mice. Immunity 3:549-560.

38. Pan, C., N. Baumgarth, and J.R. Parnes. 1999. CD72-deficient mice reveal nonredundant roles of CD72 in B cell development and activation. Immunity 11:495-506.
39. Cong, Y.Z., E. Rabin, and H.H. Wortis. 1991. Treatment of murine CD5- B cells with anti-Ig, but not LPS, induces surface CD5: two B-cell activation pathways. International Immunology 3:467-476.
40. Arnold, L.W., C.A. Pennell, S.K. McCray, and S.H. Clarke. 1994. Development of B-1 cells: segregation of phosphatidyl choline-specific B cells to the B-1 population occurs after immunoglobulin gene expression. Journal of Experimental Medicine 179:1585-1595.
41. Rabin, E., Y. Cong, T. Imanishi-Kari, and H.H. Wortis. 1992. Production of 17.2.25 mu transgenic and endogenous immunoglobulin in X-linked immune deficient mice. European Journal of Immunology 22:2237-2242.
42. Watanabe, N., S. Nisitani, K. Ikuta, M. Suzuki, T. Chiba, and T. Honjo. 1999. Expression levels of B cell surface immunoglobulin regulate efficiency of allelic exclusion and size of autoreactive B-1 cell compartment. J Exp Med 190:461-469.
43. Lam, K.P., and K. Rajewsky. 1999. B Cell Antigen Receptor Specificity and Surface Density Together Determine B-1 versus B-2 Cell Development. J Exp Med 190:471-478.
44. Hayakawa, K., M. Asano, S.A. Shinton, M. Gui, D. Allman, C.L. Stewart, J. Silver, and R.R. Hardy. 1999. Positive selection of natural autoreactive B cells. Science 285:113-116.
45. Healy, J.I., R.E. Dolmetsch, L.A. Timmerman, J.G. Cyster, M.L. Thomas, G.R. Crabtree, R.S. Lewis, and C.C. Goodnow. 1997. Different nuclear signals are activated by the B cell receptor during positive versus negative signaling. Immunity 6:419-428.
46. Bikah, G., J. Carey, J.R. Ciallella, A. Tarakhovsky, and S. Bondada. 1996. CD5-mediated negative regulation of antigen receptor-induced growth signals in B-1 B cells. Science 274:1906-1909.
47. Sen, G., G. Bikah, C. Venkataraman, and S. Bondada. 1999. Negative regulation of antigen receptor-mediated signaling by constitutive association of CD5 with the SHP-1 protein tyrosine phosphatase in B-1 B cells. Eur J Immunol 29:3319-3328.
48. Rothstein, T.L., and D.L. Kolber. 1988. Anti-Ig antibody inhibits the phorbol ester-induced stimulation of peritoneal B cells. Journal of Immunology 141:4089-4093.

The Role of Complement Receptors CD21/CD35 in Positive Selection of B-1 Cells

R. R. Reid@*†, S. Woodcock±†, A. P. Prodeus@*, J. Austen±#, L. Kobzik*±,
H. Hechtman±#, F. D.Moore, Jr. ±# and M. C. Carroll°*@.
From the Departments of °Pediatrics, *Pathology and #Surgery, Harvard Medical School and @The Center for Blood Research and ±Brigham and Women's Hospital, Boston, MA. 02115

Introduction

The complement system participates in both innate and adaptive immunity (1, 2, 3). Recent studies using mice deficient in specific components of serum complement or complement receptors (CD21/CD35) have not only confirmed many known roles for complement but shed new light on mechanisms important in natural and adaptive immunity (2). For example, one mechanism by which complement enhances the generation of memory responses to T-dependent antigens is by mediating a survival signal in germinal center (GC) B cells via the CD21/CD19/Tapa-1 co-receptor. Alternatively, the deficient mice have been important in confirming the importance of complement in models of inflammation such as ischemia reperfusion injury and identifying the classical pathway as being essential for induction of injury (4). It was while studying the mechanism for reperfusion injury, that we identified a potential novel pathway in which complement is involved in positive selection of B-1 cells. We were struck by the absence of pathogenic, natural antibody in the Cr2-deficient animals despite their normal level of circulating IgM. Thus, despite normal serum levels of IgM, the mice appeared to be missing the specificity, which mediates reperfusion injury. In this review, we focus on the studies leading up to this observation and speculate on possible mechanisms.

Ischemia reperfusion injury is mediated by classical pathway complement

Ischemia reperfusion (I/R) injury refers to the acute inflammatory response that follows reperfusion of ischemic tissues. In general, the longer the period of ischemia the more pronounced the inflammatory response. A possible mechanism was identified by Weisman et al several years ago, when they reported partial blocking of injury with a soluble inhibitor of complement C3b (sCR1) in a rat cardiac model (5). Soluble CR1 receptor is very effective in inactivation of C3 convertase, i.e. the enzyme complex that converts native C3 to its active C3b form. It serves both as a co-factor in factor I cleavage of C3b and in displacing C3b from the convertase. Interestingly, intravenous injection of mice with sCR1 prior to induction of ischemia significantly reduced injury on reperfusion of the treated animals. Histologic examination of treated animals revealed a reduction in the level of complement deposition in cardiac muscle and in the extent of tissue injury (5). Similar results were obtained on pretreatment of mice with sCR1 in a hindlimb model of injury (6). In their model, the blood flow to the lower hindlimb was blocked for approximately 2 hours by placing a tourniquet on the hindlimb. Prior to re-establishment of blood flow to the tissues, mice were administered radio-iodinated albumin intravenously as a marker for permeability. While these studies determined that injury was mediated by complement they did not address the question of mechanism of initiation of injury.

To examine the mechanism of ischemia-reperfusion injury, mice deficient in complement C3 were treated in the hindlimb model. Not too surprisingly, the C3-/- mice were partially protected from injury based on an approximate 50% reduction in permeability index (Fig. 1). Thus, complement C3 is essential for induction of full injury in this murine model.

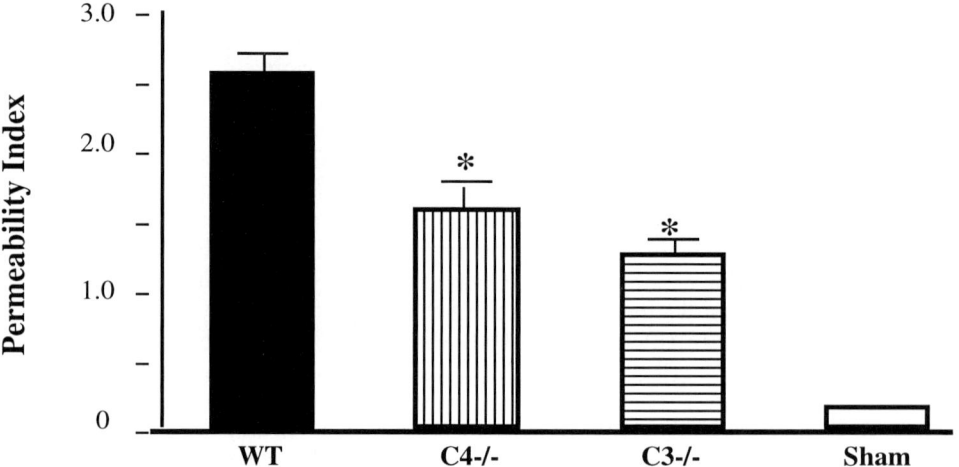

* p< 0.00001

Fig.1. Ischemia reperfusion injury is mediated by classical pathway of complement. Mice underwent 2 h of hindlimb ischemia and 3 h of reperfusion. Hindlimb P.I. was reduced significantly in C3-/- and C4-/- treated compared to complement sufficient littermates. The hindlimb permeability index (P.I.) was determined by the ratio (CPM/g dry muscle)/CPM/g blood). Error bars represent the standard error of the mean and significance is indicated by an asterisk. From Weiser et al. 1996.

However, these experiments did not identify how complement was activated. The serum complement system can be activated by at least three distinct pathways, classical, lectin or alternative. Knowing which pathway is involved, is important as it suggests a mechanism for injury. For example, the classical pathway is activated very efficiently by IgM and IgG isotypes of immunoglobulin or by the serum recognition protein C-reactive protein. Whereas, the lectin pathway is activated following recognition of specific carbohydrates such as mannan by mannan binding lectin (MBL) (7). In both pathways, complement C4 is required in forming an enzyme complex with C2 that catalyzes cleavage of the central component C3. By contrast, the alternative pathway activates spontaneously leading to conversion of C3 to its active form (C3b) and attachment to foreign- or self- tissues. The pathway is tightly regulated as all host cells express regulator proteins such as membrane cofactor or decay accelerating factor which prevent amplification of the complement pathway by inactivating or displacing the C3 convertase (1). One approach for determining the pathway involved is use of mice deficient in C4, i.e. cannot form C3 convertase via classical or lectin pathways. Comparison of mice deficient in either C3 or C4 with WT controls in the hindlimb model, revealed that C4 was also required for induction of full injury (4). This finding was important as it suggested that antibody or MBL might be involved.

Natural IgM mediates ischemia reperfusion injury

To determine if antibody was involved in mediating I/R injury, mice totally deficient in immunoglobulin, RAG2-/- (recombinase activating gene-2 deficient) were characterized along with the complement deficient animals in the hindlimb model. Significantly, the RAG-2-/- mice were protected to a similar level as observed in the complement deficient animals (4). Since the RAG2-/- animals are also missing mature lymphocytes, it was important to determine that the pathogenic effect was antibody dependent (8). To confirm that injury was mediated by serum antibody, the deficient animals were reconstituted with either normal mouse sera (4) or purified IgM (9). In both cases, the reconstituted RAG-2-/- mice were no longer protected and injury was restored. In the latter experiments, a model of intestinal injury was used as in this model, injury is thought to be mediated primarily by complement.

The interpretation of these results is that during the period of ischemia, neo-antigens are either expressed or exposed on the endothelial cell surface. It was proposed that circulating IgM recognizes the new determinant, binds and activates classical pathway of complement. While the nature of the antigen is not known, IgM rather than IgG seems to be primarily responsible for activation of complement as reconstitution of deficient mice with pooled IgG did not significantly restore injury in the mice. An alternative hypothesis is that there is another initial event such as the MBL pathway that recognizes the altered endothelial surface, induces low level complement activation which in turn exposes new antigenic sites and the pathway is amplified by binding of IgM. The finding that IgM was involved was important as it suggests possible therapeutic approaches for blocking injury in situ. Identification of specific antibody could also provide a means for isolation of the endothelial cell surface antigen.

Pathogenic IgM is a product of B-1 cells

Since a major fraction of circulating IgM is thought to represent natural antibody, i.e. product of rearranged germline genes, we reasoned that mice bearing deficiencies in the B-1 fraction of lymphocytes might also be protected. B-1 cells have a distinct phenotype from more conventional B-2 cells in that they express low levels of IgD and CD23 and a major fraction express the cell surface protein CD5 (10, 11). They are also distinguished by reduced circulation, limited frequency in

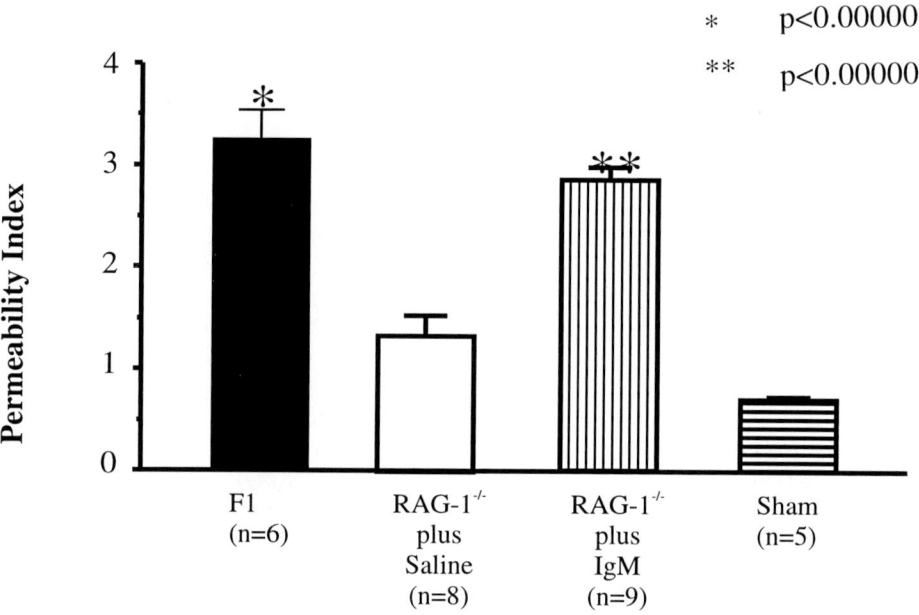

Fig 2. Intestinal permeability of inbred mice after intestinal ischemia and reperfusion or no injuri (sham). Values are means ± standard error; n= number of mice in experimental groups. F1, parent strain for recombination-activating gene-1 (RAG-1) antibody deficient mice. RAG-1 + IgM mice were reconstituted with 0.5 mg of pooled murine IgM intravenously approximately 1 hour before treatment. Results are from Williams et al. 1999.

the peripheral lymph nodes and spleen and are primarily localized within the peritoneal cavity. To examine a role for B-1 cells as a source of pathogenic IgM, antibody- deficient mice (RAG-2-/-) were reconstituted with 5×10^5 peritoneal B-1 cells and rested approximately 30 days before treatment. Ciculating IgM levels reach a near normal range within a month following adoptive transfer. Characterization of the B-1 cell reconstituted mice in the intestinal ischemia model confirmed that B-1 cells were a major source of pathogenic IgM (Fig. 2). This was an important observation because the repertoire of B-1 cell natural antibody is considerably more limited than would be expected for conventional B-2 cells. Therefore, it is possible that the pathogenic antibody represents a product of the germ-line.

Cr2-/- mice are protected from I/R injury

The initial characterization of Cr2-/- knockout mice generated in our lab revealed an approximate 50% reduction in the frequency of B1a or CD5+ B-1 cells (12). Although, characterization of another strain of Cr2-deficient mice did not identify a similar reduction (13). Whether the difference in frequency of CD5+ cells was due to variation in strain background or environmental differences is not known. Despite the reduced frequency of B-1a cells in the Cr2-/- mice, circulating levels of IgM were within the normal range. These findings suggested that the repertoire of IgM might be different in the Cr2-deficient animals. To test this hypothesis, we characterized the mice in the intestinal I/R model. Surprisingly, the Cr2-/- mice were equally protected as the complete-antibody deficient mice (Fig. 3). Comparison of survival over a five-day period following treatment in the intestinal model demonstrated a significant increase in mortality of the WT compared to Cr2-deficient animals (results not shown). Consistent with an increased mortality, a dramatic reduction in injury was observed in tissue sections harvested from treated WT or Cr2-/- deficient mice (results not shown).

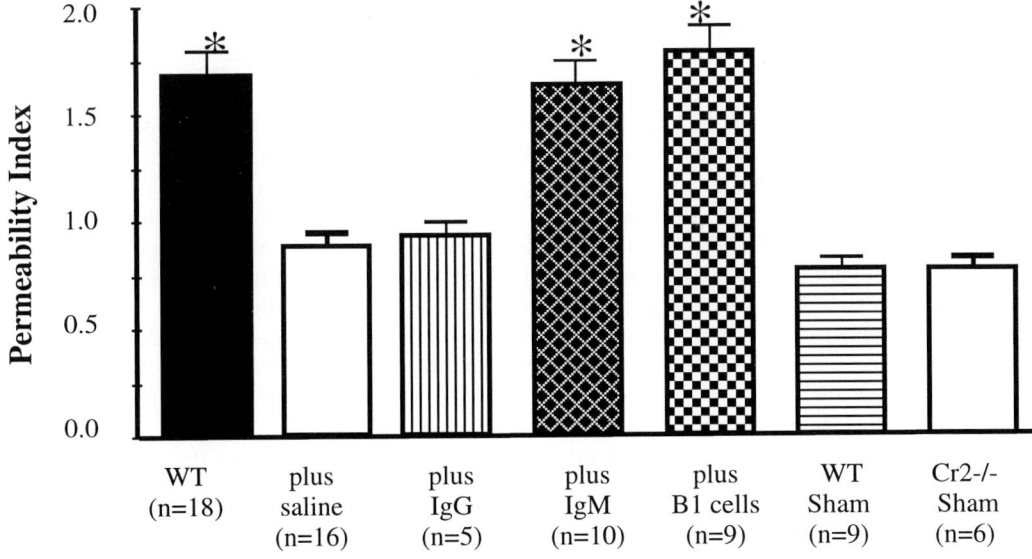

* $p < 0.00001$

Fig. 3. Intestinal permeability of inbred mice after intestinal ischemia and reperfusion or no injury (sham). WT, parent strain for Cr2-/- mice. Cr2-/- was reconstituted with pooled IgG or IgM or saline control as described in Fig. 2. Values are means ± standard error; n= number of mice in experimental groups.

Extensive injury to the mucosal layer of the intestine was observed in WT mice or Cr2-/- mice reconstituted with pooled IgM or B-1 cells. By contrast, tissue sections isolated from treated Cr2-/- mice were similar to that of sham controls. Thus, despite normal circulating levels of IgM, the Cr2-deficient mice were protected from injury. These results not only confirm the importance of B-1 cells as a source of pathogenic antibody but suggest that the complement system is somehow involved in formation or maintenance of the repertoire of natural antibody.

Complement involvement in B-1 cell selection

Two different models have been proposed to explain the development of B-1 cells. The lineage hypothesis proposes that B-1 cells develop in early fetal life as a distinct population (11). Alternatively, B-1 cells develop from the same progenitors as conventional B cells but depending on their environment, i.e. encounter with antigen, they develop into B-1 or retain the B-2 cell phenotype (14, 15). Irrespective of their origin, there is general agreement that B-1 cells are not replenished from adult bone marrow at the same frequency as B-2 cells and that their phenotype is more similar to that of early fetal liver B cells or neonatal bone marrow (BM) cells. Consistent with an early origin, their repertoire tends to be biased towards expression of more proximal VH genes and N-nucleotide addition is limited (16, 17). It seems reasonable that given the reduced replenishment by adult BM stem cells, B-1 cells are self-renewed and that antigen stimulation might be important in their renewal or even in initial selection (18). Indeed inherent to the conventional model, B-1 cells must be antigen selected.

Evidence in support of a BCR signaling requirement for positive selection of B-1 cells comes from mice bearing mutations that alter BCR signaling. For example, impairment of BCR signaling through CD19 (19, 20), vav (21) or Btk (22) dramatically affects development of B-1 cells. By contrast, loss of negative selection such as in CD22- or SHP-1 deficient mice can lead to an increase in B-1 cell frequency (23, 24). Recent, elegant studies with mice bearing two distinct Ig transgenes, V_H12 (B-1 cell phenotype) or V_H B1-8 (B-2 cell phenotype) support the hypothesis that B-1 cells are positively selected by self-antigens. For example, B cells expressing V_H12 either alone or together with B1-8 developed a B-1 cell phenotype. Whereas, few if any B cells were identified that expressed the B1-8 tg only. Thus, these results suggested that encounter of transgenic B cells with PtC resulted in expansion of those expressing V_H12 and recruitment into the B-1 pool (25). Another elegant model demonstrating antigen selection of B-1 cells was recently reported by Hardy et al (26). In their model, B cells expressing an immunoglobulin transgene specific for Thy 1.1 were selected and expanded in mice expressing the cognate antigen. By contrast, transgene + B-1 cells were not found in mice that expressed the alternative allotype Thy 1.2.

Where does complement fit into B-1 cell development? The overall reduction in B-1a cell frequency and the more specific loss of B-1 cells expressing IgM involved in I/R injury suggests a role for CD21/CD35 in either positive selection or maintenance of B-1a cells. One possible role for complement is that it enhances BCR signaling on encounter with cognate antigen. Biochemical studies and analysis of CD21/CD35 deficient mice demonstrate the importance of co-receptor signaling in activation and survival of conventional B cells (2, 3). It is very likely that B-1 cells likewise utilize co-receptor signaling to enhance the BCR signal. For example, bacteria express typical B-1 cell antigens such as phosphoryl choline and it is not unreasonable that coating of bacteria with complement ligand C3d would enhance crosslinking of the co-receptor with the BCR and enhance overall

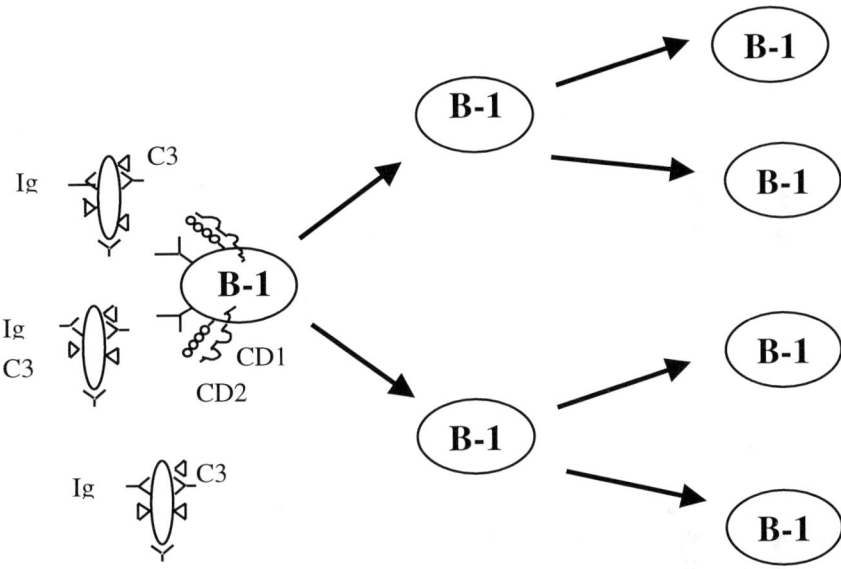

Complement-coated enteric bacteria   **Expansion of B-1 cells**

Fig. 4. Proposed role for complement and complement receptors in positive selection of peritoneal B-1 lymphocytEcounter of complement-ligand coated antigens (self- or non-self) results in co-ligation of the CD21/CD19 co-receptor and BCR on the cell surface leading to an enhanced signaling and positive selection.

signaling. Thus, antigens expressed at lower concentrations might require complement enhancement in order for the cognate B cell to recognize it and expand or be positively selected. Another role for complement receptors is in localizing antigen on follicular dendritic cells (FDC) within the lymphoid compartment. However, since the major population of B-1 cells occupy the peritoneal tissues it is not clear if they would encounter FDC within lymphoid structures. The actual site or sites in which B-1 cells undergo positive selection are not known. It is possible that they must encounter cognate antigen in early fetal development or in neonatal BM. If this is the case, it might be expected that complement receptors on stromal cells within these compartments bind antigen for presentation to B cells. It is possible that complement receptors could participate in both stages of development. First, they might enhance antigen signaling in initial positive selection. Secondly, as selected B-1 cells are replenished at peripheral sites, complement receptors might again be involved in enhancement of BCR signaling. Future studies will be important to identify the site/sites of positive selection of B-1 cells and defining the role of complement in this process.

References

1. Muller-Eberhard, H J, Molecular Organization and Function of the Complement System. Ann. Rev. Biochem. 57, 321-347, 1988.
2. Carroll, M C, The role of complement and complement receptors in induction and regulation of immunity. Ann Rev Immunol 16, 545-568, 1998.
3. Fearon, D T, Carter, R H, The CD19/CR2/TAPA-1 Complex of B Lymphocytes: Linking Natural to Acquired Immunity. Annu. Rev. Immunol. 13, 127-149, 1995.
4. Weiser, M, Williams, J, Moore, F, Kobzik, L, Ma, M, Hechtman, H, Carroll, M, Reperfusion injury of ischemic skeletal muscle is mediated by natural antibody and complement. J. Exp. Med. 1857-1864, 1996.
5. Weisman, H F, Bartow, T, Leppo, M K, Marsh, H C, Carson, G R, Concino, M F, Boyle, M P, Roux, K H, Weisfeldt, M L, Fearon, D T, Souble human complement receptor type I: In vivo inhibitor of complement suppressing post-ischemic myocardial inflammation and necrosis. Science 249, 146-151, 1990.
6. Hill, J, Lindsay, T F, Ortiz, F, Yeh, C G, Hechtman, H B, Moore, F D, Soluble complement receptor type 1 ameliorates the local and remote organ injury after intestinal ischemia-reperfusion in the rat. J. Immunol. 149, 1723-1730, 1992.
7. Epstein, J, Eichbaum, Q, Sheriff, S, Ezekowitz, R, The collectins in innate immunity. Cur Opin in Immunol 8, 29-35, 1996.
8. Shinkai, Y, Rathbun, G, Lam, K-P, Oltz, E M, Stewart, V, Mendelsohn, M, Charron, J, Datta, M, Young, F, Stall, A M, Alt, F W, Rag-2-deficient mice lack mature lymphocytes owing to inability to initiate V(D)J rearrangement. Cell 68, 855-867, 1992.
9. Williams, J P, Moore, F D, Kobzik, L, Carroll, M C, Hechtman, H B, Intestinal reperfusion injury is mediated by IgM and complement., J Appl Physiol 86; 938-42 (1999).
10. Hardy, R R, Carmack, C E, Li, Y S, Hayakawa, K, Distinctive developmental origins and specificities of murine CD5+ B cells. Immunol. Rev. 137, 91, 1994.
11. Kantor, A B, Herzenberg, L A, Origin of murine B cell lineages. Annu. Rev. Immunol. 11, 501-538, 1993.
12. Ahearn, J, Fischer, M, Croix, D, Goerg, S, Ma, M, Xia, J, Zhou, X, Howard, R, Rothstein, T, Carroll, M, Disruption of the Cr2 locus results in a reduction in B-1a cells and in an impaired B cell response to T-dependent antigen. Immunity 4, 251-262, 1996.

13. Molina, H, Holers, V, Li, B, Fung, Y, Mariathasan, S, Goellner, J, Strauss-Schoenberger, J, Karr, R, Chaplin, D, Markedly impaired humoral immune response in mice deficient in complement receptors 1 and 2. Proc. Natl. Acad. Sci. USA 93, 3357-3361, 1996.
14. Wortis, H H, Surface markers, heavy chain sequences and B cell lineages. Int. Rev. Immunol. 8, 235, 1992.
15. Clarke SH, LW, A, B-1 cell development: evidence for an uncommitted immunoglobulin (Ig)M+ B cell precursor in B-1 cell differentiation. J. Exp. med. 187, 1325-1334, 1998.
16. Gu, H, Forster, I, Rajewsky, K, Sequence homologies, N sequence insertion and JH gene utilization in VHDJH joining:implications for the joining mechanism and the ontogenic timing of Ly1 B cell and B-CLL progenitor generation. EMBO J 9, 2133, 1990.
17. Feeney, A J, Lack of N regions in fetal and neonatal mouse immunoglobulin V-D-J juntional sequences. J. Exp. Med. 172, 1377, 1990.
18. Hayakawa, K, Hardy, R R, Stall, A M, Herzenberg, L A, Herzenberg, L A, Immunoglobulin bearing B cells reconstitute and maintain the murine Ly-1 B cell lineage. Eur. J. Immunol. 16, 1313, 1986.
19. Engel, P, Zhou, L J, Ord, D C, Sata, S, Koller, B, Tedder, T F, Abnormal B lymphocyte development, activation and differentiation in mice that lack or overexpress the CD19 signal transduction molecule. Immunity 3, 39-50, 1995.
20. Rickert, R C, Rajewsky, K, Roes, J, Impairment of T-cell-dependent B cell responses and B-1 cell development in CD19-deficient mice. Nature 376, 352-355, 1995.
21. Zhang, R, Alt, F W, Davidson, L, Orkin, S H, Swat, W, Defective signaling through the T- and B-cell antigen receptors in lymphoid cells lacking the vav proto-oncogene. Nature 374, 470-473, 1995.
22. Khan, W N, Alt, F W, Gerstein, R M, Malynn, B, Larsson, I, Rathbun, G, Davidson, L, Muller, S, Kantor, A, Herzenberg, L A, Rosen, F S, Sideras, P, Defective B cell development and function in BTK 23. O'Keefe T.L., Williams G.T. , Davies S.L., Neuberger M.S., Hyperresponsive B cells in CD22-deficient mice. Science 274, 798-801, 1996.
24. Shultz, L D, Schweitzer, P A, Rajan, T V, Yi, T, Ihle, J N, Matthews, R J, Thomas, M L, Beier, D R, Mutations at the murine motheaten locus are within the hematopoietic cell protein-tyrosine phosphatase (Heph) gene. Cell 73, 1445, 1993.
25. Lam, K-P, Rajewsky, K, B cell antigen receptor specificity and surfsacce density together determine B-1 versus B-2 cell development. J. Exp. Med. 190, 471-477, 1999.
26. Hayakawa, K, Asano, M, Shinton, S A, Gui, M, Allman, D, Stewart, C L, Silver, J, Hardy, R R, Positive selection of natural autoreactive B cells. Science 285, 113-116, 1999.

Considerations on B Cell Homeostasis.

Fabien Agenes*, M. Manuela Rosado and Antonio A. Freitas.
Lymphocyte Population Biology Unit, URA CNRS 1961, Institut Pasteur, 25 Rue du Dr. Roux, 75015 Paris, France, and *Basel Institute for Immunology, Grenzacherstrasse 487, CH-4005 Basel, Switzerland.

Introduction
In adult mice the total number of lymphocytes remains constant and shows a "return tendency, due to a density dependent process to approach a stationary distribution of population densities", usually referred to as homeostasis. B cells are, however, produced continuously in either the primary lymphoid organs or by peripheral cell division: it follows that each newly formed lymphocyte can only persist if another resident lymphocyte dies. In an immune system where the total number of cells is limited, cell survival can no longer be a passive phenomenon, but rather a continuous active process where each lymphocyte must compete with other lymphocytes (Freitas and Rocha, 1997; Freitas and Rocha, 1993; Freitas et al., 1995)

B cell production and peripheral B cell numbers.
Peripheral B cell pools represent transit compartments where there is an input of newly formed cells, proliferation due to antigenic stimulation and a cellular output due to cell death and terminal differentiation (Freitas et al., 1986). The number of peripheral B cells should be a function of the rates of B cell production and B cell death. Estimates of the daily B cell production in the BM indicate that it is sufficient to replenish the peripheral B cell compartments in 3-4 days (Opstelten and Osmond, 1983; Rocha et al., 1990). It is claimed, however, that the vast majority of the newly formed B cells are counter-selected, die "in situ" or soon after leaving the marrow and never incorporate the peripheral pool (Forster and Rajewsky, 1990; Goodnow et al., 1995). The size of the B cell pool would then be limited by the insufficient "effective" production of new B cells. On the contrary it has been also claimed that a significant fraction of the peripheral B cell pool is continuously renewed from incoming newly formed cells (Freitas et al., 1986).

To study the role of B cell production in the control of peripheral B cell numbers and homeostasis we created a mouse model in which BM B cell production could be limited (Agenes et al., 1997). For this purpose we generated chimeras grafting Rag2-deficient mice with mixtures of BM cells from normal mice and from mice with a developmental block of B cell development (μMT or Rag2-deficient) (Kitamura et al., 1991; Shinkai et al., 1992). In these chimeras the number of pre-B cells can be reduced by diluting normal BM cells diluted among incompetent BM cells from the B cell-deficient donors. The number of pre-B cells was in fact found to be strictly dependent on the ratio of "normal" progenitors present in the inoculum (Agenes et al., 1997).

Chimeras containing less than 20% of the normal number of pre-B cells had reduced peripheral B cell numbers. A normal sized peripheral B cell pool, however, was generated in mice containing only 30% of the normal number of pre-B cells (Agenes et al., 1997). These results demonstrate that about 1/3 of the normal number of BM B cell precursors suffices to maintain the peripheral B cell pool size. A similar conclusion was obtained after parabiosis between one normal and two or three B cell deficient mice (Agenes et al., 1997). In these circumstances B cell production was restricted to the BM of the normal mouse since no chimerism was detectable in the BM of the different partners. In mice triads it was found that each individual mouse had physiological B cell numbers i.e. the B cell production of one mouse was sufficient the populate the peripheral pools of three mice.

In conclusion, these results demonstrate that peripheral B cells number is not determined by the rates of BM B cell production, but it is limited at the periphery. By parabiosis we show that increasing the amount of "resources" results in an expanded B cell pool in the presence of the same rate of B cell production. Thus, the selection of the primary B cell repertoires is determined not only by the direct interactions between each B cell and its ligands, the affinity or avidity of such recognition process, but also modulated by the number and nature of competitor cells.

<u>Independent homeostatic regulation of the resting and activated B cell compartments.</u>

In the mixed BM chimeras described above we also studied the size of the activated IgM-secreting effector B cell compartment. We found that regardless of the fraction of normal pre-B cells present in the BM, and the size of the mature B cell pool, the number of splenic IgM-secreting cells and the levels of serum IgM were always the same and identical to those found in normal mice or in chimeras reconstituted with 100% normal BM (Agenes et al., 1997). These findings suggest that the number of activated IgM-secreting B cells is regulated independently of the number of pre-B and mature B cells. Thus, the number of activated B cells can no longer be considered simply as a constant fraction of the number of resting B cells, but may represent an autonomous B cell compartment with different homeostatic controls. These observations also suggested that the regulation of the activated pool is a priority and only when it is replenished can peripheral mature B cells accumulate to physiological numbers.

To investigate these possibilities we studied the fate of mature lymph node B cells transferred into B cell deficient $Rag2^{-/-}$ hosts. When resting lymph node B cells were injected into B cell deficient hosts a fraction of the transferred cells expanded and constituted a highly selected population which survived for prolonged periods of time by continuous cell renewal at the periphery (Agenes and Freitas, 1999). A significant fraction of the surviving B cells expressed an activated phenotype and about 7-30% of the cells were actively engaged in IgM-secretion. By injecting different numbers of B cells, we found that when low numbers of B cells were available there was a preferential generation of activated IgM-secreting cells. With the transfer of increasing numbers of B cells there was an augmented recovery of resting B cells. Serum concentrations of IgM identical to those of control mice were readily reached in the presence of a reduced B cell number. Resting B cells accumulated at the periphery only when the activated B cells compartment was complete (Agenes and Freitas, 1999).

These results support the notion that the homeostatic regulation of the resting and activated B cell compartments is autonomous. The independent homeostatic regulation of the resting and activated B cell compartments implies an hierarchical organization of the immune system in which the first priority is the maintenance of normal serum IgM levels. It provides an efficient mechanism to ensure both a first natural barrier of protection and a maximum of repertoire diversity.

Feedback regulation of B cell differentiation.

We also investigated the possible mutual influences between established resident populations of peripheral B cells and newly coming B cell migrants. We compared the fate of a population of B cells (recent BM migrants or mature B cells) in the presence of a previously established peripheral B cell population, i.e. we followed the seeding and the persistence of a second population of B cells in B-cell deficient mice previously injected with a cohort of mature B cells (Agenes and Freitas, 1999).

We found that once established, a population of activated B cells is rather resistant to replacement and persists even in the presence of new emigrating cells generated from BM precursors or contained in a second inoculum of LN cells (Agenes and Freitas, 1999). We also demonstrated that in the presence of an established population of activated B cells there is a diminution in IgM production by a subsequently introduced B cell population. These findings indicate that there are mechanisms of feedback regulation controlling terminal B cell differentiation Feedback regulation may be exerted either through direct suppression of new cell subsets, or indirectly, by the pre-consumption of common resources necessary for the growth and/or the differentiation of the B cells (pre-emptive competition). In the latter case we must assume that the highly selected resident B cells will be more efficient in the usage of the common resources. These findings suggest that peripheral B cell selection follows the rule "first come, first served". The different possible mechanisms and the role of the final IgM product in feedback regulation are currently under investigation.

The selection of the persisting B cells and the production of IgM in the immunodeficient hosts partially mimics the development of the immune system. We would, therefore, like to extrapolate from our findings into the origin of the natural IgM secreting cells. In newborn and young mice, during the expansion of the immune system and when the mature B cells are scarce, the percentage of activated B lymphocytes is increased. In the adult mouse a proportion of the natural IgM antibodies may be produced by plasma cells derived from a pool of activated B cells selected early during ontogeny which resist replacement (Hamilton and Kearney, 1994). Substitution of the initially selected population may be only be achieved trough cellular competition based on BCR diversity and antigenic environment. Thus, a population of BM derived B cells would be more likely to enter the pool of IgM-secreting cells than a second population of LN B cells expressing a less diverse repertoire (Agenes and Freitas, 1999).

The immune system seems to be organized to ensure several alternative sources for the production of the natural antibodies which constitutes the first barrier of protection. Every newly formed B cell has the ability to differentiate into a plasma cell but this process is dependent on the nature and the number of cells already present at the periphery. During the development and expansion of the immune system an initial pool of activated B cells is selected, which can eventually be competed out by new specificities formed in the BM.

Peripheral B cell survival: the role of the BCR.

In an immune system where the total number of cells is limited, cell survival can no longer be a passive phenomenon, but rather a continuous active process (Freitas and Rocha, 1997). Maintenance of naive T cells in peripheral lymphoid organs requires continuous T cell receptor (TCR) engagement by major histocompatibility complex (MHC) molecules (Brocker, 1997; Tanchot et al., 1997). Survival of naive B cells in the peripheral pools also appears to involve interactions between the B cell receptor (BCR) and yet unidentified ligand(s). This was first suggested by experiments in which a transgenic BCR could be ablated by an inducible Cre-LoxP recombination event. It was reported that after BCR ablation, B cells rapidly disappeared from the peripheral pools (Lam and Rajewsky, 1998). This study, however, did not allow to directly correlate BCR signaling and peripheral B cell survival, as BCR ablation also leads to the arrest of new B cell production in the BM (Lam et al., 1997). In the absence of the newly formed BM migrants a significant fraction of the peripheral B cells is rapidly lost (Freitas et al., 1986; Heyman et al., 1989). The question remained on whether the mere presence of a signaling complex, e.g. IgM-Igα-Igβ or other, at the cell surface suffices to signal B cell survival, or if B cell survival requires ligand recognition. We addressed directly the role of ligand-mediated recognition in peripheral B cell survival. B cells lacking the V-region of the IgM receptor were shown to have a very short life-expectancy (Rosado and Freitas, 1998). The presence of the truncated membrane IgM transgene, lacking the V-region, provides constitutive signals that suffice to signal allelic exclusion and to promote pre-B cell development in the absence of the surrogate light chain λ5 (Corcos et al., 1995), but fails to support long-term survival of the Tg B cells (Rosado and Freitas, 1998). Differences in the antibody repertoires expressed by pre-B and peripheral B cells can also be invoked to suggest the involvement of ligand-mediated recognition in peripheral B cell positive selection and persistence (Freitas et al., 1989; Freitas et al., 1989; Gu et al., 1991).

In contrast to T cells in which TCR survival signaling seems to require the recognition of MHC class I or class II molecules, the nature of the ligand(s) that might be involved in B cell survival remain elusive.

Studies on monoclonal B mice.

The size of the peripheral B cell pool is regulated by an active homeostatic process with lymphocytes competing for survival signals. We have shown that B cell survival signals may involve antigen-specific receptors (Rosado and Freitas, 1998) implying some type of ligand recognition. In this were the case, we expect that the size of each B cell clone to be limited by the levels of exploitable "epitopes" present at the periphery, i.e. a diverse population of B cells exploiting multiple "epitopes" should occupy a larger niche than a monoclonal population.

By crossing several Ig transgenic mice with C57Bl/6-Rag2 deficient mice we obtained different lines of mice bearing homogeneous populations of truly monoclonal B cells. We found that the number of resting B cells in MoMD4 mice, containing a single B cell clone, was diminished to 60% of wild type mice. More important we found that the number of a-HEL IgM-secreting cells and the total levels of serum IgM were reduced to about 30% of controls (M. Rosado & AA. Freitas, submitted). These observations further indicate that while a significant number of resting B cells from a a-HEL clone can survive in the peripheral pools, only a minor fraction is fit to transit into the IgM-secreting pool. They show that the requirements for the maintenance of resting and activated a-HEL monoclonal B cells differ, i.e. the niche for IgM-secreting cells being proportionally much smaller that the niche for resting B cells.

If the levels of "epitopes" as survival signals, are limiting for clonal sizes then the maximal number of cells should vary for each B cell clone. Previous studies in different lines of Rag-deficient mice bearing rearranged IgH and IgL chain transgenes have reported reduced B cells numbers which apparently varied from 30-50% of control mice (Lam and Rajewsky, 1999; Young et al., 1994). Independent lines of gene targeted "quasi-monoclonal" mice also yielded different numbers of B cells according to the Tg line studied (Era et al., 1991; Sonoda et al., 1997). We found that B cell numbers in another MoSP6.L1 mice is 30% of control compared to 60% for the MoMD4 line (M. Rosado & AA. Freitas, submitted). In the same MoSP6.L1 mice the niche occupied by the activated B cells differs again from MoMD4 mice and is almost negligible (<10%). Recent observations showing that some B cells clones are more competent to fill up the CD5 compartment than others these findings indicate that the niche size for each clone varies also with tissue localization (Lam and Rajewsky, 1999). These results, also support the role of BCR specificity in the control of peripheral cell numbers, but do not exclude the involvement of other "non-specific" resources in B cell homeostasis.

Conclusion.

In adults, in steady state conditions, the number of lymphocyte populations is under homeostatic control. New lymphocytes which are continuously produced in primary and secondary lymphoid organs must compete with resident cells for survival. Lymphocyte survival must be a continuous active process which implicate cognate receptor engagement as fundamental survival signals for B lymphocytes. The conflict of survival interests between different cell types gives rise to a pattern of interactions which mimics the behavior of complex ecological systems. In response to competition, lymphocytes use different survival signals occupying different ecological niches during cell differentiation. This is the case for resting and naturally activated IgM-secreting cells.

References.

Agenes, F., and Freitas, A. A. (1999). Transfer of small resting B cells into immunodeficient hosts results in the selection of a self-renewing activated B cell population. Journal of Experimental Medicine *189*, 319-30.

Agenes, F., Rosado, M. M., and Freitas, A. A. (1997). Independent homeostatic regulation of B cell compartments. Eur. J. Immunol. *27*, 1801-1807.

Brocker, T. (1997). Survival of mature CD4 T lymphocytes is dependent on Major Histocompatibility complex Class II-expressing dendritic cells. J.Exp.Med. *186*, 1223-1232,.

Corcos, D., Dunda, O., Butor, C., Cesbron, J.-Y., Lores, P., Buchinni, D., and Jami, J. (1995). Pre-B cell development in the absence of lambda5 in transgenic mice expressing a heavy-chain disease protein. Curr.Biol. *5*, 1140-1148.

Era, T., Ogawa, M., Nishikawa, S., Okamoto, M., Honjo, T., Akagi, K., Miyazaki, J., and Yamamura, K. (1991). Differentiation of growth signal requirement of B lymphocyte precursor is directed by expression of immunoglobulin. EMBO Journal *10*, 337-42.

Forster, I., and Rajewsky, K. (1990). The bulk of the peripheral B-cell pool is stable and not rapidly renewed from the bone marrow. Proc. Nat. Acad. Sci. (USA) *87*, 4781-4785.

Freitas, A. A., Lembezat, M. P., and Coutinho, A. (1989). Expression of antibody V-regions is genetically and developmentally controled and modulated by the B lymphocyte environment. Int. Immunol. *1*, 342-354.

Freitas, A. A., Lembezat, M. P., and Rocha, B. (1989). Selection of antibody repertories: transfer of mature T lymphocytes modifies VH gene family usage in the actual and available B cell repertories of athymic mice. International Immunology *1*, 398-408.

Freitas, A. A., and Rocha, B. (1997). Lymphocyte survival: a red queen hypothesis. Science *277*, 1950.

Freitas, A. A., Rocha, B., and Coutinho, A. (1986). Lymphocyte population kinetics in the mouse. Immunol. Rev. *91*, 5-37.

Freitas, A. A., and Rocha, B. A. (1993). Lymphocyte lifespans: homeostasis, selection and competition. Immunol. Today *14*, 25-29.

Freitas, A. A., Rosado, M., Viale, A.-C., and Grandien, A. (1995). The role of cellular competition in B cell survival and selection of B cell repertoires. Eur. J. Immunol. *25*, 1729-1738.

Goodnow, C. C., Cyster, J. G., Hartley, S. B., Bell, S. E., Cooke, M. P., Healy, J. I., Akkaraju, S., Rathmell, J. C., Pogue, S. L., and Shokat, K. P. (1995).

Self-tolerance checkpoints in B lymphocyte development. Advances in Immunology *59*, 279-368.

Gu, H., Tarlinton, D., Muller, W., Rajewsky, K., and Forster, I. (1991). Most peripheral B cells are ligand selected. J. Exp. Med. *173*, 1357-1371.

Hamilton, A. M., and Kearney, J. F. (1994). Effects of IgM allotype suppression on serum IgM levels,B-1 and B-2 cells, and antibody responses in allotype heterozygousF1 mice. Developmental Immunology. *4*, 27-41.

Heyman, R. A., Borrelli, E., Lesley, J., Anderson, D., Richman, D. D., Baird, S. M., Hyman, R., and Evans, R. M. (1989). Thymidine kinase obliteration : creation of transgenic mice with controlled immune deficiency. Proc. Nat. Acad. Sci. *86*, 2698.

Kitamura, D., Roes, J., Kühn, R., and Rajewsky, K. (1991). A B-cell deficient mouse by targeted disruption of the membrane exon of the immunoglobulin mu chain gene. Nature *350*, 423-426.

Lam, K.-P., Kuhn, R., and Rajewsky, K. (1997). In vivo ablation of surface immunoglobulin on mature B cells by inducible gene targeting results in rapid cell death. Cell *90*, 1073-1083.

Lam, K.-P., and Rajewsky, K. (1999). B Cell Antigen Receptor Specificity and Surface Density Together Determine B-1 versus B-2 Cell Development. J. Exp. Med. *190*, 471-478.

Lam, K. P., and Rajewsky, K. (1998). Rapid elimination of mature autoreactive B cells demonstrated by Cre-induced change in B cell antigen receptor specificity in vivo. Proceedings of the National Academy of Sciences of the United States of America *95*, 13171-5.

Opstelten, D., and Osmond, D. G. (1983). Pre-B cells in mouse bone marrow: immunofluorescence stathmokinetic studies of the proliferation of cytoplasmic mu-chain-bearing cells in normal mice. J. Immunol. *131*, 2635-2640.

Rocha, B., Penit, C., Baron, C., Vasseur, F., Dautigny, N., and Freitas, A. A. (1990). Accumulation of bromodeoxyuridine-labeled cells in central and peripheral lymphoid organs : minimal estimates of production and turnover rates of mature lymphocytes. Eur. J. Immunol. *20*, 1697-1708.

Rosado, M. M., and Freitas, A. A. (1998). The role of the B cell receptor V region in peripheral B cell survival. Eur. J. Immunol. *28*, 2685-2693.

Shinkai, Y., Rathbun, G., Lam, K.-P., Oltz, E. M., Stewart, V., Mendensohn, M., Charron, J., Datta, M., Young, F., Stall, A. M., and Alt, F. W. (1992). RAG-2-deficient mice lack mature lymphocytes owing to inability to initiate V(D)J rearrangement. Cell *68*, 855-867.

Sonoda, E., Pewzner-Jung, Y., Schwers, S., Taki, S., Jung, S., Eilat, D., and Rajewsky, K. (1997). B cell development under the condition of allelic inclusion. Immunity 6, 225-33.

Tanchot, C., Lemonnier, F. A., Perarnau, B., Freitas, A. A., and Rocha, B. (1997). Differential requirements for survival and proliferation of CD8 naive or memory T cells. Science 276, 2057-2062 .

Young, F., Ardman, B., Shinkai, Y., Lansford, R., Blackwell, T. K., Mendelsohn, M., Rolink, A., Melchers, F., and Alt, F. W. (1994). Influence of immunoglobulin heavy- and light-chain expression on B-cell differentiation [published erratum appears in Genes Dev 1995 Dec 15;9(24):3190]. Genes & Development 8, 1043-57.

Selective Maturation of V_H12 B Cells in the Spleen Enriches for Anti-Phosphatidyl Choline B Cells: Evidence for Receptor Editing

C. Tatu and S. H. Clarke
Department of Microbiology and Immunology, University of North Carolina at Chapel Hill, Chapel Hill, North Carolina 27599

Abstract

PtC-specific B-1 cells originate from conventional B-2 (B-0) cells as a result of antigen activation. V_H12 B cells specific for PtC are enriched at two developmental checkpoints in the bone marrow; first at the pre-BI to pre-BII transition where V_H12 pre-B cells with anti-PtC V_HCDR3 are enriched, and second at the pre-BII to immature B cell transition where L chain diversity is restricted. This restriction is due to the inability of most L chains to associate with V_H12 H chains. We present evidence here of a third developmental checkpoint that enriches for PtC-specific B cells, at the transitional to mature B-2 (B-0) cell stage. Most V_H12 transitional B cells do not differentiate to a mature B-2 cell and, of those that do, most have undergone receptor editing. The Vκ4/5H L chain appears to be one of the few L chains that can support differentiation to the mature B-2 cell stage, providing an explanation for its dominance among V_H12 B cells in the spleen. Once cells reach this stage, those that bind PtC are induced to differentiate to B-1. Thus, through selection at multiple differentiative stages and the induction of extensive secondary Vκ rearrangement and receptor editing, V_H12 B cell differentiation is focused toward specificity for PtC and selection to the B-1 subset.

Introduction

B-1 cells are different from conventional or B-2 cells in a number of ways, including phenotype, tissue distribution, life span, specificities, and more (for a review see [1, 2]). Their role in the immune system is unknown, but that they are responsible for production of much of the circulating natural IgM [3, 4] suggests that they play a role in the early phases of an immune response. One of the more intriguing aspects of this population is its repertoire. Unlike the B-2 subset, the B-1 subset is enriched in polyreactive and autoreactive B cells. Among the latter are cells specific for ssDNA, IgG (rheumatoid factor), Thy-1, and phosphatidyl choline (PtC) [4-9].

We have used mice transgenic for the V_H12 rearrangement or both the V_H12 and $V\kappa4/5H$ rearrangements of an anti-PtC B cell lymphoma to follow the differentiation and segregation of PtC-specific B-1 cells [10-13]. These studies indicate that PtC-specific B-1 cells differentiate from B-2 cells as a result of antigen signaling through surface IgM. This pathway was first proposed by Wortis and colleagues based on their finding that B-2 cells stimulated *in vitro* with anti-IgM and IL-6 differentiate to B-1 [14, 15]. Thus, segregation of PtC-specific cells to the B-1 subset is achieved by antigen driven differentiation from B-2. To reflect this order of differentiation we will refer to B-2 cells as B-0 cells.

The production of PtC-binding V_H12 B cells is greatly enriched at several differentiative checkpoints during differentiation in the bone marrow [12]. The first checkpoint is at the pre-BI to pre-BII transition. Pre-B cells first produce H chains at this transition and express the pre-B cell receptor (pre-BCR). Those cells that fail to make a functional pre-BCR are eliminated [16]. We find that the overwhelming majority of V_H12 pre-B cells are eliminated at this stage in a V_HCDR3 sequence dependent manner, enriching for cells with an anti-PtC V_HCDR3 (10 amino acids in length and a Gly in the fourth position [referred to as 10/G4]). The second checkpoint is at the transition from pre-BII to immature B where L chain rearrangement occurs and IgM is first expressed. We find that most L chains are unable to associate with V_H12 (non-permissive) resulting in extensive secondary $V\kappa$ rearrangement [12]. Splenic V_H12 B cells are highly enriched in expression of the $V\kappa4/5H$ L chain, which when paired with V_H12 can bind PtC [12]. Thus, the result of selection at these two checkpoints is an enrichment of V_H12 B-0 cells that bind PtC.

We report here evidence of a third developmental checkpoint at the transition to a mature B cell. We demonstrate that V_H12 B cells expressing a Vk1A L chain transgene (encoding a permissive L chain) are impaired or blocked in their ability to differentiate to a mature B-0 cell. B cells that differentiate beyond the splenic transitional B cell stage have undergone receptor editing suggesting that the V_H12 B-0 (PtC^{neg}) and B-1 (PtC^{pos}) repertoires are highly dependent on receptor editing. We suggest that the enrichment of $V\kappa4/5H$ L chains in splenic V_H12 B-0 cells is because the $V\kappa4/5H$ L chain is one of the few L chains that can support V_H12 differentiation to the mature cell stage. Thus, as a result of selection and receptor editing the specificity of the V_H12 repertoire is focused toward PtC-binding, and ultimately toward selection to the B-1 subset.

Materials and Methods

Mice

V_H12 Tg mice (6-1) were previously described and maintained in our pathogen-free animal facility by backcrossing to C.B17 mice [10]. Vk1A Tg mice were kindly provided by Dr. Martin Weigert (Princeton University) and crossed with 6-1 mice to produce VκlA Dbl Tg mice for these studies.

Flow Cytometry and Cell Sorting
Flow cytometry analysis was performed as previously described [12] using FACScan™ (Becton Dickinson) with hardware interface and acquisition and analysis software from Cytomation, Inc. For sorting, cells were stained with antibodies to B220 and CD23 and with liposomes. The cell populations indicated were sorted on a MoFlo high speed sorter (Cytomation, Inc.).

Vκ Repertoire Determination
Total RNA was extracted from the cells of each population and used for RT-PCR using 5' RACE (Life Technologies, Inc.) and cloned as described [12]. The UNC Automated Sequencing Facility performed sequencing of randomly selected clones. Sequence analysis was performed using the BLAST sequence search facility (GenBank).

Results

Our previous analysis of B-0 cell development in 6-1 mice indicated that there is a tremendous restriction in L chain use; over half of the B-0 cells in 6-1 mice use Vκ4/5H [12]. This includes only cells that do not bind PtC or are poor binders of PtC and that do not become B-1. We attributed this restriction to the inability of a majority of L chains to associate with V_H12. However, this idea cannot fully explain the extraordinary bias in favor of Vκ4/5H use by V_H12 B-0 cells, since in the absence of a selective advantage, Vκ4/5H should not be expressed any more frequently than any other permissive L chain. We therefore speculated that Vκ4/5H L chains confers a selective advantage in driving differentiation to a mature B cell. To address this possibility we have combined the V_H12 transgene with a VκlA transgene, since VκlA L chains are V_H12 permissive [12]. This H and L chain combination was not expected to bind PtC and therefore cells expressing this combination should not drive differentiation to the B-1 cell stage. Thus, they allow analysis of V_H12 B cell differentiation to the B-0 cell stage.

6-1 mice and VκlA Tg mice were interbred to produce V_H12/VκlA (VκlA Dbl) Tg mice. The VκlA Tg mice contained an insertion of a VκlA-Jκ1 rearrangement into an endogenous Jκ locus, and were a kind gift of by Dr. Martin Weigert (Princeton University). They also contained a duplication of the germline Jκ–Cκ locus on the chromosome containing the VκlA insertion (M. Weigert, personal communication). Splenic B cells were analyzed by flow cytometry for phenotype and liposome binding. The liposomes contain PtC as a membrane constituent and encapsulate a fluorescein dye, enabling the detection of

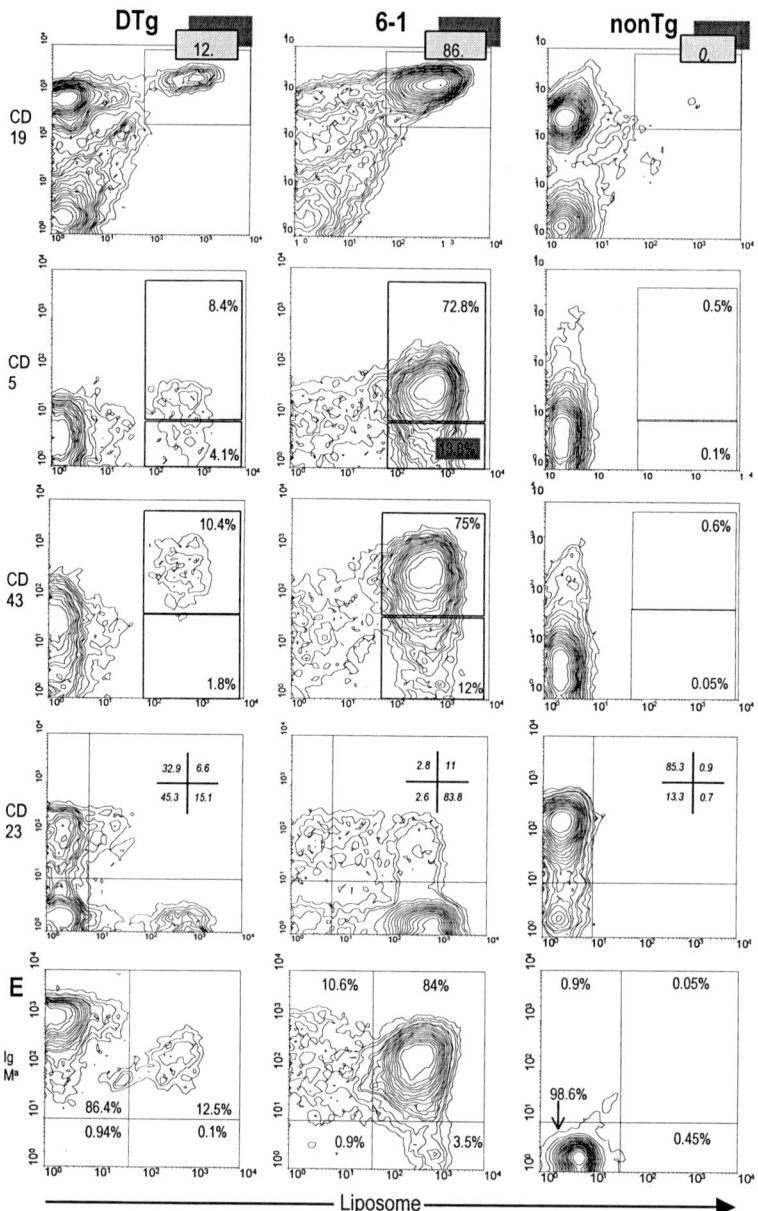

Fig. 1. Flow cytometric analysis of splenic B cells from Vk1A Dbl, 6-1, and a non-Tg littermate. Cells were stained for the markers indicated and for liposome binding. In panel A, total lymphocytes are gated. In panels B, C, D, and E, CD19$^+$ cells are gated. For each mouse, 7-8 x 10^7 cells were stained and 50,000 cells per sample were acquired on the flow cytometer. All contours are at 5% probability.

PtC-specific B cells. Unexpectedly, ~20% of the spleen cells in VκlA Dbl Tg mice bind liposomes (Fig. 1). Moreover, these cells have a B-1 phenotype. They are CD23$^-$, CD5$^+$, CD43$^+$, and CD19^{hi}, indistinguishable from the liposome binding B cells of 6-1 mice (Fig. 1).

The majority of the splenic B cells in VκlA Dbl Tg mice do not bind liposomes (Fig. 1). These cells are CD5$^-$ and CD43$^{-/low}$ indicating that they are not B-1 cells, consistent with our previous observation that differentiation of V_H12 B cells to B-1 is a function of their ability to bind PtC [12]. The reason for the low expression of CD43, which is shared by PtC^{neg} B cells of 6-1 mice, is unknown, but might indicate that they are less mature than non-Tg transitional B cells (Fig. 1). Approximately one-third of the PtC^{neg} B cells are CD23$^+$, indicating a mature B-2 cell phenotype; the remaining two-thirds are CD23$^-$, indicating a transitional B cell phenotype [17, 18]. The latter cells are also IgM^{hi} and CD19^{hi} (Fig. 1) and HSA^{lo} (data not shown) consistent with a transitional phenotype. This 1:2 ratio of mature to transitional B cells is unusual. In non-Tg mice, this ratio is 5-6:1, as seen in the non-Tg control (Fig. 1). This suggests either that transitional B cells are inhibited in their ability to become mature B-0 cells, or that mature B-0 cells are short-lived.

To understand the basis for differentiation to each of the B cell subsets (transitional, mature B-0, and B-1) in VκlA Dbl Tg mice, we determined L chain use by the cells of each. Splenic B cells were stained for B220 and CD23 expression and with liposomes and sorted. Transitional B cells were sorted as B220$^+$, CD23$^-$, PtC^{neg}; mature B-0 cells were sorted as B220$^+$, CD23$^+$, PtC^{neg}; B-1 cells were sorted as B220$^+$, CD23$^-$, and PtC^{bri}. Vκ transcripts were amplified from each population by RT-PCR using 5' RACE and cloned. Clones were selected at random and sequenced. As shown in Fig. 2 the majority of B-1 cells use Vκ4/5H, only one clone out of 22 was of the VκlA transgene rearrangement (Fig. 2C). Thus, editing of the VκlA receptor appears to have occurred to replace it with a Vκ4/5H rearrangement, conferring specificity for PtC, and driving differentiation to B-1. The junctional sequences are compatible with PtC binding (data not shown), as seen previously among B-1 cells of 6-1 mice [12].

The majority of transcripts from the mature B-0 cell population are also of endogenous origin (Fig. 2A), only 38% are VκlA. The non-VκlA rearrangements include Vκ genes from the Vκ9 (1 clone), Vκ23 (4 clones), Vκ19 (1 clone) and Vκ4/5 (2 non-H clones) families. Of the 8 Vκ4/5H rearrangements, 3 have junctional sequences incompatible with PtC-binding consistent with their B-0 phenotype, but 5 have junctional sequences compatible with PtC-binding. We have seen the latter among sorted B-0 cells of 6-1 mice [12] and speculate that they are B-1 cells that have lost IgM expression, either because they are undergoing cell division or because that have differentiated to plasmablasts.

The transitional B cells are predominantly VκlA rearrangements (Fig. 2B). The only endogenous rearrangement detected was of a Vκ4/5H rearrangement that is probably from a contaminating B-1 cell, since the junctional

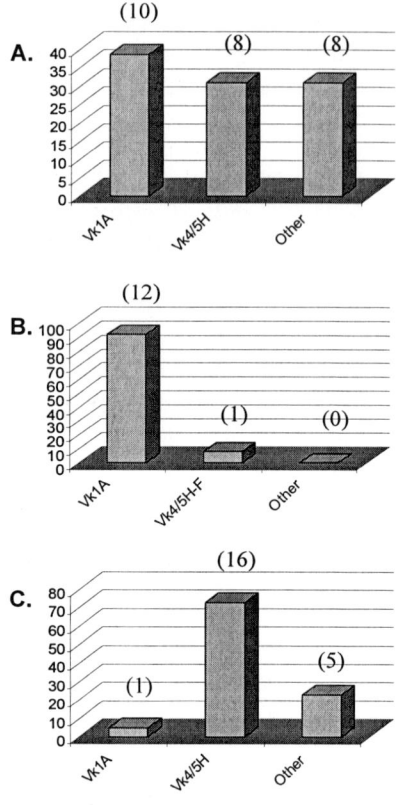

Fig. 2. The Vκ repertoire expressed by mature B-2 cells (A), transitional B cells (B) and B-1 cells (C) of Vκ1A Dbl Tg mice. The numbers in parentheses indicate the number of sequenced clones in each group. Vκ1A refers to the transgene and Vκ4/5H refers to a single member of the Vκ4/5 family. It has a F at the Vκ-Jκ junction. "Other" refers to all Vκ genes other than the transgene or Vκ4/5H, and includes other members of the Vκ4/5 family.

amino acid is typical of an anti-PtC B-1 cell. Thus, V_H12/Vκ1A B cells can differentiate to the transitional B cell stage, but their differentiation beyond this stage to the mature B-0 and B-1 cell stages appears to be either blocked or impaired. The high frequency of endogenous rearrangements in B-0 and B-1 cells suggests that continued differentiation to B-0 and, in particular, to B-1 is enhanced by editing to replace the Vκ1A-Jκ1 rearrangement with an endogenous Vκ rearrangement.

Discussion

These data suggest that V_H12/Vκ1A B cells are impaired in their ability to differentiate beyond the transitional B cell stage, despite the fact that Vκ1A L chains are V_H12 permissive. The Vκ1A rearrangements detected among mature B-0 cells may indicate that some Vκ1A-expressing cells can differentiate to this stage. However, it is also possible that these cells co-express the Vκ1A transgene and an endogenous Vκ rearrangement, and that they differentiate to the mature

B-0 stage by virtue of the endogenous L chain. This is an important consideration since the duplication of the germline Jκ–Cκ locus on the chromosome containing the Vκ1A-Jκ1 insertion makes it possible to rearrange an endogenous Vκ to a Jκ gene downstream of the insertion without stopping the expression of the Vκ1A transgene. Regardless of the explanation, because the Vκ4/5H transgene has no difficulty in driving rearrangement beyond the transitional B cell stage [10], and because Vκ1A-only transgenic mice have a near normal ratio of mature to transitional B cells (data not shown), we conclude that not all V_H12 permissive L chains are equal in their ability to drive V_H12 B cell differentiation.

These data also suggest that V_H12/Vκ1A B cells that are unable to differentiate from a transitional B cell to a mature B-0 cell undergo receptor editing. Editing is unlikely to be unique to V_H12 B cells of Vκ1A Dbl Tg mice. The impetus to edit is likely to be strong even in B cells that do not have a κ transgene. This is because the repertoire of L chains that can drive B-0 cell differentiation appears to be small. The Vκ23 and Vκ4/5 families account for 9 of 11 endogenous rearrangements detected among sorted B-0 cells of Vκ1A Dbl Tg mice (excluding the 5 Vκ4/5H rearrangements with junctional sequences compatible with PtC-binding). Such a small repertoire makes it likely that the initial Vκ rearrangement in a V_H12 B cell of any mouse (Tg or non-Tg) will encode an L chain that cannot drive B-0 differentiation. Thus, receptor editing must be a general feature of V_H12 B-0 cell differentiation. This provides an explanation for the finding that Vκ4/5H rearrangements from 6-1 B-0 cells are more skewed to downstream Jκ4 and Jκ5 gene segments than are Vκ rearrangements encoding other permissive L chains [12]. We suggest that this is the result of editing to replace permissive L chains that cannot support B-0 differentiation and forcing Vκ4/5H rearrangements to use more downstream Jκ genes.

Why V_H12/Vκ1A transitional B cell differentiation to B-0 is blocked is not known. This and other V_H12 B cells expressing non-Vκ4/5H L chains may be autoreactive and therefore eliminated by negative selection. Alternatively, such cells may not be positively selected. Surface IgM is required for B cell survival [19] and it may be that receptors composed of V_H12 H chains and non-permissive L chains are unable to provide a survival signal. However, the level of IgM on the surface of V_H12/Vκ1A transitional B cells is not significantly different from that of non-Tg mice (Fig. 1) suggesting a mechanism influenced by some aspect of the V regions, possibly the ability to bind a ligand.

Receptor editing as a means to replace autoreactive receptors on developing B cells in the bone marrow is well established [20-22]. In these models the autoreactive B cells have two possible fates, edit to acquire a non-autoreactive receptor and continue differentiation or die in the bone marrow. Autoreactive cells do not generally leave the bone marrow. That the unedited cells of Vκ1A Dbl Tg mice are in the spleen makes the induction of editing in this model unique. It may be that the signal to edit is made in the bone marrow, but that the unedited cells do not carry out cell death until later, after they reach the spleen. This difference from other receptor editing models could be related to the strength of the negative signal, as the previously described models involve B cells with high affinity anti-self receptors. Another possibility is that the signal for editing in this model is received in the spleen, not the bone marrow, allowing exit from the bone marrow. This would mean that receptor editing occurs at a later stage in differentiation than believed, although germinal center B cells that have lost specificity for the antigen driving the response can edit [23].

Significantly, the repertoire that supports B-0 differentiation includes Vκ4/5H, the gene used to encode anti-PtC. A high frequency (>50%) of V_H12 B-0 cells from 6-1 mice express Vκ4/5H [12], and 3 out 16 endogenous rearrangements (excluding those with a PtC-binding junctional sequence) of B-0 cells from Vκ1A Dbl Tg mice express this gene (Fig. 2). Thus, the differentiation of transitional V_H12 B cells to mature B-0 cells constitutes the third checkpoint in V_H12 B cell differentiation favoring the generation of PtC-specific B cells. The first limits V_HCDR3 diversity and enriches for 10/G4 V_HCDR3, and the second limits L chain diversity and favoring a subset of κ L chains that includes Vκ4/5H. The combined effect of selection at these three checkpoints is an unprecedented enrichment for B cells specific for a single antigen - PtC.

This enrichment of PtC-specific B-0 cells will enhance the production of PtC-specific B-1 cells. Our analysis of PtC-specific B-1 cell differentiation in V_H12 Tg mice indicates that they derive from B-0 cells [10-13]. The evidence is threefold: First, the segregation to B-1 occurs after immunoglobulin gene rearrangement. Second, in the absence of functional Bruton's tyrosine kinase, a kinase involved in surface IgM signaling, PtC-specific B-1 cell differentiation is impaired and these cells have a B-0 phenotype [11]. Third, intermediates of the B-0 to B-1 pathway have been identified in V_H12 Tg mice [13]. We have shown by adoptive transfer that these cells can differentiate to B-1. Moreover, in the presence of cyclosporin A, an inhibitor of lymphocyte activation, these intermediate cells acquire a more B-0-like phenotype. Wortis and colleagues originally proposed this pathway based on their finding that anti-IgM and IL-6 to induce B-0 cells to acquire a B-1 phenotype in vitro [14, 15]. Thus, selection for B-0 cells expressing the 10/G4 will be to increase the production of PtC-specific B-1 cells.

These data argue for a strong evolutionary selection promoting the generation of PtC-specific B cells. This is also suggested by the observation of Booker and Haughton [24] that the sequences of the V_H12 gene and the V_H11 gene, which also encodes PtC-specific antibodies, are more evolutionarily conserved than other V_H genes. This selection is presumably because anti-PtC provide a survival, although in what way is unknown. Interestingly, anti-PtC antibodies have been shown to provide survival value against bacterial infection in an acute peritonitis model [25]. In addition, anti-PtC are a component of natural IgM, which has recently been demonstrated to be important in focusing virus to secondary lymphoid organs and in providing resistance [26]. Thus, anti-

PtC could be a vital component in the early stages of bacterial or viral infections.

Acknowledgements

We are grateful to Larry Arnold for his many helpful discussions and for the assistance of the UNC Flow Cytometry Center and the UNC Automated Sequencing Facility. This work was supported by the National Institutes of Health grants AI29576 and AI43587 and by grant 79017 from the American Cancer Society and a grant from the Arthritis Foundation.

References

1. Herzenberg LA., Stall AM, Lalor PA, Sidman C, Moore WA, Parks D, Herzenberg LA (1986) The Ly-1 B cell lineage. Immunol Rev 93:81-102
2. Kipps TJ (1989) The CD5 B cell. Adv Immunol 47:117-185
3. Hayakawa K, Hardy R R, Parks D R, Herzenberg LA (1983) The Ly-1 B cell subpopulation in normal, immunodeficient, and autoimmune mice J Exp Med 157:202-218
4. Hayakawa K, Hardy RR, Honda M, Herzenberg LA, Steinberg AD, Herzenberg LA (1984) Ly-1 B cells: functionally distinct lymphocytes that secrete IgM autoantibodies. Proc Natl Acad Sci USA 81:2494-2498
5. Casali P, Burastero SE, Nakamura M, Inghirami G, Notkins AL (1987) Human lymphocytes making rheumatoid factor and antibody to ssDNA belong to Leu-1$^+$ B cell subset. Science 236: 77-81
6. Hardy RR, Hayakawa K, Shimizu M, Yamasaki K, Kishimoto T (1987) Rheumatoid factor secretion from human Leu-1+ B cells. Science 236; 81-83
7. Mercolino TJ, Arnold LW, Hawkins LA, Haughton G (1988) Normal mouse peritoneum contains a large number of Ly-1$^+$ (CD5) B cells that recognize phosphatidyl choline. Relationship to cells that secrete hemolytic antibody specific for autologous erythrocytes. J Exp Med 168:687-698
8. Hayakawa K, Asano M, Shinton SA, Gui M, Allman D, Stewart CL, Silver J, Hardy RR (1999) Positive selection of natural autoreactive B cells. Science 285:113-116
9. Murakami, M, Tsubata T, Okamoto M, Shimizu A, Kumagai S, Imura H, Honjo T (1992) Antigen-induced apoptotic death of Ly-1 B cells responsible for autoimmune disease in transgenic mice. Nature 357:77-80
10. Arnold LW, Pennell CA, McCray SK, Clarke SH (1994) Development of B-1 cells: Segregation of phosphatidyl choline-specific B cells to the B-1 population occurs after immunoglobulin gene expression. J Exp Med 179:1585-1595
11. Clarke SH, Arnold LW (1998) B-1 cell development: evidence for an uncommitted immunoglobulin (Ig)M$^+$ B cell precursor in B-1 cell differentiation. J Exp Med 187;1325-1334
12. Tatu C, Ye J, Arnold LW, Clarke SH (1999) Selection at multiple checkpoints focuses V_H12 B cell differentiation toward a single B-1 cell specificity. J Exp Med 190:903-914
13. Arnold LW, McCray SK, Tatu C, Clarke SH (2000) Identification of a precursor to PtC-specific B-1 cells suggesting that B-1 cells differentiate from splenic conventional B cells in vivo: cyclosporin A blocks differentiation to B-1. J Immunol in press
14. Cong Y-z, Rabin E, Wortis HH (1991) Treatment of murine CD5- B cells with anti-Ig, but not LPS, induces surface CD5: two B-cell activation pathways. Int Immunol 3:467-476

15. Rabin E, Cong Y-z, Wortis HH (1992) Loss of CD23 is a consequence of B-cell activation. In: Herzenberg LA, Haughton G, Rajewsky K (eds). CD5 B Cells in Development and Disease. Annals of the New York Academy of Sciences 651:130-142
16. Ye J, McCray SK, Clarke SH (1996) The transition of pre-BI to pre-BII cells is dependent on the structure of the m/surrogate L chain receptor. EMBO J 15:1524-1533
17. Carsetti R, Kohler G, Lamers MC (1995) Transitional B cells are the target o fnegative selection in the B cell compartment. J Exp Med 181:2129-2140
18. Loder F, Mutschler B, Ray RJ, Paige CJ, Sideras P, Torres R, Lamers MC, Carsetti R (1999) B cell development in the spleen takes place in discrete steps and is determined by the quantity of B cell receptor-derived signals J Exp Med 190:75-89
19. Lam KP, Kuhn R, Rajewsky K (1997) In vivo ablation of surface immunoglobulin on mature B cells by inducible gene targeting results in rapid cell death. Cell 90:1073-1083
20. Gay D, Saunders T, Camper S, Weigert M (1993) Receptor editing: an approach by autoreactive B cells to escape tolerance. J Exp Med 177:999
21. Tiegs SL, Russell DM, Namazee D (1993) Receptor editing in self-reactive bone marrow B cells. J Exp Med 177:1009
22. Melamed D, Benschop RJ, Cambier JC, Nemazee D (1998) Developmental regulation of B lymphocyte immune tolerance compartmentalizes clonal selection from receptor selection. Cell 92:173
23. Han S, Dillon SR, Zheng B, Shimoda M, Schlissel MS, Kelsoe G (1997) V(D)J recombinase activity in a subset of germinal center B lymphocytes. Science 278:301
24. Booker JK, Haughton G (1994) Mechanisms that limit diversity of autoantibodies. II. Evolutionary conservation of the Ig variable region genes which encode naturally occurring autoantibodies. Int Immunol 6:1427-1436
25. Boes M, Prodeus AP, Schmidt T, Carroll MC, Chen J (1998) A critical role of natural immunoglobulin M in immediate defense against systemic bacterial infection. J Exp Med 188:2381-2386
26. Ochsenbein AF, Fehr T, Lutz C, Suter M, Brombacher F, Hengartner H, Zinkernagel RM (1999) Control of early viral and bacterial distribution and disease by natural antibodies Science 286:2156-2159

CD5 and Other Superantigens May Select and Maintain Rabbit Self-renewing B-lymphocytes and Human B-CLL Cells

R. G. Mage[1] and R. Pospisil[2]
[1]Laboratory of Immunology, NIAID, NIH, Bethesda, MD 20892-1892, USA and [2]VIDIA, Videnska 1083, Prague 142 00, Czech Republic

Introduction

Although the rabbit has been the animal model that revealed many of the intricacies of the immune system over the years (reviewed in Mage 1998), it is currently not a popular species for study. Nonetheless the immune system of this species continues to reveal new insights. This report will review our studies in the rabbit which suggested that CD5 and other superantigens influence the survival and expansion of developing self-renewing B-lymphocytes. These studies led us to propose that CD5 also influences development of normal and pathologic human B-cells such as the CD5 positive cells that develop into B-CLL.

The Appendix and GALT as Central Sites of Primary Repertoire Development in Young Rabbits

The search for a mammalian bursal equivalent in the 60's (Archer et al. 1963) was abandoned when it was recognized that mammalian B-lymphocytes develop and undergo gene rearrangements in sites such as bone marrow, fetal liver, spleen or omentum (Cooper and Lawton 1972). However, even the chicken bursa is not a site of V-gene rearrangement but a site where B-lymphocytes undergo cellular expansion and diversification of their rearranged V_H and V_L genes by gene conversion (Pink et al. 1985; Reynaud et al. 1994). With the discovery that most B-lymphocytes in normal rabbits have rearranged the allotype-encoding V_H1 gene (Becker et al., 1990), and that the appendix is a site of early development and expansion of B lymphocytes, the idea of a mammalian bursal equivalent was revived (Weinstein et al. 1994a, 1994b). When we showed that gene conversion is also a major mechanism for V-gene diversification in the appendix, the parallel with the chicken bursa was particularly striking (Weinstein et al. 1994a, 1994b). Surgical removal of the newborn rabbit appendix and other GALT leads to markedly reduced peripheral B cell numbers (Cooper et al. 1968; Vajdy et al. 1998), diminished antibody responses (Cooper et al, 1968; Dasso and Howell

1997) and diminished diversification of rearranged V_H-gene sequences (Vajdy et al. 1998). The fact that diversification does occur in "GALT-less" rabbits is probably explained by recent observations that gene conversion occurs in germinal centers (GCs) of spleens and lymph nodes of adult rabbits (Sehgal et al. 1998; Schiaffella et al. 1999; Winstead et al. 1999). We and others have suggested that some new repertoire may also be developed in adult peripheral GCs as gene conversion occurs to improve specific antibody affinities [e.g.to haptens such as DNP (Sehgal et al. 1998, Schiaffella et al. 1999) or fluorescein (Winstead et al. 1999)].

CD5 and Self-renewing B-cells of Rabbits

In contrast to mouse and human where only a small proportion of B cells express CD5, in rabbits essentially all peripheral B cells express this glycoprotein (Raman and Knight 1992) and most dark zone B cells in appendix GCs express high levels of CD5 (Pospisil et al. 1996; Pospisil and Mage 1998a). CD5+ B cells appear to develop early in ontogeny and be maintained throughout life by self-renewal.

Chronic Allotype Suppression
The fact that most rabbit B-lymphocytes develop early in life and are capable of self renewal in the periphery explains early observations that fetal or neonatal exposure to antibodies to kappa light chain (b-allotypes) or to heavy chain V_Ha allotypes resulted in chronic depressed expression of the targeted allotype (reviewed in Mage, 1975). The partial recovery from suppression with time is also compatible with our recent observations that some cells with rearranged germline V_H (Sehgal et al. 1998a) and V_κ sequences (Sehgal et al. ms in preparation) are found in developing splenic GCs of adults. This suggests that although B-lymphopoiesis may be limited in adult rabbits (Crane et al. 1996), it is not totally absent.

B-Cell Survival and Expansion

Gut Flora as Superantigens, Mitogens or Antigens
The GCs that arise in primary lymphoid tissue such as young rabbit appendix may be driven to develop by superantigen, self-antigen or other mediators of proliferation but these reactions are not necessarily specific for one particular antigen. Although GCs develop in the absence of antigen in the chicken bursa and sheep ileal Peyer's patches, gut flora stimulate B-cell expansion in follicles after hatching or birth (Ekino 1993; Reynolds and Morris 1984). Studies of germ-free rabbits showed that lymphoid follicles do not develop normally in the absence of

gut flora (Stepankova et al. 1980). We suggested that in these rabbits, CD5 and other self antigens can provide B cells with survival signals and limited expansion in the absence of stimulation by environmental antigens (Pospisil and Mage 1998a).

Development of New V_H Framework Region Sequences ($V_H a2$ allotype) in Mutant *ali* Rabbits

In the mutant *Alicia* rabbit (Kelus and Weiss 1986), a small deletion at the 3' end of the V_H gene cluster (Allegrucci et al., 1990; Becker and Knight, 1990) led to loss of the $V_H 1$ gene that encodes the a2 allotype (Knight and Becker, 1990). Although homozygous mutant *ali/ali* rabbits lack the $V_H 1a2$ gene, gene-conversion-like changes lead to B cells with a2-like surface Ig. We found that alterations in FR1 and FR3 sequences could be accounted for by gene-conversion-like changes that utilized candidate donor sequences upstream of a rearranged $V_H 4$ gene, the first functional gene in the mutants' V_H region cluster (Chen et al., 1993, 1995; Sehgal et al. 1998a). Our studies of the appearance of cells bearing a2-like epitopes in the appendix of mutant rabbits led us to the conclusion that there was positive selection and expansion of B cells based on framework region structures ($V_H a$ allotypes) (Pospisil et al., 1995). We proposed that the preferential expansion and survival of B cells based on FR1 and FR3 expression involved "superantigen"-like interactions with endogenous and exogenous ligands. The exogenous ligands probably come from gut flora; one endogenous ligand appears to be CD5 (Pospisil et al., 1996; Pospisil and Mage 1998).

CD5 is a Ligand for B-cell Surface Immunoglobulin

After we showed that "positive" selection occurs during B-lymphocyte development in the rabbit appendix and that the selection favors B cells expressing surface immunoglobulins with $V_H a2$ structures (Pospisil et al., 1995), we searched for possible ligands present in the young rabbit appendix that might interact with the a2 epitopes. We were surprised to discover that in a cell-binding assay, F(ab')$_2$ fragments, especially those bearing $V_H a2$ framework region determinants, specifically interacted with IgM+ appendix B-cells. Furthermore, immobilized F(ab')$_2$ fragments could selectively bind and permit isolation of CD5 from lysates of surface-biotinylated appendix cells. We proposed that interactions of V_H framework region structures with CD5 may affect maintenance and selective expansion of particular B cells and thus contribute to autostimulatory growth (Pospisil et al.1996; Pospisil and Mage 1998a, 1998b). The amount of signaling and qualitative differences in signaling may determine B-cell negative or positive selection. CD5-V_H interaction alone may induce a signal that is sufficient to promote expansion and/or survival of B cells or may influence the fate of B-cell selection in combination with other signals

Translation of Findings Made in Rabbit to Man

As in the Rabbit, Human Ig Binds to CD5 and Permits Isolation of CD5 From Lysates of CLL Cell Lines

As previously described for rabbit F(ab')$_2$ and CD5 (Pospisil et al. 1996), the binding of lysates of surface-biotinylated EBV-CLL cells to human immunoglobulin (HIg) covalently coupled to Dynabeads likewise enabled isolation of CD5. If cell lysates from biotinylated EBV-CLL cells were first preincubated with anti-CD5-coated beads to remove CD5, the CD5 molecules were no longer bound and recovered from HIg-beads (Pospisil et al., 1999).

Human Ig Binds to Purified Recombinant CD5

To further analyze the binding interaction of human Ig to CD5, we tested the binding of biotinylated HIg to purified recombinant CD5 molecules by ELISA. These results are summarized in Fig 1.

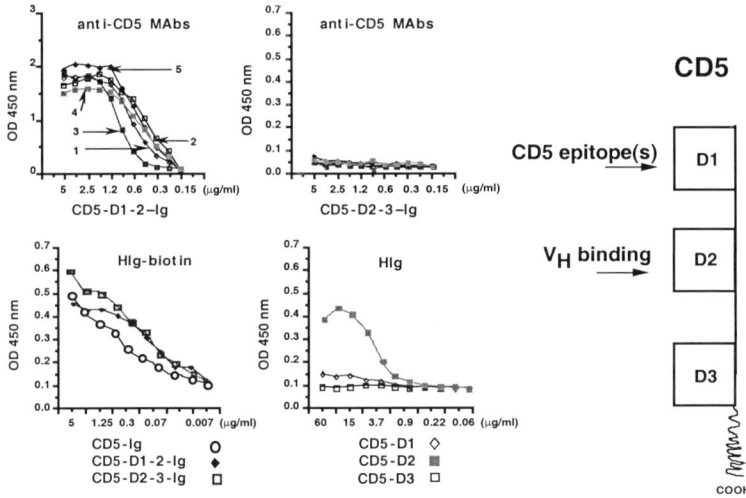

Fig. 1 (Adapted from Pospisil et al. 1999) CD5 epitopes map to the D1 domain of human CD5 and V$_H$ binding maps to the CD5-D2 domain. The materials and methods used can be found in Pospisil et al. 1999. Top panels. Anti-CD5 antibodies bound to CD5-D1-2 proteins but not to CD5-D2-3 proteins. The anti-CD5 antibodies tested were: 1-BL1a (Immunotech), 2-Leu-1 (Becton Dickinson), 3-MCA1341 (Serotech), 4-NCL-CD5-2 (Novocastra Laboratories Ltd) and 5-T1 (Coulter Corp.). Bottom panels. HIg bound to hCD5-Ig fusion proteins representing the full length extracellular portion of human CD5 (CD5Ig) and to CD5-D1-2-Ig, CD5-D2-3-Ig. Using a more sensitive sandwich ELISA assay HIg also bound to D2 protein but not to D1 or D3 proteins.

The extracellular region of CD5 is composed of three scavenger receptor cysteine-rich (SRCR) domains termed D1, D2 and D3 (Starling et al. 1997). When a fusion protein (hCD5-Ig) containing all of the extracellular domains of human CD5 was coated on ELISA plates, both anti-CD5 antibodies and HIg bound. Using CD5 fusion proteins containing CD5-D1-2 or CD5-D2-3 domains, HIg bound to both CD5-D1-2 and CD5-D2-3 proteins. Using a more sensitive sandwich ELISA assay, we were able to show that HIg also bound to the single D2 domain but not to D1 or D3 domains. When CD5 fusion proteins containing single domains were used, the murine monoclonal anti-CD5 antibodies bound detectably to D1 only, but not to D2 or D3. All five anti-CD5 antibodies tested bound to CD5-D1-2 proteins but not to CD5-D2-3. These data suggest that anti-CD5 antibodies recognize epitope(s) in the first extracellular SRCR domain of human CD5 and are consistent with reports that anti-murine CD5 monoclonal antibodies also bind to the D1 domain of murine CD5 (Starling et al. 1997). Calvo et al. (1999) reported similar results with a panel of nine different anti-CD5 monoclonal antibodies; eight reacted with epitopes assigned to D1 and one (83-C4) reacted with D3. In contrast to D1-domain specificity of the tested anti-CD5 antibodies, our data document that the V_H-mediated binding of HIg to CD5 targets the extracellular D2 domain (Pospisil et al. 1999).

Skewed Expression of V_H Genes in CLL. Correlation with V_H-CD5 Binding/Inhibition

The expression of V_H genes in CLL is skewed toward particular members of V_H families. Compared to some earlier studies (Kipps 1993) recent data evaluating expressed Ig rather that Ig gene rearrangements found a higher proportion of CLL cases expressed V_H1 family members; members of the V_H3 gene family were underrepresented and V_H1 and V_H5 family members were overrepresented. Of the V_H1 family genes expressed in CLL, 51p1-related (V_H1-69) genes predominate. The V_H4-34 (V_H4-21) gene is also overrepresented (Dighiero et al. 1996; Fais et al. 1998). In addition, the presence of mutations varies depending upon the V_H family expressed by the B-CLL cell. A hierarchy was seen with mutations in $V_H3 > V_H4 > V_H1$ (Fais et al. 1998). We asked whether the interaction of CD5 with surface Ig is V_H family restricted. We tested a panel of human monoclonal Igs of different V_H families for their capacity to inhibit anti-CD5 staining of CLL cells as measured by flow microfluorometry. Igs encoded by members of the V_H4 and V_H1 families showed the highest inhibition. Among V_H4 monoclonal Igs, one is known to use the V_H4-34 (V_H4-21) gene and two of the V_H1 monoclonal Igs are known to use 51p1-related (V_H1-69) genes. Although the binding of CD5 to V_H does not appear to be restricted to one family, there is a correlation of higher inhibition with V_H families that are over-represented in B-CLL. There is no apparent correlation with V_L usage. Variable inhibitions were observed with different members of a given V_H family possibly because of different V_H sequences (Pospisil et al. 1999). Interestingly, in a recent study of V_H genes expressed in familial CLL patients, both peripheral white blood cell counts and the levels of CD5 on the surface of the patient's cells were significantly lower when the V_H

gene sequences were diversified rather than germline in sequence (Sakai et al. 2000). These inverse correlations could reflect the influence of CD5-Ig interaction on proliferative capacity of the leukemic clone.

B-CLL cells : malignant equivalent to follicular mantle-zone B-cells

The continual "tickling" of the BCR by diverse low-affinity self-antigens was suggested to provide a form of continuously acting positive selection on the primary B-cell pool as well as on secondary B cells (Neuberger 1997). Recently it was shown that signals induced via the transmembrane BCR, can override the death signal delivered by CD95. Signaling rescued cells directly in a very rapid fashion without up-regulation of new anti-apoptotic proteins (Catlett and Bishop 1999). In addition, the development of B-1 cells greatly depends on the density of specific BCRs at the cell surface (Lam and Rajewsky 1999) and on the strength of signals through BCR triggered by self-antigens (Watanabe et al. 1999). The normal human B cell counterpart to CLL may be found in mantle zones of the secondary B cell follicles (Lankester et al. 1995). B-CLL are clonal populations of membrane IgM$^+$ or IgM$^+$/mIgD$^+$ CD5$^+$ B cells that appear to be arrested in a differentiation stage phenotypically resembling follicular mantle-zone B-cells. When B cells first enter the lymphoid tissue they are unable to directly migrate into follicles. Instead, they accumulate in the interfollicular regions, the outer T cell zone, red pulp and in the marginal zone (Lortan et al., 1987, Forster et al., 1996). Signaling by the BCR was shown to promote follicular migration of B cells entering lymphoid tissue. The persistence signal needed for B-cell maintenance in the interfollicular regions of the appendix, tonsils and other lymphoid organs may be provided by the interaction of surface Ig with CD5 on the same or other B cells (**Fig**. 2).

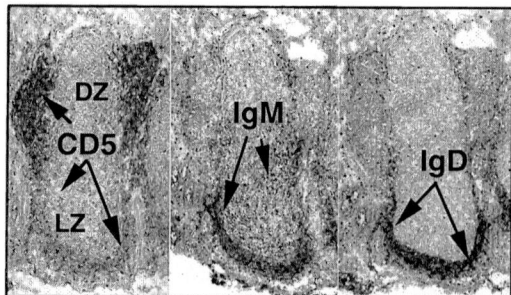

Fig. 2 Anti-CD5 staining of cryostat sections (7 µm) of human appendix from a 9-year-old defines T-cell areas, mantle zone IgM-rich and IgD-rich areas, germinal center (GC) cells and some cells in interfollicular regions. BCR-CD5 interaction in these locations may signal persistence.

Although prior appendectomy may constitute a risk factor for some leukemias and other neoplastic diseases (Howson, 1983), in one study (Bierman, 1968), CLL patients had a somewhat lower incidence of previous appendectomies than patients who did not have neoplastic disease. The appendix may be an important site of early events in human B cell development and selection of pathologic B-cell clones. As many as 50% of CLL patients have leukemia cells that use V_H1-69, V_H4-34 (V_H4-21), V_H4-39 (V_H4-18) and V_H5-51 (V_H5-251) genes. These genes are also often found expressed by follicular mantle zone B cells. The presence of these genes may constitute a risk factor for developing B-CLL (Dighiero et al., 1996). CD5 may stimulate and select B cells expressing these genes. The selection process acting on certain expressed V_H may be CLL-related, providing B cells with survival signals.

Concluding Remarks and Summary

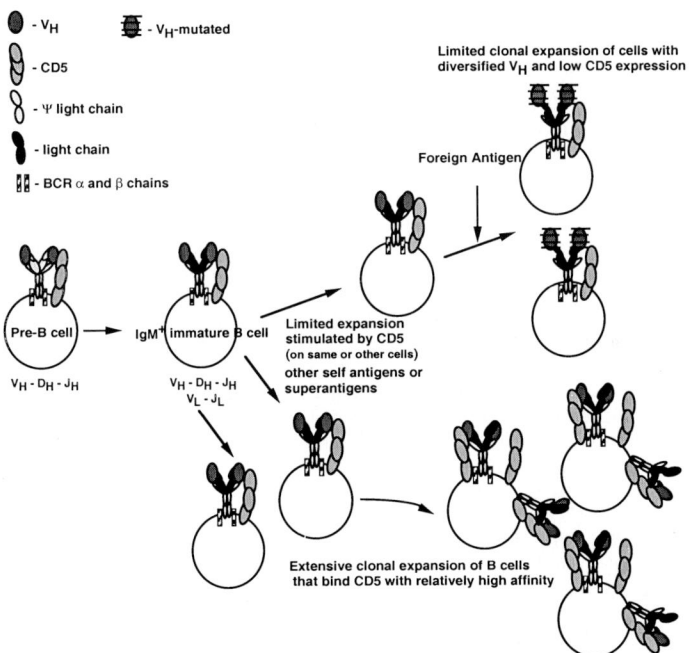

Fig. 3 Possible role(s) of CD5-V_H interaction in the evolution of normal and pathologic B cell clones. B cell repertoires are influenced by combinatorial, expression, and selection biases at various stages of differentiation. The persistence signals needed for B cell maintenance and survival may be provided by the interaction of surface Ig with CD5. B-1 cell development depends on the density of BCRs of certain specificities at the cell surface. Mutations of V_H gene sequences could affect CD5 binding and the strength of signals through the BCR triggered by self antigens.

Together, our previous studies in the rabbit model (Pospisil et al, 1996, 1998a, 1998b) and the data on human V_H (Pospisil et al. 1999) indicate that the CD5-V_H interaction does not represent conventional antibody-ligand binding but is likely to be associated with framework regions as changes in V_H sequences may alter or eliminate binding. Two recent reports suggest that B-CLL represents transformation of B cells at two distinct differentiation stages and with different clinical courses (Damle et al. 1999; Hamblin et al. 1999). Notably, those with unmutated V_H sequences and high frequencies of occurrence of a few V_H genes such as V_H1-69, may have higher CD5 expression and clearly have a more aggressive form of CLL. CD5 as a ligand for V_H framework regions may maintain, select or expand normal, autoimmune or transformed B cells and contribute to skewing of the V_H repertoires in normal B-lymphocytes and B-CLL cells.

Acknowledgements

The contributions of numerous colleagues whose materials, efforts and ideas contributed to the work presented here are gratefully acknowledged. They include: C. Alexander, A.O. Anderson, A. Aruffo, M.A. Bowen, H. Chen, G. Cooper, J. Dasso, M. Fitts, P. Fuschiotti, G.E. Marti, H. Obiakor, E. Schiaffella, D. Sehgal, G. Silverman, and P. Weinstein.

References

Allegrucci M , Newman BA , Young-Cooper GO , Alexander CB , Meier D , Kelus AS , Mage RG (1990) Altered phenotypic expression of immunoglobulin heavy-chain variable-region (V_H) genes in Alicia rabbits probably reflects a small deletion in the V_H genes closest to the joining region. Proc Natl Acad Sci USA 87:5444-5448

Allegrucci M , Young-Cooper GO , Alexander CB , Newman BA , Mage RG (1991) Preferential rearrangement in normal rabbits of the 3' V_{Ha} allotype gene that is deleted in Alicia mutants; somatic hypermutation/conversion may play a major role in generating the heterogeneity of rabbit heavy chain variable region sequences. Eur J Immunol 21:411-417

Archer OK, Sutherland DER, Good RA (1963) Appendix of the rabbit: a homologue of the bursa in the chicken? Nature 200:337-339

Becker RS, Suter M, Knight KL (1990) Restricted utilization of V_H and C_H genes in leukemic rabbit B cells. Eur J Immunol 20:397-402

Becker RS, Knight KL (1990) Somatic diversification of immunoglobulin heavy chain VDJ genes: evidence for somatic gene conversion in rabbits. Cell 63: 987-997

Bierman HR (1968) Human appendix and neoplasia. Cancer 21:109-118

Calvo J, Padilla O, Places L, Vigorito E, Vila JM, Vilella R, Mila J., Vives J, Bowen MA Lozano F (1999) Relevance of individual CD5 extracellular domains on antibody recognition, glycosylation and comitogenic signaling. Tissue Antigens. 54:16-26

Catlett IM, Bishop GA (1999) A novel mechanism for rescue of B cells from CD95/Fas-mediated apoptosis. J.Immunol. 163:2378-2381

Cooper MD, Perey DY, Gabrielsen AE, Sutherland DE, McKneally MF, Good RA. (1968) Production of an antibody deficiency syndrome in rabbits by neonatal removal of organized intestinal lymphoid tissues. Int. Arch. Allergy. Appl. Immunol. 33:65-88

Cooper MD and Lawton AR (1972) The mammalian "bursa equivalent": Does lymphoid differentiation along plasma cell lines begin in gut-associated lymphoepithelial tissues (GALT) of mammals? Contemp Topics Immunobiol 1:49-68

Crane MA, Kingzette M, Knight KL (1996) Evidence for limited B-lymphopoiesis in adult rabbits. J. Exp. Med. 183:2119-2121

Damle RN, Wasil T, Fais F, Ghiotto F, Valetto A, Allen SL, Buchbinder A, Budman D, Dittmar K, Kolitz J, Lichtman SM, Schulman P, Vinciguerra VP, Rai KR, Ferrarini M, Chiorazzi N (1999) Ig V gene mutation status and CD38 expression as novel prognostic indicators in chronic lymphocytic leukemia. Blood 94: 1840-1847

Dasso JF, Howell MD (1997) Neonatal appendectomy impairs mucosal immunity in rabbits. Cellular Immunol 182: 29-37

Dighiero G, Kipps T, Schroeder HW, Chiorazzi N, Stevenson F, Silberstein, LE, Caligaris-Cappio F, Ferrarini M (1996) What is the CLL B-lymphocyte? Leukemia and Lymphoma (Suppl 2) 22, 13-39.

Ekino S (1993) Role of environmental antigens in B cell proliferation in the bursa of Fabricius at neonatal stage. Eur J Immunol 23:772-775

Fais F, Ghiotto F, Hashimoto S, Sellers B, Valetto A, Allen SL, Schulman P, Vinciguerra VP, Rai K, Rassenti LZ, et al.: (1998) Chronic lymphocytic leukemia B cells express restricted sets of mutated and unmutated antigen receptors. J Clin Invest 102: 1515-1525

Forster R, Mattis AE, Kremmer E, Wolf E, Brem G, Lipp M (1996) A putative chemokine receptor, BLR1, directs B cell migration to defined lymphoid organs and specific anatomic compartments of the spleen. Cell 87: 1037-1047

Hamblin TJ, Davis Z, Gardiner A, Oscier DG, Stevenson FK (1999) Unmutated Ig V_H genes are associated with a more aggressive form of chronic lymphocytic leukemia. Blood 94: 1848-1854

Howson CP (1983) Appendectomy and subsequent cancer risk. J Chronic Dis 36: 391-396

Kelus AS, Weiss S (1986) Mutation affecting the expression of immunoglobulin variable regions in the rabbit.Proc Natl Acad Sci U S A 83:4883-4886

Kipps TJ (1993) Immunoglobulin genes in chronic lymphocytic leukaemia. Blood Cells 19: 615-625.

Knight KL, Becker RS (1990) Molecular basis of the allelic inheritance of rabbit immunoglobulin VH allotypes: implications for the generation of antibody diversity. Cell 60: 963-970

Lam K-P Rajewsky K (1999) B cell antigen receptor specificity and surface density together determine B-1 versus B-2 cell development. J.Exp.Med. 190: 471-477

Lankester AC, van Schijndel GM, van der Schoot CE, van Oers MH, van Noesel CJ, van Lier R A (1995) Antigen receptor nonresponsiveness in chronic lymphocytic leukemia B cells. Blood 86: 1090-1097

Lortan JE, Roobottom CA, Oldfield S, MacLennan IC (1987) Newly produced virgin B cells migrate to secondary lymphoid organs but their capacity to enter follicles is restricted. Eur J Immunol 17:1311-1316

Mage RG (1975) Allotype suppression in rabbits: effects of anti-allotype antisera upon expression of immunoglobulin genes.Transplant. Rev. 27: 84-99 1-152

Mage RG (ed) (1998) Immunology of Lagomorphs. In: Pastoret PP, Bazin, H, Griebel P, Govaerts H (eds) Handbook of Vertebrate Immunology . Academic Press, London, pp 223-260

Neuberger MS (1997) Antigen receptor signaling gives lymphocytes a long life. Cell 90: 971-973

Pink JR, Ratcliffe MJ, Vainio O (1985). Immunoglobulin-bearing stem cells for clones of B (bursa-derived) lymphocytes. Eur. J. Immunol. 15:617-620.

Pospisil R, Young-Cooper GO, Mage RG (1995) Preferential expansion and survival of B lymphocytes based on V_H framework 1 and framework 3 expression: "Positive" selection in appendix of normal and V_H-mutant rabbits. Proc Natl Acad Sci USA 92:6961-6965

Pospisil R, Fitts MG, Mage RG (1996) CD5 is a potential selecting ligand for B cell surface immunoglobulin framework region sequences. J Exp Med 184:1279-1284

Pospisil R, Mage, RG (1998a) Rabbit Appendix— A Site of Development and Selection of the B-Cell Repertoire. In: Kelsoe G, Flajnik M (eds) Somatic Diversification of the Immune Response. Springer-Verlag, Heidelberg, pp 59-70. (Contemporary Topics in Microbiology and Immunology, vol 229)

Pospisil R, Mage RG (1998b) CD5 and other superantigens as "ticklers" of the B cell receptor. Immunology Today 19:106-108

Pospisil R, Silverman GJ, Marti GE, Aruffo A, Bowen MA, Mage RG (1999) CD5 is a Potential Selecting Ligand for B-Cell Surface Immunoglobulin: A Possible Role in Maintenance and Selective Expansion of Normal and Malignant B Cells. Leukemia & Lymphoma In press

Raman C, Knight KL (1992) CD5+ B cells predominate in peripheral tissues of rabbit. J

Immunol 149:3858-3864

Reynaud CA, Bertocci B, Dahan A, Weill JC (1994) Formation of the chicken B-cell repertoire: ontogenesis, regulation of Ig gene rearrangement, and diversification by gene conversion. Adv Immunol 57:353-378

Reynolds JD, Morris B (1984) The effect of antigen on the development of Peyer's patches in sheep. Eur J Immunol 14:1-6

Sakai A, Marti GE, Caporaso N, Pittaluga S, Touchman JW, Fend F, Raffeld M (2000) Analysis of expressed immunoglobulin heavy chain genes in familial B-CLL. Blood In press

Schiaffella E, Sehgal D, Anderson AO, Mage RG (1999) Gene conversion and hypermutation during diversification of V_H sequences in developing splenic germinal centers of immunized rabbits. J Immunol 162: 3984-3995

Sehgal D, Mage RG, Schiaffella E (1998a) V_H mutant rabbits lacking the V_H1a2 gene develop a2+ B cells in the appendix by gene conversion-like alteration of a rearranged V_H4 gene. J Immunol. 160: 1246-1255

Sehgal D, Schiaffella E, Anderson AO, Mage RG(1998) Analyses of single B cells by PCR reveals rearranged V_H with germline sequences in spleens of immunized rabbits: Implications for B cell repertoire maintenance and renewal. J Immunol 161: 5347-5356

Starling SGC, Llewellyn MB, Whitney GS, Aruffo A (1997) The Ly-1.1 and Ly-1.2 epitopes of murine CD5 map to the membrane distal scavenger receptor cysteine-rich domain. Tissue Antigens 49: 1-6

Vajdy M, Sethupathi P, Knight K L (1998) Dependence of antibody somatic diversification on gut-associated lymphoid tissue in rabbits. J Immunol 160: 2725-2729

Watanabe N, Nisitani S, Ikuta K, Suzuki M, Chiba T, Honjo T (1999) Expression levels of B cell surface immunoglobulin regulate efficiency of allelic exclusion and size of autoreactive B-1 cell compartment. J Exp Med 190: 461-469

Weinstein PD, Anderson AO, Mage RG (1994a) Rabbit IgH sequences in appendix germinal centers: VH diversification by gene conversion-like and hypermutation mechanisms. Immunity 1: 647-659

Weinstein PD, Mage RG, Anderson AO (1994b) The appendix functions as a mammalian bursal equivalent in the developing rabbit. Adv Exp Med Biol 355: 249-253

Winstead CR, Zhai S-K, Sethupathi P, Knight K L (1999) Antigen-induced somatic diversification of rabbit IgH genes: gene conversion and point mutation. J Immunol 162: 6602-6612

Selection in the Mature B Cell Repertoire

F. Martin and J.F. Kearney
Department of Microbiology, Division of Developmental and Clinical Immunology, University of Alabama at Birmingham, AL, 35294-3300, USA

Mature B lymphocytes have passed through filters of negative and positive selection and have acquired the ability to generate effector cells that secrete antibodies (plasma cells) or present antigen to T cells (antigen presenting cells) upon antigenic encounter. These B cells are derived through transitional stages from immature B cells that are generated throughout life in the mouse bone marrow. Mature B cells have an *in vivo* half-life in the range of weeks, while the immature B cells live only for a few days unless selected to enter the long-lived repertoire (Schittek et al., 1991; Hartley et al., 1993; Cyster and Goodnow, 1995; Cyster, 1997). In the immune response to a foreign antigen, there are precursors within the mature B cell repertoire that have never experienced an acute antigenic encounter (primary repertoire) and others whose phenotypic and functional characteristics have been conditioned by a previous acute antigenic exposure (secondary repertoire) represented by memory B cells. The requirement for an "acute" encounter, defined as an antigenic experience that includes the generation of antigen specific effector cells, is mandatory for the above definition. However these criteria are "blurred" by the accumulating evidence for positive selection signals in the generation of the mature primary B cell repertoire, i.e. to what extent is a BCR mediated signal required early in the development of the mature B cell repertoire. In the case of T cells, there are certain similarities between signals involved in positive selection in the thymus and acute antigenic activation in the periphery. Most importantly the signal transmitted through the TCR-MHC/peptide interaction is similar to a peripheral encounter except perhaps for the affinity of the interaction, while still little is known about the requirement for co-signals. Although future experiments will dissect the similarities and differences between developmentally associated BCR signals and those involving acute antigen encounter, it is clear that BCR mediated signaling is essential for both the positive and negative selection and response to foreign antigen exposure.

Mature B lymphocytes have been classified based on surface phenotypes, topographical location and functional characteristics, into subsets with different generation and maintenance requirements (Kantor et al., 1992; Stall et al., 1996). B1 cells, enriched in the peritoneal and pleural cavities, are generated and maintained through continuous IgM signaling, and as a result, the B1 repertoire is enriched with certain antibody specificities (Bhat et al., 1992; Hardy et al., 1994; Arnold et al., 1994). Inactivation of several molecules associated with signaling through surface IgM-ligand interactions (*btk*, Igα tail, *vav*, PKCβ, PI3 kinase p85α etc.) decreases the B1 significantly more than the B2 compartment (Tarakhovsky, 1997; Khan et al., 1995; Kerner et al., 1995; Torres et al., 1996; Inaoki et al., 1997; Tarakhovsky et al., 1995; Leitges et al., 1996; Fruman et al., 1999; Suzuki et al., 1999; Hayakawa et al., 1999). The mechanisms involved in each of the above cases are not entirely clear and involve both insufficient signaling when a positive regulator is deficient, as well as excessive signaling, when a negative signal is missing. For example, an unexpected negative regulatory function for Igα tail in IgM signaling has been recently described (Torres and Hafen, 1999; Kraus et al., 1999).

The mature B2 population seems to require also a low level of IgM receptor engagement for their development and maintenance (Cyster et al., 1996; Lam et al., 1997). Phenotypic and topographical heterogeneity exists within the mature B2 cell population with IgM^{lo} IgD^{hi} $CD21^{int}$ $CD23^{hi}$ B cells recirculating between B lymphoid follicles (FO) in spleen and lymph nodes and IgM^{hi} IgD^{lo} $CD21^{hi}$ $CD23^{lo}$ non-recirculating cells enriched primarily in the marginal zone (MZ) of the spleen (Gray et al., 1982; Oliver et al., 1997; Martin and Kearney, 2000). Studies in rats have shown that thoracic duct recirculating cells contain MZ B cell precursors and also that T-dependent memory cells colonize the marginal zone and are mobilized from this site by a new antigen encounter (Kumararatne and MacLennan, 1982; Liu et al., 1991).

In addition to a bona-fide IgM signal, several co-signaling molecules that modify the threshold of the main signal have been shown to amplify or inhibit the outcome of the selection process. These molecules include the CD19/CD21/CD81 co-signal(Engel et al., 1995; Rickert et al., 1995; Ahearn et al., 1996; Molina et al., 1996; Miyazaki et al., 1997; Tsitsikov et al., 1997), the CD22/SHP-1/*lyn* pathway (Cornall et al., 1998), CD5 (Bikah et al., 1996) and probably other molecules less well characterized at this time. It was also shown that signals mediated through CD40 participate in B cell maintenance/maturation particularly when associated with deficiencies in *btk* signaling (Khan et al., 1997; Oka et al., 1996).The majority of newly-formed B lymphocytes produced daily in the bone marrow do not enter the mature B cell repertoire (MacLennan, 1998; Cyster, 1997). However, the critical molecular interactions and microenvironments that select which clones will be successfully recruited are still poorly understood.

In the 81x transgenic mouse model we have shown that the mature B cell repertoire is biased by the expression of the VH transgene and the mice have larger marginal zones, smaller follicles and reduced numbers of B1 cells (Martin et al., 1997; Chen et al., 1997). More recently the process of B cell selection into the marginal zone selection was analyzed in detail. Using clonal markers this process was shown to be the result of a selective enrichment from newly formed bone marrow cells through splenic transitional B cells and finally the acquisition of a marginal zone B cell phenotype. As seen in Table 1, when $CD21^{int}CD23^{high}$ and $CD21^{high}CD23^{low}$ B cells were sorted from adult 81x transgenic spleen as representatives of follicular (FO) and marginal zone (MZ) cells, and their immunoglobulin light chain genes were sequenced, their repertoire was quite different.

Table 1. Vκ – Jκ sequences in sorted splenic B-cell subpopulations from 81x transgenic mice

CD21^{int} CD23^{high}	CD21^{high} CD23^{low}	CD21^{high} CD23^{low} 35-1$^+$	CD21^{high} CD23^{low} 35-1$^-$
Vk1C – Jk5^a (1)^b	**Vk1C – Jk5 (6)**	**Vk1C – Jk5 (10)**	Vk1C – Jk1 (1)
Vk1C – Jk1 (1)	Vk1 II – Jk1 (1)		Vk1A – Jk5 (1)
Vk1C – Jk2 (1)	Vk8 I a – Jk5 (1)		Vk1 II – Jk1 (1)
Vk2 II – Jk5 (1)	Vk10 V b – Jk5 (1)		Vk2 II – Jk1 (1)
Vk4/5 IV a^c – Jk1 (1)			Vk4/5 IV a – Jk2 (1)
Vk4/5 VI – Jk4 (1)			Vk4/5 IV b – Jk1 (1)
Vk8 I a – Jk1 (1)			Vk4/5 VI a – Jk2 (1)
Vk10 V a – Jk1 (1)			Vk4/5 VI a – Jk5 (1)
Vk23 V – Jk1 (1)			Vk4/5 VI b – Jk5 (1)
VkRF V a – Jk1 (1)			Vk8 I a – Jk5 (1)
			Vk8 I b – Jk2 (1)
			Vk19/28 V – Jk1 (1)

^{a} all Vk1C – JK5 have the canonical junction recognized by the 35-1 anti-Id
^{b} numbers in parenthesis represent the number of times a given sequence was found
^{c} individual genes within the same family and group are designated by lower case

Six of nine sequences derived from the MZ were identical and presented a canonical joint (VκlC-Jκ5 recognized by the anti-Id antibody 35-1) while no repeats appeared in 10 FO sequences. Second we determined if there are other dominant clones beside the VκlC-Jκ5 canonical rearrangement inside the CD21^{high}CD23^{low} population enriched in the MZ of the spleen. As expected, all 10 clones sequenced from the CD21^{high}CD23^{low}35-1$^+$ sorted cells contained the canonical composition while from 12 clones derived from the CD21^{high}CD23^{low}35-1$^-$ cells no repeats were found (Table 1). This shows that clones beside the Id$^+$ (35-1$^+$ clone) do not go through such an extreme enrichment process in the 81x TG mice. These data confirm at a molecular level the enrichment of the idiotype bearing B cells expressing the canonical VκlC-Jκ5 light chain and its validity as a marker for these B cells.

Altering the composition of the B-cell receptor through N region additions decreases the rate of clonal production of this particular rearrangement and its MZ enrichment. This process can be recapitulated *in vitro* by purified CD21^{low} immature bone marrow B cells and is due to preferential clonal survival (Martin and Kearney, 2000). We also showed that generation of the entire marginal zone depends on an intact CD19 molecule while the enrichment process requires a functional *btk* tyrosine kinase. As seen in Fig.1, the percentage of B cells is reduced in TG xid mice in comparison to TG (81x) mice and that the Id enrichment in the MZ is ablated. We tried to rescue these defects by crossing the TG xid on a bcl-xL transgenic background (Fang et al., 1996). In these mice the bcl-xL transgene is expressed in the early immature bone marrow B cell development stages as well as in immature splenic B cells but is downregulated in mature B lineage stages (T. Behrens – personal communication).

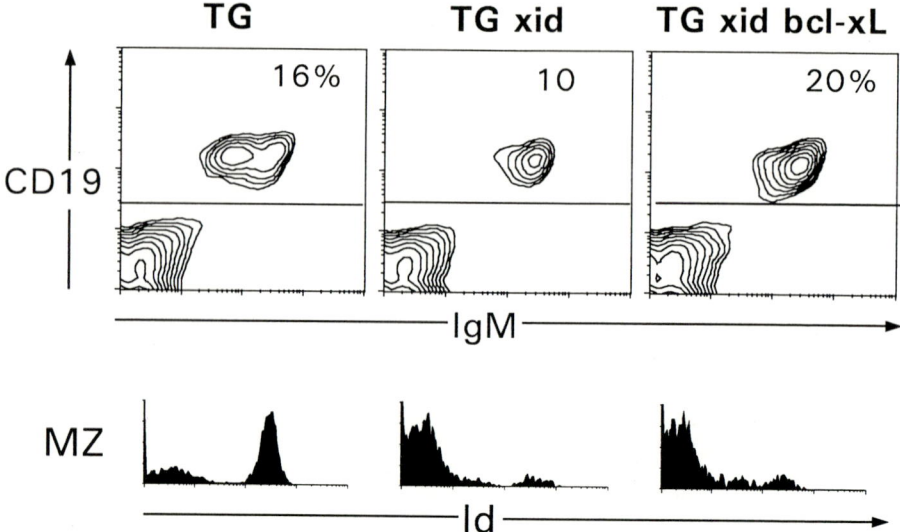

Fig. 1. Splenic B cell compartments in TG (81x), TG xid and TG xid bcl-xL mice. Spleen cells were stained with antibodies against IgM, CD19, CD21 and Id (35-1) and analyzed by flowcytometry. Lymphoid cells are displayed as two color contour plots (*top*) and cells of the marginal zone (MZ) – $IgM^{high}CD21^{high}$ are gated for the Id expression (*bottom*).

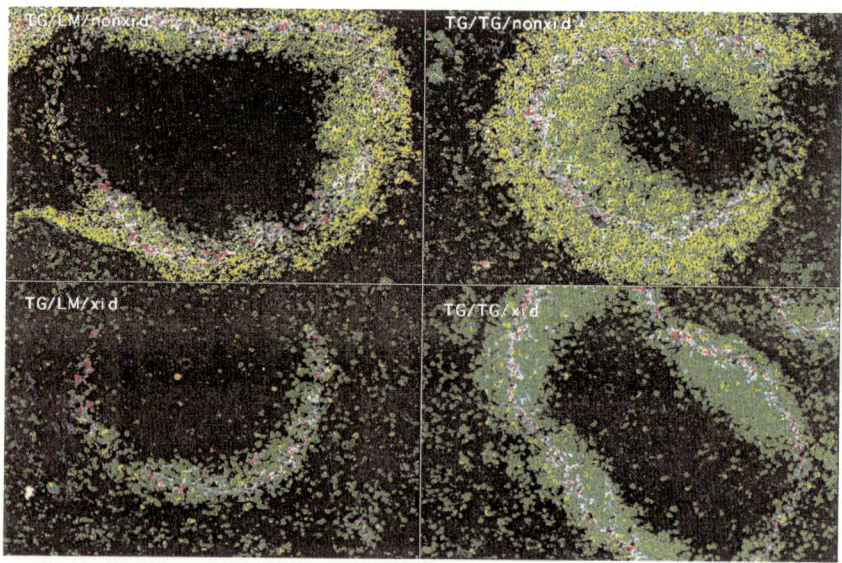

Fig. 2. Fluorescent microphotographs of spleen sections from TG (81x) (*top left*), TG bcl-xL (*top right*), TG xid (*bottom left*) and TG xid bcl-xL (*bottom right*). Spleen sections are stained with anti-IgM (green), MOMA-1 (blue) and anti-Id (35-1, orange).

This transgene should permit rescue of bcl-xL dependent steps in the early stages up to the splenic transitional cells but not later in the maintenance of mature B cells. Indeed the presence of the bcl-xL transgene rescues the defect in total B cell numbers, but does not rescue the process of Id enrichment in the MZ. This rescue in cell number but not in Id$^+$ B cells can also be seen in fluorescent photomicrographs of the spleen from the mice described in Fig. 2. The introduction of the bcl-xL transgene in 81x mice increases the size of both the FO and MZ compartments (compare the top panels - TG/LM/nonxid with TG/TG/nonxid) while in the absence of the *btk* function, both compartments are decreased (compare top and bottom left panels). In 81x xid mice (bottom left) not only the B cell follicles and marginal zones are very small but, in addition large numbers of B cells (green) are scattered in an unorganized pattern throughout the red pulp and T cell zones. Introduction of bcl-xL in 81x xid, restores the B cell follicles and marginal zones to sizes comparable to the regular trangenic mice (compare bottom right with top left). Idiotype positive cells (orange) are abundant in 81x and 81x bcl-xL transgenic mice (top panels), but not 81x xid and 81x xid bcl-xL (bottom panels).

These data taken together suggest a scenario similar to that modeled in Fig. 3. Normally B cells mature from newly formed (NF) through transitional (TR) stages and into the marginal zone (MZ). As showed by us and others, B cells undergo a great deal of selection before they reach the MZ repertoire, with some clones being at an advantage (striped cells – Id$^+$ cells in our case) and other at a disadvantage (the filled cells) in going through these processes. It seems that *btk* is playing a key role in these transitions, since in its absence the entire size of the compartments is reduced, and the selection of certain clones over others is lost (see *btk* ko).

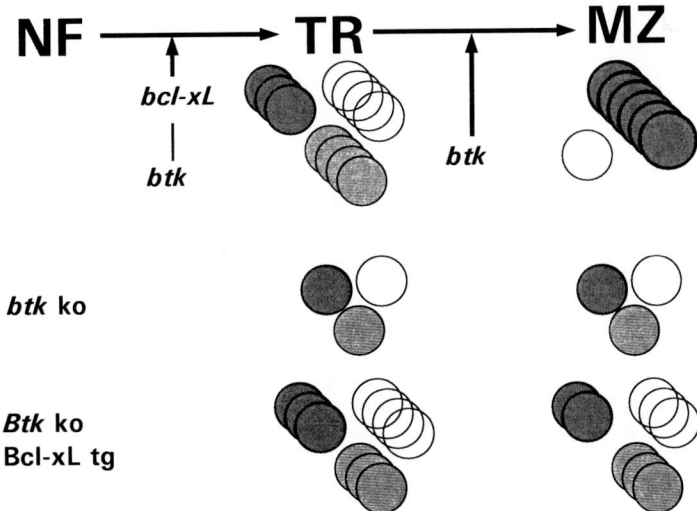

Fig. 3. Developmental model showing the involvement of *btk* and bcl-xL in the marginal zone positive selection

Certain stages of the first step in this developmental process are rescued by a bcl-xL transgene and the size of the peripheral pool becomes similar to that of 81x mice. However there is a second phase of the selection into the marginal zone repertoire that is not rescued in our model. At this time we do not know if this is simply due to the downregulation of the bcl-xL transgene at these late stages or represents indeed a bcl-xL independent function of *btk*.

In conclusion we have identified two stages in B cell maturation that are *btk* dependent, one at the progression from NF into TR that is rescued by the overexpression of bcl-xL and the other from TR into MZ that is not rescued in our model system and may be less bcl-xL dependent.

The result of the above-described mechanisms is a powerful positive selection that transforms a random repertoire derived from multiple Ig gene rearrangements and reshaped by central negative selection in order to function optimally when encountering pathogens. The positive selection seems to be dependent and is controlled by (i) specificity of the BCR, (ii) BCR receptor density and (iii) multiple co-receptor signals which may be stage-specific. Positive selection results in the establishment of different repertoires within each of the microenvironments where they develop under the locally exerted selective pressures. It is now clear that we can compartmentalize the mature long lived B cell repertoire into a follicular recirculating repertoire, a static marginal zone repertoire and a B1 repertoire enriched in the peritoneal and pleural cavities. Recent data on the functional abilities of B1 (Karras et al., 1997; Tanguay et al., 1999; Zimecki et al., 1994) and marginal zone B cells (Oliver et al., 1997; Oliver et al., 1999), suggest that these two B cell subsets have evolved because they provide the first line responses for gut/peritoneum and blood borne antigens. On the other hand the large repertoire of mature recirculating follicular B cells represents a diverse pool from which antigen specific cells are recruited for long-term T-dependent antigen responses and memory generation.

Acknowledgements

We wish to thank Lisa Jia for invaluable technical help and Ann Brookshire for editorial help. We are grateful to Dr. Tim Behrens for the bcl-xL transgenic mice. This work was supported by NIH grants AI 14782 and CA13148.

References

Ahearn, J.M., Fischer, M.B., Croix, D., Goerg, S., Ma, M., Xia, J., Zhou, X., Howard, R.G., Rothstein, T.L., and Carroll, M.C. (1996). Disruption of the Cr2 locus results in a reduction in B-1a cells and in an impaired B cell response to T-dependent antigen. Immunity *4*, 251-262.

Arnold, L.W., Pennell, C.A., McCray, S.K., and Clarke, S.H. (1994). Development of B-1 cells: segregation of phosphatidyl choline-specific B cells to the B-1 population occurs after immunoglobulin gene expression. J.Exp.Med. *179*, 1585-1595.

Bhat, N.M., Kantor, A.B., Bieber, M.M., Stall, A.M., Herzenberg, L.A., and Teng, N.N. (1992). The ontogeny and functional characteristics of human B-1 (CD5+ B) cells. Int.Immunol. 4, 243-252.

Bikah, G., Carey, J., Ciallella, J.R., Tarakhovsky, A., and Bondada (1996). CD5-mediated negative regulation of antigen receptor-induced growth signals in B-1 B cells. Science 274, 1906-1909.

Chen, X., Martin, F., Forbush, K.A., Perlmutter, R.M., and Kearney, J.F. (1997). Evidence for selection of a population of multi-reactive B cells into the splenic marginal zone. Int.Immunol. 9, 27-41.

Cornall, R.J., Cyster, J.G., Hibbs, M.L., Dunn, A.R., Otipoby, K.L., Clark, E.A., and Goodnow, C.C. (1998). Polygenic autoimmune traits: Lyn, CD22, and SHP-1 are limiting elements of a biochemical pathway regulating BCR signaling and selection. Immunity 8, 497-508.

Cyster, J.G. (1997). Signaling thresholds and interclonal competition in preimmune B-cell selection. Immunol.Rev. 156, 87-101.

Cyster, J.G. and Goodnow, C.C. (1995). Antigen-induced exclusion from follicles and anergy are separate and complementary processes that influence peripheral B cell fate. Immunity 3, 691-701.

Cyster, J.G., Healy, J.I., Kishihara, K., Mak, T.W., Thomas, M.L., and Goodnow, C.C. (1996). Regulation of B-lymphocyte negative and positive selection by tyrosine phosphatase CD45. Nature 381, 325-328.

Engel, P., Zhou, L.J., Ord, D.C., Sato, S., Koller, B., and Tedder, T.F. (1995). Abnormal B lymphocyte development, activation, and differentiation in mice that lack or overexpress the CD19 signal transduction molecule. Immunity 3, 39-50.

Fang, W., Mueller, D.L., Pennell, C.A., Rivard, J.J., Li, Y.S., Hardy, R.R., Schlissel, M.S., and Behrens, T.W. (1996). Frequent aberrant immunoglobulin gene rearrangements in pro-B cells revealed by a bcl-xL transgene. Immunity 4, 291-299.

Fruman, D.A., Snapper, S.B., Yballe, C.M., Davidson, L., Yu, J.Y., Alt, F.W., Cantley, and LC (1999). Impaired B cell development and proliferation in absence of phosphoinositide 3-kinase p85alpha. Science 283, 393-397.

Gray, D., MacLennan, I.C., Bazin, H., and Khan, M. (1982). Migrant mu+ delta+ and static mu+ delta- B lymphocyte subsets. Eur.J.Immunol. 12, 564-569.

Hardy, R.R., Carmack, C.E., Li, Y.S., and Hayakawa, K. (1994). Distinctive developmental origins and specificities of murine CD5+ B cells. Immunol.Rev. 137, 91-118.

Hartley, S.B., Cooke, M.P., Fulcher, D.A., Harris, A.W., Cory, S., Basten, A., and Goodnow, C.C. (1993). Elimination of self-reactive B lymphocytes proceeds in two stages: arrested development and cell death. Cell 72, 325-335.

Hayakawa, K., Asano, M., Shinton, S.A., Gui, M., Allman, D., Stewart, C.L., Silver, J., and Hardy, R.R. (1999). Positive selection of natural autoreactive B cells. Science 285, 113-116.

Inaoki, M., Sato, S., Weintraub, B.C., Goodnow, C.C., and Tedder, T.F. (1997). CD19-regulated signaling thresholds control peripheral tolerance and autoantibody production in B lymphocytes. J.Exp.Med. 186, 1923-1931.

Kantor, A.B., Stall, A.M., Adams, S., and Herzenberg, L.A. (1992). Adoptive transfer of murine B-cell lineages. Ann.N.Y.Acad.Sci. 651, 168-169.

Karras, J.G., Wang, Z., Huo, L., Howard, R.G., Frank, D.A., and Rothstein, T.L. (1997). Signal transducer and activator of transcription-3 (STAT3) is constitutively activated in normal, self-renewing B-1 cells but only inducibly expressed in conventional B lymphocytes. J.Exp.Med. 185, 1035-1042.

Kerner, J.D., Appleby, M.W., Mohr, R.N., Chien, S., Rawlings, D.J., Maliszewski, C.R., Witte, O.N., and Perlmutter, R.M. (1995). Impaired expansion of mouse B cell progenitors lacking Btk. Immunity 3, 301-312.

Khan, W.N., Alt, F.W., Gerstein, R.M., Malynn, B.A., Larsson, I., Rathbun, G., Davidson, L., Muller, S., Kantor, A.B., and Herzenberg, L.A. (1995). Defective B cell development and function in Btk-deficient mice. Immunity 3, 283-299.

Khan, W.N., Nilsson, A., Mizoguchi, E., Castigli, E., Forsell, J., Bhan, A.K., Geha, R., Sideras, P., and Alt, F.W. (1997). Impaired B cell maturation in mice lacking Bruton's tyrosine kinase (Btk) and CD40. Int.Immunol. *9*, 395-405.

Kraus, M., Saijo, K., Torres, R.M., and Rajewsky, K. (1999). Ig-alpha cytoplasmic truncation renders immature B cells more sensitive to antigen contact. Immunity *11*, 537-545.

Kumararatne, D.S. and MacLennan, I.C. (1982). The origin of marginal-zone cells. Adv.Exp.Med.Biol. *149*, 83-90.

Lam, K.P., Kuhn, R., and Rajewsky, K. (1997). In vivo ablation of surface immunoglobulin on mature B cells by inducible gene targeting results in rapid cell death. Cell *90*, 1073-1083.

Leitges, M., Schmedt, C., Guinamard, R., Davoust, J., Schaal, S., Stabel, S., and Tarakhovsky, A. (1996). Immunodeficiency in protein kinase cbeta-deficient mice. Science *273*, 788-791.

Liu, Y.J., Zhang, J., Lane, P.J., Chan, E.Y., and MacLennan, I.C. (1991). Sites of specific B cell activation in primary and secondary responses to T cell-dependent and T cell-independent antigens. Eur.J.Immunol. *21*, 2951-2962.

MacLennan, I.C. (1998). B-cell receptor regulation of peripheral B cells. Curr.Opin.Immunol. *10*, 220-225.

Martin, F., Chen, X., and Kearney, J.F. (1997). Development of VH81X transgene-bearing B cells in fetus and adult: sites for expansion and deletion in conventional and CD5/B1 cells. Int.Immunol. *9*, 493-505.

Martin, F. and Kearney, J. F. Positive Selection from Newly Formed to Marginal Zone B Cells Depends on the Rate of Clonal Production, CD19 and *btk*. Immunity 12(1), 39-49.

Miyazaki, T., Muller, U., and Campbell, K.S. (1997). Normal development but differentially altered proliferative responses of lymphocytes in mice lacking CD81. EMBO J *16*, 4217-4225.

Molina, H., Holers, V.M., Li, B., Fung, Y., Mariathasan, S., Goellner, J., Strauss-Schoenberger, J., Karr, R.W., and Chaplin, D.D. (1996). Markedly impaired humoral immune response in mice deficient in complement receptors 1 and 2. Proc.Natl.Acad.Sci.U.S.A. *93*, 3357-3361.

Oka, Y., Rolink, A.G., Andersson, J., Kamanaka, M., Uchida, J., Yasui, T., Kishimoto, T., Kikutani, H., and Melchers, F. (1996). Profound reduction of mature B cell numbers, reactivities and serum Ig levels in mice which simultaneously carry the XID and CD40 deficiency genes. Int Immunol *8*, 1675-1685.

Oliver, A.M., Martin, F., Gartland, G.L., Carter, R.H., and Kearney, J.F. (1997). Marginal zone B cells exhibit unique activation, proliferative and immunoglobulin secretory responses. Eur.J.Immunol. *27*, 2366-2374.

Oliver, A.M., Martin, F., and Kearney, J.F. (1999). IgMhighCD21high lymphocytes enriched in the splenic marginal zone generate effector cells more rapidly than the bulk of follicular B cells. J.Immunol. *162*, 7198-7207.

Rickert, R.C., Rajewsky, K., and Roes, J. (1995). Impairment of T-cell-dependent B-cell responses and B-1 cell development in CD19-deficient mice. Nature *376*, 352-355.

Schittek, B., Rajewsky, K., and Forster, I. (1991). Dividing cells in bone marrow and spleen incorporate bromodeoxyuridine with high efficiency. Eur.J.Immunol. *21*, 235-238.

Stall, A.M., Wells, S.M., and Lam, K.P. (1996). B-1 cells: unique origins and functions. Semin.Immunol. *8*, 45-59.

Suzuki, H., Terauchi, Y., Fujiwara, M., Aizawa, S., Yazaki, Y., Kadowaki, T., and Koyasu, S. (1999). Xid-like immunodeficiency in mice with disruption of the p85alpha subunit of phosphoinositide 3-kinase. Science *283*, 390-392.

Tanguay, D.A., Colarusso, T.P., Pavlovic, S., Irigoyen, M., Howard, R.G., Bartek, J., Chiles, T.C., and Rothstein, T.L. (1999). Early induction of cyclin D2 expression in phorbol ester-responsive B-1 lymphocytes. J.Exp.Med. *189*, 1685-1690.

Tarakhovsky, A. (1997). Bar Mitzvah for B-1 cells: how will they grow up? J.Exp.Med. *185*, 981-984.

Tarakhovsky, A., Turner, M., Schaal, S., Mee, P.J., Duddy, L.P., Rajewsky, K., and Tybulewicz, V.L. (1995). Defective antigen receptor-mediated proliferation of B and T cells in the absence of Vav. Nature *374*, 467-470.

Torres, R.M., Flaswinkel, H., Reth, M., and Rajewsky, K. (1996). Aberrant B cell development and immune response in mice with a compromised BCR complex. Science 272, 1804-1808.

Torres, R.M. and Hafen, K. (1999). A negative regulatory role for Ig-alpha during B cell development. Immunity 11, 527-536.

Tsitsikov, E.N., Gutierrez-Ramos, J.C., and Geha, R.S. (1997). Impaired CD19 expression and signaling, enhanced antibody response to type II T independent antigen and reduction of B-1 cells in CD81- deficient mice. Proc Natl Acad Sci U S A 94, 10844-10849.

Zimecki, M., Whiteley, P.J., Pierce, C.W., and Kapp, J.A. (1994). Presentation of antigen by B cells subsets. I. Lyb-5+ and Lyb-5- B cells differ in ability to stimulate antigen specific T cells. Arch.Immunol.Ther.Exp.(Warsz.) 42, 115-123.

Immunoglobulin V_H Gene Analysis in Rat: Most Marginal Zone B Cells Express Germline Encoded V_H Genes and Are Ligand Selected

P.M. Dammers[1], A. Visser[1], E.R. Popa[2], P. Nieuwenhuis[1], N.A. Bos[1] and F.G.M. Kroese[1].
[1]Department of Histology and Cell Biology and [2]Department of Clinical Immunology, University of Groningen, Groningen, The Netherlands

Introduction

Most peripheral B cells in man and rodents are small recirculating cells expressing relatively low levels of surface (s)IgM and high levels of sIgD. In lymphoid tissues, these recirculating B cells are predominantly located in follicular structures (B cell follicles) and therefore called recirculating follicular B (RF-B) cells [1]. Marginal zone B cells (MZ-B cells) are a distinct B cell subset. In contrast to RF-B cells, MZ-B cells do not recirculate and are exclusively found in spleen, where they are located in a distinct area surrounding the B cell follicles and PALS [2-4]. MZ-B cells differ from RF-B cells in many respects. MZ-B cells are slightly larger with lightly pyroninophilic cytoplasm and less condensed nuclear chromatin [5, 6]. They express high levels of sIgM and high levels of complement receptors CR1 (CD35) and CR2 (CD21), and low levels of sIgD [7-11]. In addition, MZ-B cells express higher basal levels of the T cell co-stimulatory molecules B7-1 (CD80) and B7-2 (CD86) [12]. These characteristics suggest that MZ-B cells are in a somewhat activated state, which could be the result of previous antigenic experience.

There is evidence that MZ-B cells are functionally involved in T cell-independent type 2 (TI-2) responses to (lipo-)polysaccharide antigens. These antigens form the major constituent of cell walls of encapsulated bacteria like *Streptococcus pneumoniae*, *Neisseria meningitidis* and *Haemophilus influenzae* [13]. The TI-2 antigens Ficoll and pneumococcal capsular polysaccharides localize on MZ-B cells very shortly after injection [14, 15]. In addition, these antigens were shown to react preferentially with MZ-B cells on tissue sections of human spleen [16]. Binding of TI-2 antigens to MZ-B cells appears to be dependent on the presence of complement rather than on (antigen-specific) serum immunoglobulin [15]. The anatomical location together with the activated phenotype, presumably allows MZ-B cells to respond more rapidly than other B cells to bloodborn infections with encapsulated bacteria. Experiments in mice, using *ex-vivo* stimulation of purified B cell subsets, revealed that MZ-B cells respond indeed differently to mitogens like LPS, dextran-conjugated anti-IgM or anti-IgD antibodies or CD40 liga-

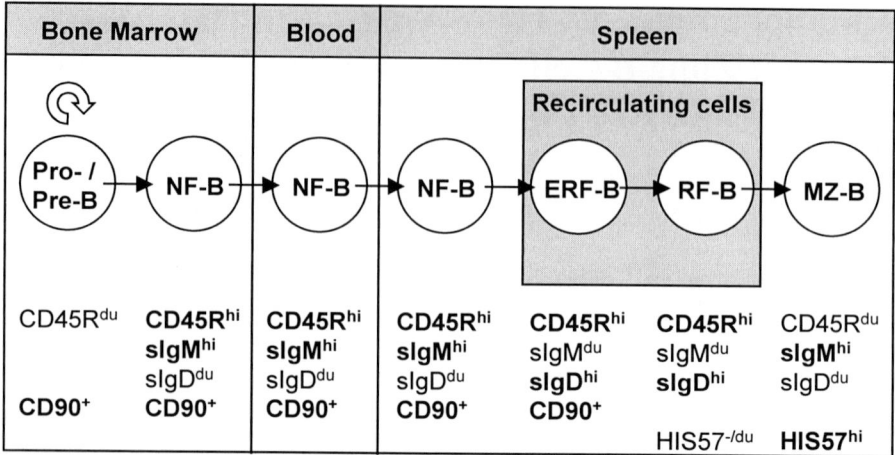

Fig. 1. Model of virgin B cell development in the rat. Cell devisions occur only at the stage of Pro/pre-B cells. Molecules that are highly expressed are in bold typeface (see text for further explanation).

tion than RF-B cells [12, 17, 18]. Oliver et al. [12] showed that MZ-B cells proliferate more rapidly in response to LPS and anti-CD40 ligation than RF-B cells. In addition, they showed that MZ-B cells are able to upregulate the T cell-costimulatory molecules B7-1 and B7-2 much faster after LPS and anti-CD40 stimulation than RF-B cells.

B Cell Development in Rat

Peripheral Virgin B Cell Differentiation

In rat, various stages in B cell development can be distinguished [19]. A model of virgin B cell development in rat is shown in Fig. 1. CD45R (B220) recognized by the HIS24 mAb is expressed in all B-lymphoid stages, albeit that lower levels of CD45R are detected on MZ-B cells [1, 20]. In rat, immature B cells highly express CD90 (Thy-1 antigen), whereas mature B cells are indispensably CD90$^-$ [21, 22]. In rat spleen, two subsets of immature CD90$^+$ B cells can be distinguished on basis of relative expression levels of sIgM and sIgD: newly formed B (NF-B) cells (sIgM^{hi}sIgD^{lo}CD90$^+$) and early recirculating follicular B (ERF-B) cells (sIgM^{lo}sIgD^{hi}CD90$^+$). In addition to these two immature CD90$^+$ B cell stages, also two mature CD90$^-$ B cell stages can be distinguished in rat spleen: RF-B cells (sIgM^{lo}sIgD^{hi}CD90$^-$) and MZ-B cells (sIgM^{hi}sIgD^{lo}CD90$^-$) [23]. In Fig. 2 we show the four B cell subsets in rat spleen as defined by flow cytometry using anti-IgM and anti-CD90 mAbs. In bone marrow of rats, in addition to NF-B cells, large numbers of Pro/pre-B cells are present which can be detected by their absence of sIgM (and sIgD) and their high CD90 expression (Fig. 2). In contrast to ERF-B cells and RF-B cells, NF-B cells do not have full recirculating capacity,

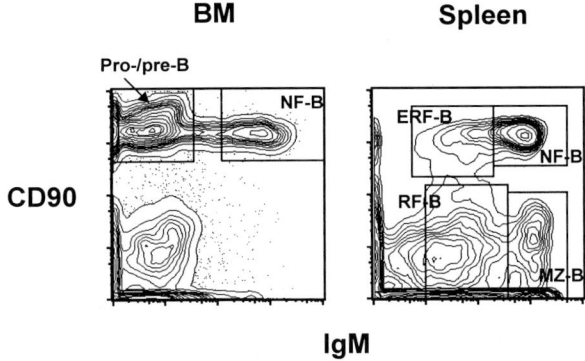

Fig. 2. B cell subpopulations in bone marrow (BM) and spleen of normal untreated adult PVG rat, as revealed by CD90 and sIgM expression.

illustrated by their absence from thoracic duct lymph and lymph nodes [23]. Presumably NF-B cells lack molecules like CD62L (L-selectin) necessary for adherence to high endothelial venules of lymph nodes. NF-B cells and ERF-B cells probably represent two sequential stages in B cell development towards the mature $CD90^-$ B cells (Fig. 1) [23]. Recently, we have characterized a mAb (HIS57) recognizing a determinant that is highly expressed on MZ-B cells and to a lower extend also on some RF-B cells (Fig.1; HIS57) [24]. Tissue sections of normal spleen stained with mAb HIS57 show similar staining pattern as observed with mAbs to CD1d in mice [25], however, no reactivity was observed with a cell line (T9CD1) transfected with rat CD1d (in collaboration with Dr. A. Matsuura, Sapporo Medical University, Sapporo, Japan).

Origin of MZ-B Cells

MZ-B cells reappear in spleen of lethally irradiated animals after transfer of syngeneic bone marrow or fetal liver cells, indicating that progenitors of MZ-B cells are ultimately found among bone marrow and fetal liver cells [23, 26-28]. Although some experiments suggested that MZ-B cells in rat belong to a distinct B cell lineage [4, 29, 30], transfer of thoracic duct lymphocytes (TDL) into cyclophosphamide-treated or irradiated rats provided evidence that MZ-B cells originate actually from recirculating B cells [30-32]. More recently, by using adriamycin (doxorubicin) induced B cell depletion, we determined the exact phenotype of these B cells, and established that the mature $CD90^-$ RF-B cells rather than the immature $CD90^+$ NF-B cells and ERF-B cells are the immediate MZ-B precursor cells [24]. In rats, two days after a single i.v. injection of adriamycin, MZ-B cells and all immature $CD90^+$ B cells are severly depleted (including all precursor B cells in bone marrow), while appreciable numbers of RF-B cells ($sIgM^{lo-}sIgD^{hi}CD90^-$) remain present (Fig. 3, day 2 after injection). In these adriamycin treated rats, MZ-B cells reappear at a time point at which newly generated (immature $CD90^+$) B cells are still absent (Fig. 3, day 15). This indicates that MZ-B cells that reappear in these adriamycin-treated rats originate from adriamycin-resistant, mature $CD90^-$ RF-B cells.

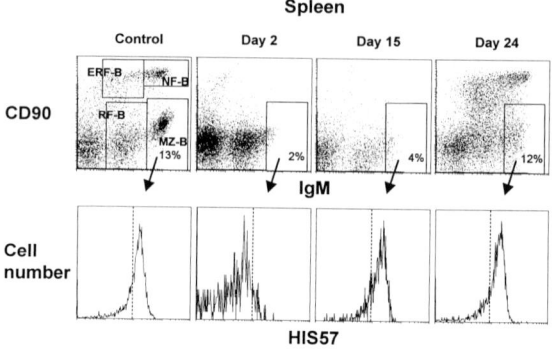

Fig. 3. Three-color flow cytometry analysis of the B cell subpopulations in spleen, as revealed by staining for sIgM and CD90, of rats taken 2, 15 and 24 days after i.v. injection with adriamycin (control = untreated). For each time point, the relative expression levels of HIS57 antigen is shown for sIgM^{hi}CD90$^-$ cells, to monitor reappearing MZ-B cells. Numbers are percentage of cells within the lymphoid gate.

MZ-B Cells: Naive versus Memory B Cells

The splenic marginal zone probably consists of a population of B cells of mixed origin. The presence of MZ-B cells in spleen of antigen-free [33], germfree [34, 35], nude [29, 36] and TCR$^{-/-}$ animals [12], implies that MZ-B cells can develop independently from exogenous antigens and T cell help. MZ-B cells in rat develop in ontogeny before germinal centers can be detected [37]. Together, these observations indicate that MZ-B cells can be naive (i.e. non-memory) B cells. However, there is also evidence that the splenic marginal zone contains antigen-experienced memory B cells. In rats, specific hapten-binding (memory) B cells were localized in the marginal zone of the spleen after immunization with T cell-dependent (TD) or TI-1 antigens [38]. More recently, Dunn-Walters et al. [39] and Tierens et al. [40] found by mutation analysis of microdissected MZ-B cells that most MZ-B cells in the human spleen carry somatically mutated Ig V_H genes. In addition to these somatic mutations, virtually all MZ-B cells in man express CD27 [41], which is selectively expressed by memory B cells, but not by naive B cells [42].

Are MZ-B Cells a Selected Population of Cells?

Kinetic studies of B cell differentiation reveals that only a small proportion of RF-B cells eventually becomes MZ-B cell. The flow and kinetics of cells within the mature CD90$^-$ B cell compartment in rat is depicted in Fig. 4. The mature CD90$^-$ B cell compartment (RF-B and MZ-B together) in rat consists of about 800 million cells. Deenen et al. [22] determined that 1% of the CD90$^-$ B cells are renewed each day (8 million cells/day). Given the renewal rate of MZ-B cells in rat of 0.5% per day [43], approximately 0.2 million RF-B cells per day enter the MZ-B cell stage, which is only 0.03% of all RF-B cells within the animal. Most cells that leave the RF-B cell compartment each day (1% of the total number of RF-B cells) either die by (presumably) apoptosis or differentiate to e.g. Ig secreting plasma cells after antigenic stimulation. There are some indications that the small numbers of RF-B cells that eventually become incorporated into the pool of MZ-B cells are selected on basis of their Ig receptor specificity. Firstly, circumstantial evidence indicates that MZ-B cells function in response to TI-2 antigens and

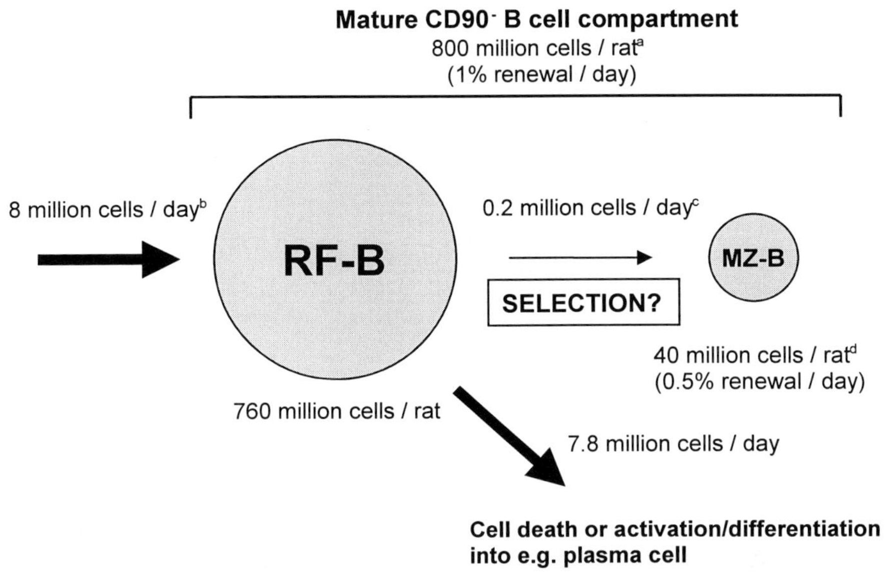

Mature CD90⁻ B cell compartment
800 million cells / rat[a]
(1% renewal / day)

8 million cells / day[b] → RF-B

RF-B → 0.2 million cells / day[c] → MZ-B

SELECTION?

760 million cells / rat

40 million cells / rat[d]
(0.5% renewal / day)

7.8 million cells / day

→ Cell death or activation/differentiation into e.g. plasma cell

Fig. 4. Fate of RF-B cells. [a] Absolute numbers of cells and renewal rate of the mature CD90⁻ B cell compartment are obtained from [22]. [b] In steady state conditions the number of B cells entering the mature CD90⁻ B cell compartment per day is 1% of the total number of CD90⁻ B cells (800 million). [c] The renewal rate of MZ-B cells is derived from [43]. [d] Total number of MZ-B cells per rat spleen is the average of five 3-months-old male PVG rats and determined by three-color flow cytometry using sIgM, sIgD and CD90 staining.

therefore MZ-B cells most likely express Ig receptors specific for this type of antigens. Secondly, Mason et al. [44] have shown that the majority of peripheral B cells in (soluble-)lysozyme/anti-lysozyme double-transgenic mice are anergic. Immunohistological staining of spleen sections of these lysozyme-tolerant mice clearly demonstrated the presence of B cell follicles, but the absence of marginal zones. Apparently, some kind of activation process is necessary for B cells to enter the MZ-B cell stage, which is blocked by the anergic state of the B cells in these double-transgenic mice. Finally, Chen et al. [45] have previously demonstrated that the splenic marginal zone in V_H81X Ig transgenic mice (81xtg model) contain a selected population of multi-reactive B cells.

Rat Ig V_H Gene Region Sequences Reveal that a Similar V_H Gene Family Relationship Exists in Rat as in Mouse

As in other mammalian species, diversity in the Ig heavy chain and light chain variable regions in rat is created by recombination of variable (V), diversity (D) (heavy chain only) and joining (J) gene segments [19]. As a consequence of

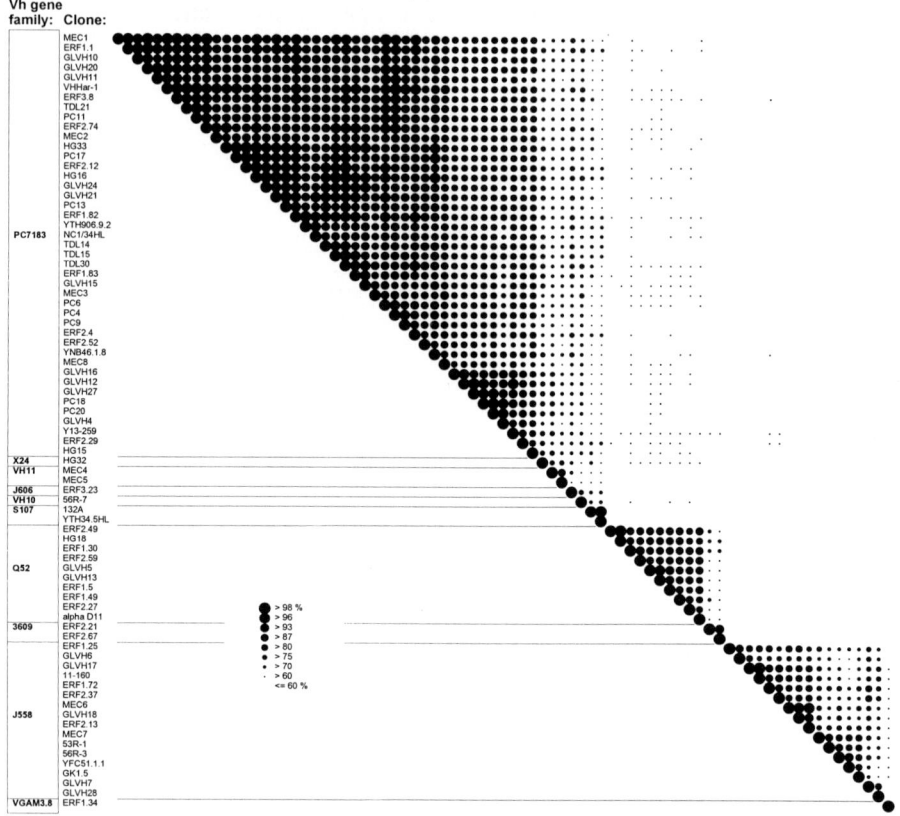

Fig. 5. Homology dot plot of rat Ig V_H gene sequences described in [19]. The dot plot was established with V_H gene sequences originating from both cDNA (rearranged $V_H D J_H$) and genomic DNA. V_H gene family subdivisions were made according to the >80% sequence identity rule [46]. Mouse V_H gene family nomenclature is used, which results from comparison of rat V_H genes with members of mouse V_H gene families.

the evolutionary process of gene duplication and diversification, Ig heavy chain V-region genes (V_H genes) can be grouped into genes with highly homologous sequences (V_H gene families). Brodeur et al. [46] determined that in mice members of a V_H gene family have generally more than 80% sequence identity, whereas the sequence identity between members of different V_H gene families is in general less than 70%. In previous years, we and others have obtained a large number of V_H gene sequences of rat origin [19, 47]. In order to determine whether rat V_H genes can be subdivided into V_H gene families similar to mouse, we made a pair-wise comparison of all known rat V_H gene sequences (listed in [19]). From this comparison a homology matrix was compiled, which is depicted in Fig. 5. The sequences used in Fig. 5 are shown in descending order of identity, with clone MEC1 arbitrarily put on top. The percentage of identity between two individual V_H genes is represented by a dot. The size of these dots is proportional to the identity between the sequences (large dots represent higher sequence identity). As can be withdrawn from the triangular patterns formed by the large dots, V_H genes of rat can also be subdivided into families based on sequence identity. Similar to mouse, V_H gene family members in rat generally share more than 80% sequence identity, whereas the identity between members of different families is most frequently less 70%. Furthermore, rat and mouse V_H gene sequences reveal high similarity when compared to each other. So far, each rat V_H gene sequence could be assigned, according to the V_H gene family definition mentioned above, to a particular mouse V_H gene family. Because of this high similarity between mouse and rat, we use the mouse V_H gene family nomenclature also for designation of V_H gene families in rat. V_H gene sequences analyzed thus far reveal the existence of at least 10 V_H gene families in rat (PC7183, X24, Vh11, S107, J606, Vh10, Q52, 3609, J558 and VGAM3.8). We assume that the number of V_H gene families in rat is more than 10, because Southern blot hybridization analysis of *Eco*RI digested genomic DNA, using radiolabeled mouse V_H gene family specific DNA probes, suggest the existence of at least 14 V_H gene families in rat [19, 47]. These data also indicate that the V_H gene family subdivision in mouse and rat was already established before the two species diverged during evolution, which is also in agreement with the homologous organization of the J_H locus in both species [48].

Ig V_H Gene Analysis of MZ-B Cells in Rat

In order to investigate whether the marginal zone in rodents contain memory B cells similar in extend as seen in man and to determine whether MZ-B cells are selected on basis of Ig specificity, we analysed V_H genes expressed by both flow cytometry- and immunohistologically-defined MZ-B cells in rat. To this end, we investigated the repertoire of productive PC7183 V_H genes in four stages of B cell development (splenic RF-B and MZ-B cells, and bone marrow Pro/pre-B and NF-B cells), in two individual 6-months-old untreated PVG rats (raised under conventional conditions). B cell subsets were sorted by flow cytometry on basis of sIgM, sIgD and CD90 expression (Fig. 2).

Germline PC7183 V_H gene sequences were established by comparing the PC7183 V_H gene sequences obtained in the present study with each other and with other known PC7183 V_H gene sequences [19]. The relative numbers of germline and mutated PC7183 V_H genes expressed in the four analysed B cell subsets are shown in Fig. 6. $V_H DJ_H$-μ transcripts expressed by the immature $CD90^+$ Pro/pre-B cells and NF-B cells, and by the mature $CD90^-$ RF-B cells are essentially all encoded by germline V_H genes. In contrast, MZ-B cells exhibit a higher frequency of cells carrying mutated V_H genes (about 20%). In addition, the replacement over silent mutation ratios (R/S ratio) observed in FR and CDR regions of some V_H gene sequences from MZ-B cells, indicate that the mutations seen in V_H genes of

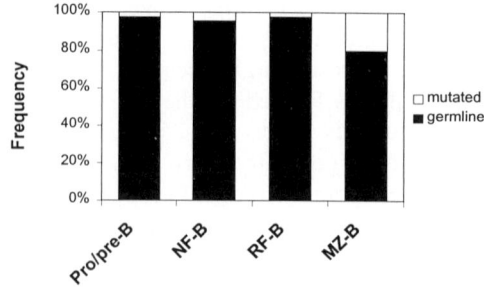

Fig. 6. Ratio of germline and mutated Ig V_H genes expressed in bone marrow Pro/pre-B cells (n=40) and NF-B cells (n=44), and splenic RF-B (n=43) and MZ-B cells (n=40). Data are obained from two 6-months-old PVG rats, which were raised under conventional conditions. V_H genes were considered mutated when more than 2 mutations were revealed upon comparison with the germline sequence (see text for further explanation).

MZ-B cells are not randomly distributed and therefore likely the result of selection (affinity-maturation). Moreover, from one rat we obtained four MZ-B cell V_HDJ_H region sequences that are clonally related to each other. This clone was identified by their identical V_H-D-J_H junctional sequence (CDR3 region). Mutations are observed in the V_H region of two of these clonally related sequences, which strongly suggests that this MZ-B cell clone presumably originates from a germinal center.

Similar results were obtained from MZ-B cells microdissected from spleen sections. Presently, we have analysed 17 productively rearranged V_HDJ_H region sequences from follicular areas and 10 from marginal zone areas of the spleen. Somatic mutations were apparent in only 1 out of 10 PC7183 V_H gene sequences from B cells located in the marginal zone, whereas no somatic mutations were observed in the 17 V_H genes isolated from follicular B cells. This indicates that most MZ-B cells in adult conventional rats, defined by either flow cytometry or anatomy, display no signs of somatic mutations and therefore these cells are considered to be naive (non-memory) B cells.

Whether MZ-B cells are selected on basis of their Ig specificity was revealed by analysing the usage of individual members of the PC1783 V_H gene family that were expressed in the flow cytometry-defined B cell subsets. Although the overall PC1783 V_H gene family usage is very similar between RF-B cells and MZ-B cells, difference is apparent in usage of certain PC7183 members. For instance, PC7183 family members PC-1 and PC-4 are frequently expressed by RF-B cells, whereas these genes are not or rarely expressed among MZ-B cells. Difference in PC7183 V_H gene usage is also observed between the immature $CD90^+$ NF-B cells from bone marrow and the mature $CD90^-$ B cells from spleen (RF-B cells and MZ-B cells). This is in agreement with several experiments that revealed that selection of B cells occurs between the stages of NF-B cells in bone marrow and mature RF-B cells in the periphery [22, 49]. In addition, our study reveals that little selection takes place between Pro/pre-B cells and NF-B cells in bone marrow on basis of Ig specificity.

Additional evidence that MZ-B cells are a ligand-selected population of cells derives from the V_H-D-J_H junctional sequences (CDR3 region). By comparison of the mean CDR3 length in V_HDJ_H-μ transcripts obtained from the flow cytometry-defined B cell subsets, we found that the CDR3 length in V_HDJ_H regions expressed by MZ-B cells is on average 2-3 amino acids shorter than the CDR3 length in V_HDJ_H regions expressed in the other three B cell subsets (Fig. 7). The shorter CDR3 regions observed in MZ-B cells is not due to completely lack of N nucleotide additions or excess in exonuclease activity at V_H-D-J_H junctions in these cells (Dammers et al., manuscript in preparation). Consequently, the use of shorter CDR3 regions in $V_H DJ_H$ transcripts of MZ-B cells, compared to Pro/pre-B cells, NF-B cells and RF-B cells, is most likely the result of a selection event rather than a fundamental difference in the mechanisms of recombination.

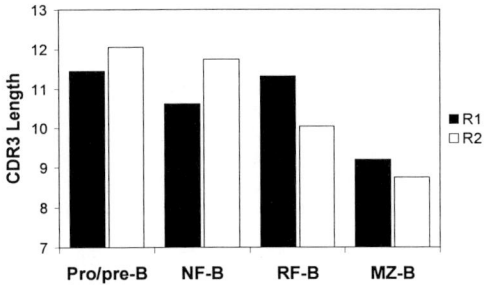

Fig. 7. Mean CDR3 length (in codons, according to Kabat et al. [50]) of $V_H DJ_H$ regions expressed in Pro/pre-B cells and NF-B cells from bone marrow, and RF-B cells and MZ-B cells from spleen. Data are obained from two untreated 6-months-old PVG rats (R1 and R2). For each rat, mean CDR3 lengths were calculated from about 20 $V_H DJ_H$ region sequences per B cell subset. (see text for further explanation).

The present data indicate that most MZ-B cells in rat appear to be naive (non-memory) B cells. In addition, based on the Ig V_H gene repertoire expressed by MZ-B cells and the shorter CDR3 length encoded by the V_H-D-J_H junction of these cells, we conclude that although these B cells are "naive" they are, however, ligand selected. The mechanism of this selection process is still unknown.

References

1. Kroese FGM, Butcher EC, Lalor PA, Stall AM, Herzenberg LA (1990) The rat B cell system: the anatomical localization of flow cytometry-defined B cell subpopulations. Eur J Immunol 20:1527-1534
2. Gray D, MacLennan ICM, Bazin H, Khan M (1982) Migrant $\mu^+\delta^+$ and static $\mu^+\delta^-$ B lymphocytes subsets. Eur J Immunol 12:564-569
3. Kraal G (1992) Cells in the marginal zone of the spleen. Int Rev Cytol 132:31-74
4. MacLennan ICM, Gray D, Kumararatne DS, Bazin H (1982) The lymphocytes of splenic marginal zones: a distinct B-cell lineage. Immunol Today 3:305-307
5. Kearney JF, Cooper MD, Klein J, Abney ER, Parkhouse RME, Lawton AR (1977) Ontogeny of Ia and IgD on IgM-bearing B lymphocytes in mice. J Exp Med 146:297-301
6. Nieuwenhuis P (1971) On the origin and fate of immunologically competent cells. Thesis, University of Groningen, Groningen
7. Bourgois A, Kitajima K, Hunter IR, Askonas BA (1977) Surface immunoglobulins of lipopolysaccharide-stimulated spleen cells: the behaviour of IgM, IgD and IgG. Eur J Immunol 7:151-153
8. Monroe JG, Havran WL, Cambier J (1983) B lymphocyte activation: entry into cell cycle is accompanied by decreased expression of IgD but not IgM. Eur J Immunol 13:208-213
9. Gray D, McConnell I, Kumararatne DS, MacLennan ICM, Humphrey JH, Bazin H (1984) Marginal zone B cells express CR1 and CR2 receptors. Eur J Immunol 14:47-52

10. Timens W, Boes A, Poppema S (1989) Human marginal zone B cells are not an activated B cell subset: strong expression of CD21 as a putative mediator for rapid B cell activation. Eur J Immunol 19:2163-2166
11. Bohnsack JF, Cooper NR (1988) CR2 ligands modulate human B cell activation. J Immunol 141:2569-2576
12. Oliver AM, Martin F, Kearney JF (1999) IgM^{high}CD21^{high} lymphocytes enriched in the splenic marginal zone generate effector cells more rapidly than the bulk of follicular B cells. J Immunol 162:7198-7207
13. Spencer J, Perry ME, Dunn-Walters DK (1998) Human marginal-zone B cells. Immunol Today 19:421-426
14. Van den Eertwegh AJM, Laman JD, Schellekens M, Boersma WJA, Claassen E (1992) Complement-mediated follicular localization of T-independent type-2 antigens: the role of marginal zone macrophages revisited. Eur J Immunol 22:719-726
15. Harms G, Hardonk MJ, Timens W (1996) In vitro complement-dependent binding and in vivo kinetics of pneumococcal polysaccharide TI-2 antigens in the rat spleen marginal zone and follicle. Infect Immunity 64:4220-4225
16. Peset Llopis M-J, Harms G, Hardonk MJ, Timens W (1996) Human immune response to pneumococcal polysaccharides: complement-mediated localization preferentially on CD21-positive splenic marginal zone B cells and follicular dendritic cells. J Allergy Clin Immunol 97:1015-1024
17. Snapper CM, Yamada H, Smoot D, Sneed R, Lees A, Mond JJ (1993) Comparative in vitro analysis of proliferation, Ig secretion, and Ig class switching by murine marginal zone and follicular B cells. J Immunol 150:2737-2745
18. Oliver AM, Martin F, Gartland GL, Carter RH, Kearney JF (1997) Marginal zone B cells exhibit unique activation, proliferative and immunoglobulin secretory responses. Eur J Immunol 27:2366-2374
19. Kroese FGM (ed) (1998) Immunology of the rat. In: Pastoret P-P, Bazin H, Govaerts A, Griebel PJ (eds) Handbook of vertebrate immunology. Academic Press, London, pp 137-222
20. Kroese FGM, Wubbena AS, Opstelten D, Deenen GJ, Schwander EH, de Leij L, Vos H, Poppema S, Volberda J, Nieuwenhuis P (1987) B lymphocyte differentiation in the rat: production and characterization of monoclonal antibodies to B lineage associated antigens. Eur J Immunol 17:921-928
21. Crawford JM, Goldschneider I (1980) Thy1 antigen and B lymphocyte differentiation in the rat. J Immunol 124:969-976
22. Deenen GJ, Kroese FGM (1993) Kinetics of B cell subpopulations in peripheral lymphoid tissues: evidence for the presence of phenotypically distinct short-lived and long-lived B cell subsets. Int Immunol 5:735-741
23. Kroese FGM, de Boer NK, de Boer T, Nieuwenhuis P, Kantor AB, Deenen GJ (1995) Identification and kinetics of two recently bone marrow-derived B cell populations in peripheral lymphoid tissues. Cellular Immunol 162:185-193
24. Dammers PM, de Boer NK, Deenen GJ, Nieuwenhuis P, Kroese FGM (1999) The origin of marginal zone B cells in the rat. Eur J Immunol 29:1522-1531
25. Amano M, Baumgarth N, Dick MD, Brossay L, Kronenberg M, Herzenberg LA, Strober S (1998) CD1 expression defines subsets of follicular and marginal zone B cells in the spleen: β_2-microglobulin-dependent and independent forms. J Immunol 161:1710-1717
26. Owen JJT, Wright DE, Habu S, Raff MC, Cooper MD (1977) Studies on the generation of B lymphocytes in fetal liver and bone marrow. J Immunol 118:2067-2072
27. Rozing J, Brons NHC, van Ewijk W, Benner R (1978) B Lymphocyte differentiation in lethally irradiated and reconstituted mice. Cell Tiss Res 189:19-30
28. Kroese FGM, Wubbena AS, Nieuwenhuis P (1986) Germinal centre formation and follicular antigen trapping in the spleen of lethally X-irradiated and reconstituted rats. Immunology 57:99-104
29. Bazin H, Gray D, Platteau B, MacLennan ICM (1982) Distinct δ^+ and δ^- B lymphocyte lineages in the rat. Ann NY Acad Sci 399:157-173
30. MacLennan ICM, Gray D (1986) Antigen-driven selection of virgin and memory B cells. Immunol Rev 91:61-85

31. Kumararatne DS, MacLennan ICM (1981) Cells of the marginal zone of the spleen are lymphocytes derived from recirculating precursors. Eur J Immunol 11:865-869
32. Lane PJL, Gray D, Oldfield S, MacLennan ICM (1986) Differences in the recruitment of virgin B cells into antibody responses to thymus-dependent and thymus-independent type-2 antigens. Eur J Immunol 16:1569-1575
33. Bos NA, Bun JCAM, Meedendorp B, Wubbena AS, Kroese FGM, Ploplis VA, Cebra JJ (2000) B cell populations in antigen-free mice. Old Herborn University Seminar Monograph (in press)
34. Iijima S, Yamane T (1968) The spleen of germ free animals. In: Miyakawa M, Luckey TD (eds) Advances in germfree research and gnotobiology. Iliffe Books Ltd, London, pp 139-148
35. Pollard M (1967) Germinal centers in germfree animals. In: Cottier H, Odartschenko N, Schindler R, Congdon CC (eds) Germinal centers in immune responses. Springer-Verlag, Berlin, New York, pp 343-348
36. Kumararatne DS, Bazin H, MacLennan ICM (1981) Marginal zones: the major B cell compartment of the rat spleens. Eur J Immunol 11:858-864
37. Kroese FGM, Wubbena AS, Kuijpers KC, Nieuwenhuis P (1987) The ontogeny of germinal centre forming capacity of neonatal rat spleen. Immunol 60:597-602
38. MacLennan ICM, Liu Y-J (1991) Marginal zone B cells respond to both polysaccharide antigens and protein antigens. Res Immunol 142:346-351
39. Dunn-Walters DK, Isaacson PG, Spencer J (1995) Analysis of mutations in immunoglobulin heavy chain variable region genes of microdissected marginal zone (MGZ) B cells suggests that the MGZ of human spleen is a reservoir of memory B cells. J Exp Med 182:559-566
40. Tierens A, Delabie J, Michiels L, Vandenberghe P, De Wolf-Peters C (1999) Marginal-zone B cells in the human lymph node and spleen show somatic hypermutations and display clonal expansion. Blood 93:226-234
41. Tangye SG, Liu Y-J, Aversa G, Phillips JH, de Vries JE (1998) Identification of functional human splenic memory B cells by expression of CD148 and CD27. J Exp Med 188:1691-1703
42. Klein U, Rajewsky K, Küppers R (1998) Human immunoglobulin $(Ig)M^+IgD^+$ peripheral blood B cells expressing the CD27 cell surface antigen carry somatically mutated variable region genes: CD27 as a general marker for somatically mutated (memory) B cells. J Exp Med 188:1679-1689
43. Gray D (1988) Population kinetics of rat peripheral B cells. J Exp Med 167:805-816
44. Mason DY, Jones M, Goodnow CC (1992) Development and follicular localization of tolerant B lymphocytes in lysozyme/anti-lysozyme IgM/IgD transgenic mice. Int Immunol 4:163-175
45. Chen X, Martin F, Forbush KA, Perlmutter RM, Kearney JF (1997) Evidence for selection of a population of multi-reactive B cells into the splenic marginal zone. Int Immunol 9:27-41
46. Brodeur PH, Riblet R (1984) The immunoglobulin heavy chain variable region (Igh-V) locus in the mouse I. One hundred Igh-V genes comprise seven families of homologous genes. Eur J Immunol 14:922-930
47. Vermeer LA (1995) Molecular analysis of rat B cell differentiation. Thesis, University of Groningen, Groningen
48. Lang P, Mocikat R (1991) Immunoglobulin heavy-chain joining genes in the rat: comparison with mouse and human. Gene 102:261-264
49. Gu H, Tarlinton D, Müller W, Rajewsky K, Föster I (1991) Most peripheral B cells in mice are ligand selected. J Exp Med 173:1357-1371
50. Kabat EA, Wu TT, Perry HM, Gottesman KS, Foeller C (1991) Sequences of proteins of immunological interest. NIH Publication, Bethesda MD

IV

CD5

STAT3 Activation, Chemokine Receptor Expression, and Cyclin-Cdk Function in B-1 Cells

T. L. Rothstein[1], G. M. Fischer[1,2], D. A. Tanguay[3], S. Pavlovic[3], T. P. Colarusso[1], R. M. Gerstein[4], S. H. Clarke[5], and T. C. Chiles[3]

[1]Immunobiology Unit, Department of Medicine, Boston University Medical Center, Boston, MA. USA
[2] Department of Microbiology, Boston University Medical Center, Boston, MA, USA
[3] Department of Biology, Boston College, Chestnut Hill, MA, USA
[4] Department of Molecular Genetics and Microbiology, University of Massachusetts Medical School, Worcester, MA, USA
[5] Department of Microbiology and Immunology, University of North Carolina Chapel Hill, NC, USA

Introduction

B-1 cells constitute a specific subset of B lymphocytes, first distinguished from B-2 (conventional B) cells by relatively low level expression of the pan-T cell surface glycoprotein, CD5, but presently known to manifest many additional characteristic phenotypic, functional, biochemical and molecular features (reviewed in more detail elsewhere, [1-3]). B-1 cells appear early in development and initially dominate the B cell pool but subsequently decline in relative number with B-2 cell production. In mice, coelomic tissues, especially the peritoneal cavity, are relatively enriched for B-1 cells, whereas the spleen contains a very low proportion (but perhaps equal number) of B-1 cells, and peripheral lymph nodes are virtually devoid of this population. Peritoneal and splenic B-1 cells are similar to each other, and different from B-2 cells, in expressing sIgM at high levels, and sIgD and B220 at low levels; however, peritoneal and splenic B-1 cells are distinguished by exclusively peritoneal expression of Mac-1. Peritoneal B-1 cells also contain a small "sister" population that is identifiable by the absence of CD-5 (but presence of Mac-1) expression. B-1 cells contribute the major share of nonimmune (resting) serum IgM and IgA, which plays a key role in limiting microbial dissemination [4]. B-1 cell-derived immunoglobulin contains little somatic mutation and N-insertion and is thus germline-like and repertoire-restricted. The over-representation of particular antigen specificities and V_H families suggests that B-1 cells constitute a positively selected population.

The origin of B-1 cells remains uncertain. Adoptive transfer experiments early on suggested that B-1 cells represent a separate lymphocyte lineage, the precursors for which are not present in adult bone marrow. Rather, B-1 cell repopulation requires concurrent transfer of sIg-positive B-1 cells, thereby defining the capacity of B-1 cells for self-renewal [5-7]. More recent studies have suggested that B-1 cell differentiation results from a particular form of sIg signaling. This is supported by the demonstration that B-2 cells acquire CD5 expression within days of sIg crosslinking in vitro, and the observation that mice transgenic for certain (B-1-derived) immunoglobulin receptors contain an overabundance of B-1 cells [8, 9].

B-1 cells appear to manifest a predisposition for abnormal growth. B-1 cells represent the cell of origin for the human malignant disease, chronic lymphocytic leukemia, and in the mouse, clonal expansions of B-1 cells occur frequently with increasing age [10-13]. Further, B-1 cells are readily immortalized in culture in the absence of exogenous agents [14]. The origin of these unusual growth characteristics remains unknown.

B-1 cells differ dramatically from B-2 cells in the signals required to induce cell cycle progression to S phase [15-18]. Whereas sIg crosslinking by anti-immunoglobulin antibody (anti-Ig) drives B-2 cells to enter S phase and incorporate thymidine, B-1 cells do not enter S phase after anti-Ig treatment and are thus hyporesponsive to sIg signaling; this may reflect a negative influence of sIg signaling [19]. On the other hand, treatment with phorbol esters alone, such as phorbol myristate acetate (PMA) drives B-1 cells to enter S phase, whereas B-2 cells do not respond similarly. Instead B-2 cells are induced to enter S phase by phorbol ester only in conjunction with a calcium ionophore. Further, the B-1 cell response to phorbol ester treatment is quite rapid, with the peak of S phase occurring at 24-30 hours (as opposed to 54-60 hours for B-2 cells stimulated with PMA plus ionomycin). These results, along with the propensity of B-1 cells for self-renewal, clonal expansion, malignant transformation, and in vitro immortalization, mentioned above, suggest the possibility that B-1 cells differ from B-2 cells by constitutively expressing an endogenous mediator that promotes growth and abbreviates the signaling requirements for cell cycle progression. The present work was undertaken to address this issue.

Results

To identify growth related mediators that are differentially expressed by B-1 and B-2 cells, peritoneal B-1 cells were purified by negative selection as previously described [20] and compared with B-1-depleted splenic B-2 cells. Early work failed to reveal a substantial difference in several activation and growth related characteristics, including resting levels of intracellular Ca^{++}, expression of the growth related genes c-*myc*, and *egr-1*, and levels of the nuclear, DNA-binding factors NF-κB, and AP-1, as well as STAT5 and STAT6 ([20-22] and unpublished observations).

Constitutive Expression of Nuclear Activated STAT3 by B-1 Cells

In examining transcription factor binding to the SIE site from the c-*fos* promoter, however, a marked difference between B-1 and B-2 cells was found--untreated B-1 cells expressed nuclear SIE-binding activity corresponding to SIF-A, whereas untreated B-2 cells did not. By a number of criteria this material was shown to be authentic, active STAT3 phosphorylated on tyrosine705, including immunoinhibition, shift-western, and western blot analyses using anti-STAT3 and ptyr705STAT3-specific antibodies [23]. Not only is B-1 cell STAT3 constitutively activated by tyrosine705 phosphorylation, it is also constitutively phosphorylated on serine727 (which has been reported to enhance STAT3 DNA binding activity, [24]). Although cytokines such as IL-6 and IL-10 produce STAT3 activation in B-2 cells within 15 minutes, this is unlikely to be the mechanism by which B-1 cells acquire nuclear ptyr705STAT3, because cytokine-induced STAT3 tyrosine phosphorylation is typically transient and does not last as long as the duration of ex vivo B-1 cell purification (about 30 hours) and B-1 cells are washed many times during this procedure. Along the same lines, although crosslinking sIg on B-2 cells induces activation of STAT3 within 3 hours, this is unlikely to be the mechanism by which B-1 cells acquire nuclear ptyr705STAT3, because the duration of STAT3 activation is brief and STAT3 SIE-binding activity is no longer detected after the 2 day treatment required to induce CD5 expression in vitro [23].

The faithfulness with which nuclear ptyr705STAT3 expression marks the B-1 population was tested by examining B-1 cells that reside in the spleen. Here the situation was found to be quite different. Neither FACS-sorted (B220+,CD23-) splenic B-1 cells from normal mice, nor FACS-sorted (B220+,CD23-) splenic B-1 cells from $V_{H}12$ and $V_{H}12V_{\kappa}4$ immunoglobulin transgenic mice (that contain large numbers of splenic B-1 cells, [9]) were found to contain nuclear SIE-binding activity corresponding to SIF-A. And in companion control experiments, peritoneal B-1 cells from $V_{H}12V_{\kappa}4$ transgenic mice expressed constitutively activated STAT3 just like peritoneal B-1 cells from normal mice. Thus peritoneal and splenic B-1 cells differ completely with respect to constitutive expression of nuclear activated STAT3. To the extent that sIg triggering is thought to be responsible for B-1 cell differentiation, this further suggests that antigen receptor signaling is not responsible for constitutive expression of activated STAT3. Moreover, these results make it less likely that an autocrine mechanism accounts for B-1 expression of activated STAT3, because the splenic density of B-1 cells in $V_{H}12V_{\kappa}4$ transgenic mice is high without provoking STAT3 phosphorylation. These considerations focus attention on endogenous mechanisms and unique features of the peritoneal environment as causative agents leading to expression of activated STAT3 by peritoneal B-1 cells. This in turn raises the question of how peritoneal B-1 cells come to be directed to the peritoneum.

Chemokine Receptor Expression by B-1 and B-2 Cells

To address this issue, purified peritoneal B-1 and splenic B-2 cells were examined for the presence of chemokine receptor transcripts by RNase protection assay [25-27]. Although no chemokine receptors present on B-1 and not present on B-2 cells were identified, two receptors, CCR2 and CCR3, were found to be expressed by splenic B-2 cells and not by peritoneal B-1 cells. The expression of chemokine receptor transcripts by splenic B-1 cells from immunoglobulin transgenic mice paralleled the situation observed with splenic B-2 cells. Thus, splenic B-1 cells expressed CCR2 and CCR3, just like splenic B-2 cells, but unlike peritoneal B-1 cells. Here again, as with activated STAT3, peritoneal and splenic B-1 cells differ in a constitutively expressed molecular feature.

Distinctive Features of Peritoneal and Splenic B-1 Populations

There are then several known characteristics that distinguish peritoneal and splenic B-1 cells; these are: surface Mac-1 expression, nuclear expression of ptyr705STAT3, and expression of CCR2 and CCR3 transcripts. These results suggest that the B-1 population is not uniform, but may be subdivided into distinct B-1 populations as shown in Table 1. The nature of the events leading to the separate localization of B-1 cells with distinctive characteristics remains uncertain, and could reflect differences in the origins of these B-1 cells as well as local environmental influences on B-1 cells that are directed to different locations.

Distinguishing Features of Peritoneal And Splenic B-1 Cells

	B-1-S	B-1-P
Mac-1	−	+
ptyrSTAT3	−	+
CCR2/CCR3	+	−
Trafficking	?	?
Ontogeny	?	?

Table 1

Cyclin D2 Expression in PMA-Stimulated B-1 Cells

The rapidity of the proliferative response to phorbol ester treatment suggests that induction of cell cycle mediators is uniquely regulated in B-1 cells, perhaps in turn reflecting the presence of a growth related mediator like ptyrSTAT3. To delineate differences in cell cycle control, peritoneal B-1 and splenic B-2 cells were stimulated with PMA at 300 ng/ml, and analyzed for cyclin and cdk expression and function. Initial experiments focused on cyclin D2, previously reported to be the most abundant D-type cyclin in stimulated murine B cells [28, 29] and which has been shown to be required for lymphocyte G1 progression [30]. Immunoblotting of B-1 cell lysates showed that cyclin D2 expression was upregulated 2-4 hours after stimulation with PMA [31]. In contrast there was no induction of cyclin D2 expression in PMA-treated B-2 cells over a 24 hour period; cyclin D2 only appeared in B2 cells 24 hours after mitogenic stimulation with anti-Ig. Despite the rapidity with which cyclin D2 was expressed in PMA-stimulated B-1 cells, induction was transcriptionally regulated as described in other systems [32], and as demonstrated here by sensitivity to cycloheximide and to actinomycin D, and by correlation with upregulation of cyclin D2 mRNA. To examine the participation of cyclin D2 in complexes with cyclin-dependent kinases (cdk), cdk4 and cdk6 were immunoprecipitated with kinase-specific antisera and, after size separation by

SDS-PAGE, immunoblotted with anti-cyclin D2. By this measure PMA-stimulated B-1 cells contained cyclin D2-cdk4 and, to a lesser extent, cyclin D2-cdk6, complexes, whereas such complexes were detected in B-2 cells only after stimulation with anti-Ig, and not after treatment with PMA. To address the activity of cyclin D2-cdk complexes, the capacity of immunoprecipitated cyclin D2 to phosphorylate a truncated retinoblastoma (Rb) substrate protein was determined. PMA-induced cyclin D2 (complexes) immunoprecipitated from B-1, but not from B-2, cells phosphorylated Rb, as compared to untreated controls. These results strongly suggest that early cyclin D2 expression, preferentially induced in B-1 cells by PMA, is an important determinant of B-1-specific phorbol ester-induced cell cycle progression, and focuses attention on the presently unknown mechanism by which cyclin D2 induction is upregulated by PMA in the absence of calcium ionophore.

Cyclin D3-Cdk4/6 Complexes in PMA-Stimulated B-1 and B-2 Cells

Rb phosphorylating activity was examined further in PMA-stimulated B-1 cells by immunoblotting endogenous Rb with an antibody specific for the phosphoserine780 form of Rb. Although an increase in pser780Rb was observed after 4 hours stimulation (the time of peak cyclin D2 induction) serine780 phosphorylation increased progressively at 12 and 24 hours of stimulation. This suggested that additional cyclin-cdk complexes play a role in Rb phosphorylation and cell cycle progression produced by PMA in B-1 cells. To identify these mediators, cyclin D3 was examined (as cyclin D1 is not expressed in murine B cells [28, 29]).

Immunoblotting of B-1 cell lysates showed that cyclin D3 expression was upregulated at 16-24 hours after stimulation by PMA. Unlike the situation with cyclin D2, however, cyclin D3 was also upregulated in PMA-stimulated B-2 cells. To examine the participation of cyclin D3 in cdk-containing complexes, cdk4 and cdk6 were immunoprecipitated with kinase-specific antibody as before and then immunoblotted with anti-cyclin D3. This demonstrated that 24 hours after PMA stimulation both B-1 and B-2 cells contained cyclin D3-cdk4 complexes at similar levels. In addition, both B-1 and B-2 cells contained cyclin D3-cdk6 complexes, although the level in PMA-stimulated B-2 cells was somewhat lower than the level in PMA-stimulated B-1 cells.

To address the activity of cyclin D3-cdk complexes, the capacity of immunoprecipitated cdk4 to phosphorylate a retinoblastoma substrate protein was determined; it was found that cdk4 immunoprecipitated from B-1 cells stimulated with PMA for 16-24 hours phosphorylated Rb. This represents a time when cyclin D2 is no longer present, and expression of cyclin D3 and D3-cdk complexes is maximal. In direct contrast, cdk4 immunoprecipitated from B-2 cells at various times up to 24 hours following the addition of PMA failed to phosphorylate Rb, although cdk4 from control B-2 cells stimulated with the mitogenic combination of PMA plus ionomycin did so. Thus cyclin D3-cdk complexes are similarly induced by PMA in B-1 and B-2 cells, but the Rb phosphorylating activity of these complexes is expressed only in B-1 cells. These results strongly suggest that negative regulation of cyclin D3-cdk complexes in B-

2 cells plays a role in the differential responses of B-1 and B-2 cells to PMA stimulation, adding another layer to the unique induction of cyclin D2-cdk complexes in B-1 cells that correlates with PMA-responsiveness (see Table 2). This in turn raises the possibility that inhibitors which target D-type cyclin holoenzyme complexes, cdk-activating kinase (CAK), or cdc25 phosphatase, may be differentially regulated in B-1 and B-2 cells [33].

It is interesting to note that the promoter for p21, a cdk2 inhibitor, contains 3 STAT binding sites arranged in tandem order, and that the activated STAT3 constitutively expressed in B-1 cells binds to one of these sites in isolation [34]. It may be that the bulk of Rb phosphorylation is carried out by cyclin E-cdk2 complexes, the activity of which is enhanced in B-1 cells by cyclin D2-cdk4-mediated sequestration of [35], or lower expressed levels of, p21 and/or p27. In keeping with this, Rb phosphorylating activity immunoprecipitated with anti-cdk2 from B-1 cells was greatly increased by PMA stimulation, whereas PMA produced no increase in cdk2-associated kinase activity in B-2 cells.

Cell Cycle Mediators Stimulated by PMA in B-1 and B-2 Cells		
	Cyclin D2 + Cdk4/6	Cyclin D3 + Cdk4/6
B-1 Cells	Present Active	Present Active
B-2 Cells	Absent -	Present Inactive

Table 2

DISCUSSION

The STAT3 transcription factor is strongly associated with cell growth. In particular, Darnell and colleagues have recently reported that STAT3 is an oncogene. This conclusion rests on several pieces of evidence. 1) Human tumor samples and tumor-derived cell lines constitutively express activated STAT3 [36]; 2) Src oncogene transformed cell lines contain activated STAT3 and are growth inhibited by dominant negative STAT3 [37, 38]; and, 3) Mutant, constitutively activated STAT3 transforms NIH 3T3 and rat 3Y1 cells in vitro and transformed 3Y1 cells are tumorigenic in vivo [39]. The oncogenicity of STAT3 suggests that constitutively expressed ptyr705STAT3 may be responsible, all or in part, for the unusual growth-related properties of murine B-1 cells, including self-renewal, clonal expansion, and neoplastic transformation in vivo, and immortalization in vitro. This places emphasis on identifying the constitutively active kinase or constitutively suppressed phosphatase responsible for baseline STAT3 phosphorylation in B-1 cells. The observation that B-1 cells constitutively express small amounts of tyrosine phosphorylated STAT1 suggests that the substrate specificity is not restricted to STAT3 [23]; however, the absence of constitutively activated STAT5 and STAT6 in B-1 cells suggests that constitutive phosphorylation of STAT3 is not part of a general induction of STAT phosphorylating mechanisms or of cellular activation. Although JAK kinases are responsible for cytokine-induced STAT phosphorylation [40, 41], recent reports indicate that other molecules are capable of producing STAT activation including *src* and *abl* kinases [37, 38, 42-45], which may be candidate mediators of constitutive B-1 STAT3 phosphorylation. It should be noted that Frank and colleagues did not find evidence of constitutive STAT3 tyrosine phosphorylation in malignant lymphocytes obtained from patients with chronic lymphocytic leukemia [46]; however, elevated levels of (transcriptionally enhancing) serine phosphorylation were observed [24].

The study of activated STAT3 led to the observation that peritoneal and splenic B-1 cells differ in this characteristic. If activated STAT3 is indeed responsible for the unusual growth related properties of B-1 cells, this would suggest that neoplastic clonal expansions are derived primarily from peritoneal rather than splenic B-1 cells, and would further imply that self-renewal is primarily a property of the B-1 cell subset. The distinction between B-1 and B-2 cells currently rests on differences in the expression of surface Mac-1, tyrosine phosphorylation of STAT3 and transcriptional activation of CCR2 and CCR3. The significance and origin of these distinctive pools of B-1 cells is uncertain. The distinguishable characteristics of these two populations suggest that they may originate from B-1 cells produced in different ways, or directed at an early stage to different locations. Alternatively, the suggestion that B-1 cells circulate between the peritoneal cavity and the spleen would infer that local environmental influences are responsible for upregulating and downmodulating the characteristics noted above, in which case it would be important to determine the nature of such local factors. Other differences between peritoneal and splenic B-1 cells, as well as between B-1 and B-2 cells, are presently being sought through general screens of gene expression using DNA macro-array and gene chip technology.

The hyper-responsiveness of B-1 cells to PMA treatment is reflected in the early induction of functional cyclin D2, and later induction of active cyclin D3, complexes with cyclin-dependent kinases. It is not surprising in a general sense that PMA induces functional D-type cyclin-cdk complexes, inasmuch as PMA pushes B-1 cells through the Rb checkpoint into S phase. Of particular interest, however, is the nature of the regulatory controls present in B-2 cells that block cell cycle progression in response to PMA. These clearly include, at least, the absence of cyclin D2 induction and the presence of one or more inhibitors of cyclin D3-cdk4 activity. Further study of these mediators is likely to reveal the manner in which PMA differentially affects B-1 and B-2 cells, which will in turn reflect, and help identify, additional molecular differences between these B cell subsets that may well be related to the potential for regulated and unregulated growth. Future experiments will also be directed toward further elucidating the relationship, if any, between phorbol ester responsiveness, activated STAT3 expression, and cyclin-cdk formation and function.

Acknowledgements

This work was supported by United States Public Health Service grant AI29690 (TLR) awarded by the National Institutes of Health and grant MCB9603784 (TCC) awarded by the National Science Foundation.

References

1. Morris, D.L., and T.L. Rothstein. 1994. CD5+ B (B-1) cells and immunity. In Handbook of B and T Lymphocytes. E.C. Snow, editor. Academic Press, Inc., San Diego. 421-45.
2. Kantor, A.B., and L.A. Herzenberg. 1993. Origin of murine B cell lineages. Annual Review of Immunology 11:501-38.
3. Haughton, G., L.W. Arnold, A.C. Whitmore, and S.H. Clarke. 1993. B-1 cells are made, not born. Immunol Today 14, 2:84-7; discussion 87-91.
4. Ochsenbein, A.F., T. Fehr, C. Lutz, M. Suter, F. Brombacher, H. Hengartner, and R.M. Zinkernagel. 1999. Control of Early Viral and Bacterial Distribution and Disease by Natural Antibodies. Science 286, 5447:2156-2159.
5. Hardy, R.R., K. Hayakawa, D.R. Parks, and L.A. Herzenberg. 1984. Murine B cell differentiation lineages. J Exp Med 159, 4:1169-88.
6. Hayakawa, K., R.R. Hardy, and L.A. Herzenberg. 1985. Progenitors for Ly-1 B cells are distinct from progenitors for other B cells. J Exp Med 161, 6:1554-68.
7. Hayakawa, K., R.R. Hardy, A.M. Stall, and L.A. Herzenberg. 1986. Immunoglobulin-bearing B cells reconstitute and maintain the murine Ly-1 B cell lineage. Eur J Immunol 16, 10:1313-6.
8. Cong, Y.Z., E. Rabin, and H.H. Wortis. 1991. Treatment of murine CD5- B cells with anti-Ig, but not LPS, induces surface CD5: two B-cell activation pathways. Int Immunol 3, 5:467-76.
9. Arnold, L.W., C.A. Pennell, S.K. McCray, and S.H. Clarke. 1994. Development of B-1 cells: segregation of phosphatidyl choline-specific B cells to the B-1 population occurs after immunoglobulin gene expression. J Exp Med 179, 5:1585-95.
10. Stall, A.M., M.C. Farinas, D.M. Tarlinton, P.A. Lalor, L.A. Herzenberg, and S. Strober. 1988. Ly-1 B-cell clones similar to human chronic lymphocytic leukemias routinely develop in older normal mice and young autoimmune (New Zealand Black-related) animals. Proc Natl Acad Sci U S A 85, 19:7312-6.
11. Raveche, E.S., P. Lalor, A. Stall, and J. Conroy. 1988. In vivo effects of hyperdiploid Ly-1+ B cells of NZB origin. J Immunol 141, 12:4133-9.
12. Martin, P.J., J.A. Hansen, R.C. Nowinski, and M.A. Brown. 1980. A new human T-cell differentiation antigen: unexpected expression on chronic lymphocytic leukemia cells. Immunogenetics 11, 5:429-39.
13. Martin, P.J., J.A. Hansen, A.W. Siadak, and R.C. Nowinski. 1981. Monoclonal antibodies recognizing normal human T lymphocytes and malignant human B lymphocytes: a comparative study. J Immunol 127, 5:1920-3.

14. Braun, J., Y. Citri, D. Baltimore, F. Forouzanpour, L. King, K. Teheranizadeh, M. Bray, and S. Kliewer. 1986. B-Ly1 cells: immortal Ly-1+ B lymphocyte cell lines spontaneously arising in murine splenic cultures. Immunol Rev 93:5-21.
15. Rothstein, T.L., T.R. Baeker, R.A. Miller, and D.L. Kolber. 1986. Stimulation of murine B cells by the combination of calcium ionophore plus phorbol ester. Cell Immunol 102, 2:364-73.
16. Rothstein, T.L., and D.L. Kolber. 1988. Peritoneal B cells respond to phorbol esters in the absence of co-mitogen. J Immunol 140, 9:2880-5.
17. Klaus, G.G., A. O'Garra, M.K. Bijsterbosch, and M. Holman. 1986. Activation and proliferation signals in mouse B cells. VIII. Induction of DNA synthesis in B cells by a combination of calcium ionophores and phorbol myristate acetate. Eur J Immunol 16, 1:92-7.
18. Rothstein, T.L., and D.L. Kolber. 1988. Anti-Ig antibody inhibits the phorbol ester-induced stimulation of peritoneal B cells. J Immunol 141, 12:4089-93.
19. Bikah, G., J. Carey, J.R. Ciallella, A. Tarakhovsky, and S. Bondada. 1996. CD5-mediated negative regulation of antigen receptor-induced growth signals in B-1 B cells. Science 274, 5294:1906-9.
20. Morris, D.L., and T.L. Rothstein. 1993. Abnormal transcription factor induction through the surface immunoglobulin M receptor of B-1 lymphocytes. J Exp Med 177, 3:857-61.
21. Wang, Z., D.L. Morris, and T.L. Rothstein. 1995. Constitutive and inducible levels of egr-1 and c-myc early growth response gene expression in self-renewing B-1 lymphocytes. Cell Immunol 162, 2:309-14.
22. Cohen, D.P., and T.L. Rothstein. 1991. Elevated levels of protein kinase C activity and alpha-isoenzyme expression in murine peritoneal B cells. J Immunol 146, 9:2921-7.
23. Karras, J.G., Z. Wang, L. Huo, R.G. Howard, D.A. Frank, and T.L. Rothstein. 1997. Signal transducer and activator of transcription-3 (STAT3) is constitutively activated in normal, self-renewing B-1 cells but only inducibly expressed in conventional B lymphocytes. J Exp Med 185, 6:1035-42.
24. Wen, Z., Z. Zhong, and J.E. Darnell, Jr. 1995. Maximal activation of transcription by Stat1 and Stat3 requires both tyrosine and serine phosphorylation. Cell 82, 2:241-50.
25. Kim, C.H., and H.E. Broxmeyer. 1999. Chemokines: signal lamps for trafficking of T and B cells for development and effector function. J Leukoc Biol 65, 1:6-15.
26. Schall, T.J., and K.B. Bacon. 1994. Chemokines, leukocyte trafficking, and inflammation. Curr Opin Immunol 6, 6:865-73.
27. Frade, J.M., M. Mellado, G. del Real, J.C. Gutierrez-Ramos, P. Lind, and A.C. Martinez. 1997. Characterization of the CCR2 chemokine receptor: functional CCR2 receptor expression in B cells. J Immunol 159, 11:5576-84.
28. Tanguay, D.A., and T.C. Chiles. 1996. Regulation of the catalytic subunit (p34PSK-J3/cdk4) for the major D-type cyclin in mature B lymphocytes. J Immunol 156, 2:539-48.
29. Solvason, N., W.W. Wu, N. Kabra, X. Wu, E. Lees, and M.C. Howard. 1996. Induction of cell cycle regulatory proteins in anti-immunoglobulin-stimulated mature B lymphocytes. J Exp Med 184, 2:407-17.
30. Lukas, J., J. Bartkova, M. Welcker, O.W. Petersen, G. Peters, M. Strauss, and J. Bartek. 1995. Cyclin D2 is a moderately oscillating nucleoprotein required for G1 phase progression in specific cell types. Oncogene 10, 11:2125-34.
31. Tanguay, D.A., T.P. Colarusso, S. Pavlovic, M. Irigoyen, R.G. Howard, J. Bartek, T.C. Chiles, and T.L. Rothstein. 1999. Early induction of cyclin D2 expression in phorbol ester-responsive B-1 lymphocytes. J Exp Med 189, 11:1685-90.
32. Vadiveloo, P.K., G. Vairo, A.K. Royston, U. Novak, and J.A. Hamilton. 1998. Proliferation-independent induction of macrophage cyclin D2, and repression of cyclin D1, by lipopolysaccharide. J Biol Chem 273, 36:23104-9.
33. Morgan, D.O. 1995. Principles of CDK regulation. Nature 374, 6518:131-4.
34. Karras, K.G., R. McKay, R. Lu, J. Pych, D.A. Frank, T.L. Rothstein, and B.P. Monia. 2000. STAT3 regulates cell growth, immunoglobulin production, and chemokine gene expression in a B lymphoma model of B-1 cells. Submitted.
35. Sherr, C.J., and J.M. Roberts. 1999. CDK inhibitors: positive and negative regulators of G1-phase progression. Genes Dev 13, 12:1501-12.

36. Garcia, R., and R. Jove. 1998. Activation of STAT transcription factors in oncogenic tyrosine kinase signaling. J Biomed Sci 5, 2:79-85.
37. Cao, X., A. Tay, G.R. Guy, and Y.H. Tan. 1996. Activation and association of Stat3 with Src in v-Src-transformed cell lines. Mol Cell Biol 16, 4:1595-603.
38. Bromberg, J.F., C.M. Horvath, D. Besser, W.W. Lathem, and J.E. Darnell, Jr. 1998. Stat3 activation is required for cellular transformation by v-src. Mol Cell Biol 18, 5:2553-8.
39. Bromberg, J.F., M.H. Wrzeszczynska, G. Devgan, Y. Zhao, R.G. Pestell, C. Albanese, and J.E. Darnell, Jr. 1999. Stat3 as an oncogene. Cell 98, 3:295-303.
40. Leonard, W.J., and J.J. O'Shea. 1998. Jaks and STATs: biological implications. Annual Review Of Immunology 16:293-322.
41. Schindler, C., and J.E. Darnell, Jr. 1995. Transcriptional responses to polypeptide ligands: the JAK-STAT pathway. Annu Rev Biochem 64:621-51.
42. Chaturvedi, P., M.V. Reddy, and E.P. Reddy. 1998. Src kinases and not JAKs activate STATs during IL-3 induced myeloid cell proliferation. Oncogene 16, 13:1749-58.
43. Frank, D.A., and L. Varticovski. 1996. BCR/abl leads to the constitutive activation of Stat proteins, and shares an epitope with tyrosine phosphorylated Stats. Leukemia 10, 11:1724-30.
44. Nieborowska-Skorska, M., M.A. Wasik, A. Slupianek, P. Salomoni, T. Kitamura, B. Calabretta, and T. Skorski. 1999. Signal transducer and activator of transcription (STAT)5 activation by BCR/ABL is dependent on intact Src homology (SH)3 and SH2 domains of BCR/ABL and is required for leukemogenesis. J Exp Med 189, 8:1229-42.
45. Ilaria, R.L., Jr., and R.A. Van Etten. 1996. P210 and P190(BCR/ABL) induce the tyrosine phosphorylation and DNA binding activity of multiple specific STAT family members. J Biol Chem 271, 49:31704-10.
46. Frank, D.A., S. Mahajan, and J. Ritz. 1997. B lymphocytes from patients with chronic lymphocytic leukemia contain signal transducer and activator of transcription (STAT) 1 and STAT3 constitutively phosphorylated on serine residues. J Clin Invest 100, 12:3140-8.

Role of NFAT in the Regulation of B-1 Cells

R. Berland and H.H. Wortis
Department of Pathology and Graduate Program in Immunology, Tufts University School of Medicine, Boston, MA 02111

Introduction

The B-1 subset of B cells is enriched in cells expressing autoreactive and multireactive antibodies encoded by Ig genes rearranged without insertion of N-nucleotides. A substantial body of evidence indicates that the generation of B-1 cells is dependent upon ligation of and signaling by the B cell receptor (BCR). That BCR ligation is critical is indicated by the importance of BCR specificity in B-1 cell development. In Ig-transgenic mice expressing a specificity typical of B-1 cells, transgene positive B cells are predominantly of the B-1 phenotype (Arnold et al., 1994; Okamoto et al., 1992). Conversely, introduction of a transgene derived from a B-2 cell results in transgene positive B cells that are almost exclusively B-2 (Goodnow et al., 1988; Nemazee and Burki, 1989). The importance of BCR specificity presumably indicates that B-1 development is dependent upon positive selection by (self-) antigen. The strongest evidence for this comes from a transgenic system in which a B-1 cell derived Ig heavy chain was introduced that, when paired with an appropriate light chain, is specific for thy-1 antigen (Hayakawa et al., 1999 and this volume). In wild type mice a high frequency of transgene expressing anti-thy-1 B cells was observed only in the B-1 subset. Strikingly, when the same transgene was expressed on a thy-1 negative background, transgene expressing B-1 cells failed to appear. Consistent with a critical role for BCR-mediated positive selection in B-1 development, the targeted disruption of a number of genes encoding signaling molecules crucial for BCR signaling results in decreased numbers of B-1 cells while having more modest effects on B-2 cells. In addition, two recent reports demonstrate that B-1 development is critically dependent on the cell surface density of autoreactive BCR (Lam and Rajewsky, 1999; Watanabe et al., 1999). In this meeting, Behrens and colleagues reported that in the anti-HEL/soluble HEL transgenic system of Goodnow (Goodnow et. al., 1988), B cells expressing a B-2 specificificity and driven to anergy by the presence of the appropriate antigen, also express CD5.

In order to further understand the molecular basis of B-1 development, we have made use of our earlier observation that, *ex vivo*, splenic B cells can be induced to express CD5 by BCR crosslinking (Bandyopadhyay et al., 1995; Cong et al., 1991; Teutsch et al., 1995; Wortis et al., 1995). CD5 expression is a hallmark of B-1a cells. In addition to expressing CD5 these cells exhibit three other properties of B-1 cells, altered response to PMA treatment (Rothstein et al., 1991) and (if IL-6 is included during BCR crosslinking) decreased IgD expression (Cong et al., 1991)and constitutively nuclear STAT-3 (Karras et al., 1997). We reasoned that defining the molecular basis of CD5 induction in this *ex vivo* system would lead us to molecules involved in the induction (and perhaps maintenance) of the B-1 phenotype generally.

As a starting point in this investigation we chose to look at the mechanism of induction of transcription of the CD5 gene in response to sIgM crosslinking.

Results

Identification of an anti-IgM-inducible CD5 enhancer.

In order to identify anti-IgM responsive *cis*-regulatory elements of the CD5 gene, reporter constructs were made in which CD5 5'-flanking sequences were used to drive expression of the firefly luciferase gene. These constructs were transiently transfected, along with an internal control renilla luciferase construct, into primary splenic B cells. Cells were subsequently incubated in medium alone or in medium plus F(ab')2 anti-IgM and then luciferase activity assayed (Fig. 1). A construct containing 2.2 kb of CD5 upstream sequence was induced about 15 fold by anti-IgM (Berland and Wortis, 1998). Analysis of 5'- and internal deletion constructs led to the conclusion that a minimal 122bp element located at about −1919 to −2040 upstream of the CD5 gene was necessary for anti-IgM induction (Berland and Wortis, 1998).

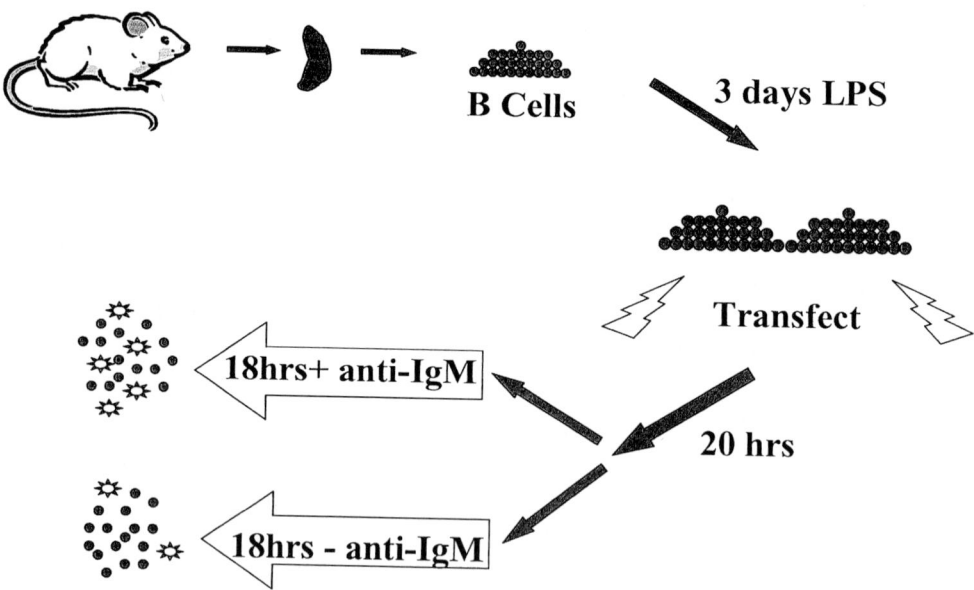

Fig. 1. Scheme of transient transfection experiments. Primary splenic B cells from BALB/cByJ mice were transfected as described (Berland and Wortis, 1998).

```
                    H4TF-1
-2040   gggctcatc..ccccctccccacctt.tgcctccaccaagcggcg..gc
        |||||||  |  ||||  ||||||||||||  |||||  ||     ||  |||    ||
        gggctcagcaacccct tccccaccttcagcctctactgggccgcgctgc


          NFAT                      Ebox      Ets              NFAT
        atggaaaacccagcc..agctcagctgacttcctcttgtgtcatggaaag
        ||||||||||||||       |||||||||||||||||||||||||||||||||||
        ctggaaaacccagcctgggctcagctgacttcctcttgtgtcatggaaag


        tgcgggtgat.aggacaggaaactgtcgcc  -1919
        |  |  |||||   |||||||||||||  ||||
        t..gagtgatggggacaggaaactcccgcc
```

Fig. 2. The CD5 enhancer. Top sequence is the murine element located from approximately –1919 to –2040 relative to the CD5 translation start site. Transcription factor binding sites are underlined. The bottom sequence is a segment of human DNA located approximately 45 kb from the CD5 gene. Vertical lines indicate nucleotide identities.

This minimal element functioned when placed downstream of the luciferase gene and in the opposite orientation (Berland and Wortis, 1998) thus qualifying as an enhancer.

The murine CD5 enhancer contains several functional transcription factor binding sites.
The CD5 enhancer contains several potential transcription factor binding sites (Fig 2, top sequence). Electrophoretic mobility shift assay (EMSA) analysis indicated that complexes formed *in vitro* on the potential ETS and Ebox sites in extracts from unstimulated splenic B cells (Berland and Wortis, 1998). In extracts from anti-IgM treated B cells four additional complexes formed, one requiring an intact H4TF-1 site and two requiring at least one intact NFAT site. The binding site of the fourth induced complex was not identified. The NFAT complexes were completely supershifted with anti-NFATc antibody (Berland and Wortis, 1998).

The role of these transcription factor binding sites in enhancer function *in vivo* was assessed by transfection experiments with constructs containing wild-type (WT) and mutant enhancers driving luciferase expression from the CD5 promoter. Mutation of the distal NFAT binding site reduced enhancer activity in the transfection assay to about 50% of the WT activity. Mutation of both NFAT sites reduced it further to about 20% of WT activity (Berland and Wortis, 1998). Mutation of the three other sites (in the context of WT NFAT sites) reduced enhancer activity to about 25% of WT activity. Mutation of any pair of these sites had only minimal effects on enhancer activity (Berland and Wortis, 1998).

The conclusion of these studies is that NFAT binding is critical to CD5 enhancer function. In addition, at least one of the other identified sites is necessary for enhancer activity.

A sequence with substantial homology to the murine CD5 enhancer is located within about 45 kb of the human CD5 gene.

To gain further evidence that the enhancer identified in the above studies is functionally significant, we searched the GenBank data base for sequences homologous to the enhancer. An element 89% identical to the murine enhancer was identified about 45 kb from the human CD5 gene (Fig. 2, bottom sequence). All transcription factor binding sites identified in the murine enhancer are conserved in the human element. This strongly suggests that the CD5 enhancer is physiologically important.

Peritoneal B-1 cells contain constitutively high nuclear and total NFATc.

We decided next to focus our efforts on the role of NFAT in enhancer activity and B-1 development. NFAT is actually a family of four related proteins encoded by four different genes (reviewed in Rao et al., 1997). Three of these family members are expressed at significant levels in peripheral lymphocytes; NFATp, NFATc, and NFAT4. All NFAT family members are reported to be localized to the cytoplasm of resting cells. Upon activation, they are dephosphorylated by the Ca2+ dependent, calcineurin (CSA) sensitive phosphatase calcineurin and translocated to the nucleus (Rao et al., 1997). In our EMSAs all of the NFAT binding activity could be supershifted by anti-NFATc antibody and none by anti-NFATp antibody (Berland and Wortis, 1998). This suggests that NFATc is the family member responsible for CD5 induction. However we cannot rule out a role for NFATp or 4 *in vivo*, particularly early in the induction, at which time NFATc levels (at least in T cells) are low (Loh et al., 1996).

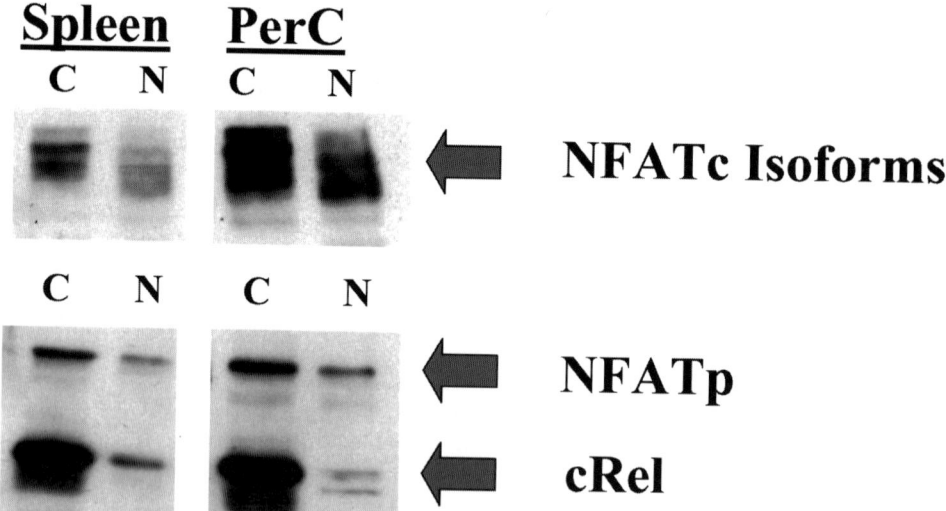

Fig. 3. Peritoneal B cells have constitutively elevated NFATc in both the nucleus and cytoplasm. Cytoplasmic (C) and nuclear (N) extracts were made from purified splenic and PerC B cells using the method of Timmerman et. al., 1997 except that 0.05% NP-40 was used and 1.8 mg/ml iodoacetamide was included in the lysis buffer. Equal cell equivalents were loaded per lane in a 10% SDS-PAGE gel. Samples were blotted to nitrocellulose. Top panels; filter probed with the monoclonal anti-NFATc antibody 7A6 (gift of G.Crabtree, (Timmerman et al., 1997)). Bottom panels; the same filter reprobed

with polyclonal rabbit anti-NFATp (gift of A. Rao, Harvard University) and rabbit anti-cRel (Santa Cruz, Santa Cruz, CA).

Given the role of NFAT in CD5 induction, we wondered if NFAT might be involved in the maintenance of CD5 expression in B-1 cells as well. The fact that many B-1 cells are known to be autoreactive suggested that perhaps in these cells there is constitutive signaling through the BCR which could maintain NFAT activation. In the anti-HEL/sHEL transgenic system of Goodnow, anergic anti-HEL B cells constitutively signal at a low level and contain nuclear NFAT (Healy et al., 1997). To address this issue, we made nuclear and cytoplasmic extracts from PerC B cells (about 70% B-1) and splenic B cells (about 5% B-1) and determined NFAT expression by Western blot. The result is shown in Fig 3. Equal cell equivalents were loaded per lane. It is apparent that B-1 cells have greatly elevated levels of NFATc in both the nucleus and cytoplasm. Reprobe of the filter with antibodies to NFATp or cREL shows no similar difference between B-1 and B-2 cells in levels of expression of these proteins. This suggests that NFATc may play a continuing role in maintaining the B-1 phenotype. Unfortunately, we were unable to transfect B-1 cells to directly assess this. We therefore turned to a cell culture system.

CH12 B cells have constitutively nuclear NFAT which contributes to CD5 expression.

CH12 is a murine B cell lymphoma line that expresses CD5 (Arnold et al., 1983). We first looked at NFAT expression in nuclear and cytoplasmic extracts of these cells. The result is shown in Fig.4. As can be seen in lanes 1 and 2, untreated CH12 cells have almost exclusively nuclear NFATc. To see if nuclear localization was calcineurin and calcium dependent, cells were treated with CSA or EGTA respectively. Lane 3 and 4 indicate that CSA treatment for 15 minutes drives most of the NFATc into the cytoplasm.

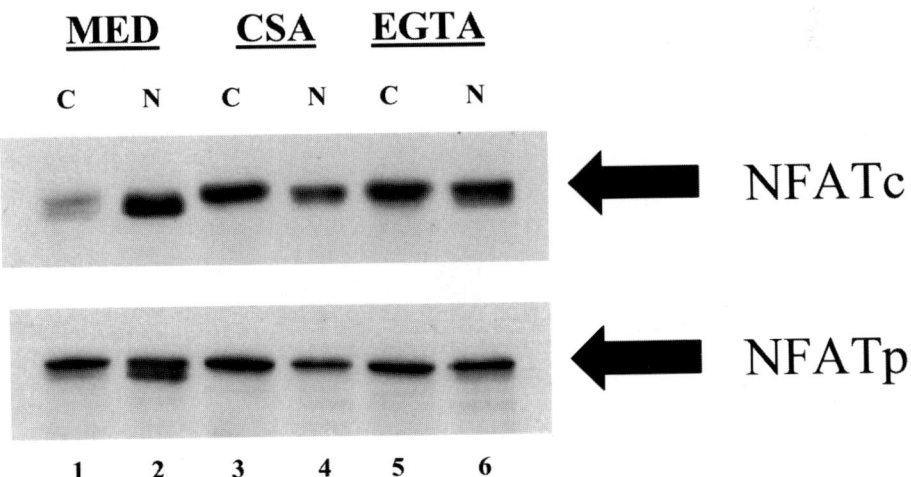

Fig 4. CH12 cells have constitutively elevated nuclear NFATc. CH12 cells were cultured either in B cell medium alone (RPMI based) or in medium plus 500 ng/ml CSA or 2.25mM EGTA. Cells were harvested and lysed as described in the legend to Fig. 3 except that NP-40 was used at 0.1%. Equal cell equivalents

of cytoplasmic (lanes 1,3, and 5) or nuclear (lanes 2,4, and 6) extract were loaded and a Western blot probed with 7A6 anti NFAT and reprobed with rabbit anti-NFATp.

Treatment with EGTA had a similar although less dramatic effect. Reprobe of this filter with anti-NFATp indicated that there may be less dramatic nuclear localization of this factor as well in CH12 cells. There exists one published study showing constitutively active NFATp in a number of human CD5 positive CLL lines (Schuh et al., 1996). In this study, NFATc activity was not examined.

We next used transient transfection to examine whether or not NFAT contributed to CD5 expression in CH12 cells. Transfection of cells with a luciferase reporter construct containing the WT CD5 enhancer resulted in a 2-3 fold increase in luciferase activity compared to transfection with a construct containing the CD5 promoter alone (Fig 5). In contrast to primary splenic B cells, the promoter alone conferred significant activity compared to a promoterless luciferase construct. To see if this enhancer activity was NFAT dependent we transfected CH12 cells with constructs containing point mutations in the distal or in both NFAT binding sites. In either case, enhancer activity was abolished (Fig 5). Thus NFAT-dependent CD5 enhancer activity is constitutively observed in CH12 cells.

Effect of targeted disruptions of the NFATp and NFATc genes (preliminary results).

The above results strongly suggest that NFAT, and particularly NFATc, plays a role in B-1 development and maintenance. In order to more firmly establish the role of NFAT in B-1 cell development we have begun studying mice with targeted disruptions of NFAT genes (Hodge et al., 1996; Oukka et al., 1998; Ranger et al., 1998; Ranger et al., 1998). To date we have performed a limited number of experiments and our results are

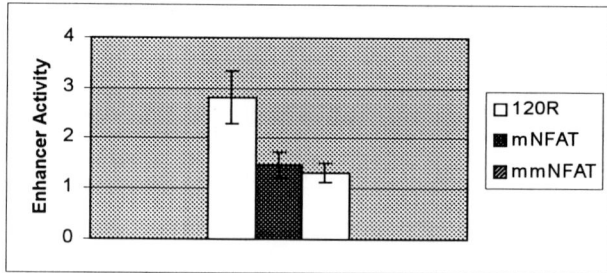

Fig 5. NFAT-dependent CD5 enhancer activity in unstimulated CH12 cells. CH12 cells were electroporated (450 V, 960μF in a Biorad Gene Pulser) with an internal control plasmid pRLTK (Promega, Madison, WI) and constructs containing the CD5 promoter (-6 to –277 relative to the ATG) with or without the enhancer element (-277Luc and pCD5Luc122R respectively in Berland and Wortis, 1998). In addition, derivatives of pCD5Luc122R (described in Berland and Wortis, 1998) were tested which contained point mutations in both NFAT sites (mmNFAT) or in only the distal NFAT site

(mNFAT). For each construct the ratio of the normalized luciferase activity (relative to pRLTK) to the normaized activity of -277 is shown.

preliminary. Disruption of the NFATp gene has little if any effect on B-1 development. However in a Rag2-/- mouse reconstituted with fetal liver from an NFATc ko embryo, peritoneal B-1 cells had about 5 fold lower levels of CD5 expression. A Rag2-/- reconstituted with fetal liver from an NFATp/NFATc double ko had few if any PerC B-1 cells. These results, although in need of repetition and extension, support the conclusion that NFATc plays a crucial role in B-1 CD5 expression.

Discussion

We initiated these studies with the aim of defining the transcription factors necessary for B-1 development . We have so far identified an anti-IgM responsive enhancer element upstream of the murine CD5 gene. This element is essential for induction of transcription from the CD5 promoter in transient transfection studies. Based on sequence homology, there appears to be a nearly identical element 45 kb from the human CD5 gene. The CD5 gene is homologous to the CD6 gene and the two genes are believed to be the result of a duplication event (Aruffo et al., 1991; Lecomte et al., 1996). In both humans and mice these genes have remained in close proximity and the CD5 enhancer is located between them. The possibility should be considered that this enhancer may also play a role in CD6 expression as this gene is induced in T cells by TCR crosslinking.

In transient transfection studies, enhancer function was dependent on the presence of two intact NFAT sites and at least one additional transcription factor binding site. This suggested to us that NFAT may be critical to B-1 cell development. In fact we had previously shown that *in vitro* induction of CD5 by sIgM crosslinking is inhibited by CSA (Teutsch et al., 1995). More recently, Clarke and colleagues have shown in cell transfer experiments that *in vivo* differentiation of CD5 negative transgenic B cells to B-1 cells is inhibited by CSA (this volume).

A significant result of the present study is that in addition to a role in induction of the B-1 phenotype, NFAT may play a role in the maintenance of this phenotype. This is suggested by the fact that B-1 cells contain extremely elevated nuclear NFATc levels. It is still necessary to demonstrate that maintenance of these elevated levels is necessary to maintenance of the B-1 phenotype. We are currently approaching this by determining if treatment of B-1 cells with CSA alters their phenotype. If it does, we will attempt to use retroviral gene transduction to overexpress a dominant negative calcinerin gene in B-1 cells.

We have shown that in the CD5+ lymphoma CH12 there is constitutive nuclear NFATc and that its binding is necessary for CD5 enhancer activity. We would next like to know if this constitutive NFAT activity is dependent on signaling through the BCR. This will be examined by isolating Ig-heavy chain loss variants of CH12 to see if NFAT activity is lost. We would then attempt to restore NFAT activity by ectopic expression of IgM. If this works we will use this system to determine if IgM-specificity is important in maintaining active NFAT in CH12 cells.

It has been known for some time that NFAT is present in B cells and inducible by both BCR and CD40 signaling (Choi et al., 1994; Venkataraman et al., 1994). However the significance of this finding has been unknown. The demonstration of a role for NFAT transcription factors in B-1 development is the first demonstration of a significant function for this family of transcription factors in B cells.

Acknowledgements

We would like to thank James Tung, Leonard Herzenberg, and Leonore Herzenberg for providing us with the construct pGL2ly1, which was the source of CD5 5'-flanking sequences used in our studies, as well as for communicating unpublished results to us. We thank Anjana Rao for rabbit anti-NFATp antibody and Gerald Crabtree for 7A6. We also thank Laurie Glimcher, Ann Ranger and Andrea Gerth for allowing us to look and peritoneal washouts from reconstituted NFAT knockout mice and for providing NFAT knockout mice for breeding at Tufts University. This work was supported in part by NIH grant AI15803 to Henry Wortis.

References

Arnold, L. W., LoCascio, N. J., Lutz, P. M., Pennell, C. A., Klapper, D., and Haughton, G. (1983). Antigen-induced lymphomagenesis: identification of a murine B cell lymphoma with known antigen specificity. J Immunol 131:2064-2068.

Arnold, L. W., Pennell, C. A., McCray, S. K., and Clarke, S. H. (1994). Development of B-1 cells: segregation of phosphatidyl choline-specific B cells to the B-1 population occurs after immunoglobulin gene expression. J Exp Med 179:1585-1595.

Aruffo, A., Melnick, M. B., Linsley, P. S., and Seed, B. (1991). The lymphocyte glycoprotein CD6 contains a repeated domain structure characteristic of a new family of cell surface and secreted proteins. J Exp Med 174:949-952.

Bandyopadhyay, R. S., Teutsch, M. R., and Wortis, H. H. (1995). Activation of B-cells by sIgM cross-linking induces accumulation of CD5 mRNA. Curr Top Microbiol Immunol 194:219-228.

Berland, R., and Wortis, H. H. (1998). An NFAT-dependent enhancer is necessary for anti-IgM-mediated induction of murine CD5 expression in primary splenic B cells. J Immunol 161:277-285.

Choi, M. S., Brines, R. D., Holman, M. J., and Klaus, G. G. (1994). Induction of NF-AT in normal B lymphocytes by anti-immunoglobulin or CD40 ligand in conjunction with IL-4. Immunity 1:179-187.

Cong, Y. Z., Rabin, E., and Wortis, H. H. (1991). Treatment of murine CD5- B cells with anti-Ig, but not LPS, induces surface CD5: two B-cell activation pathways. Int Immunol 3:467-476.

Goodnow, C. C., Crosbie, J., Adelstein, S., Lavoie, T. B., Smith-Gill, S. J., Brink, R. A., Pritchard-Briscoe, H., Wotherspoon, J. S., Loblay, R. H., Raphael, K., and et al. (1988). Altered immunoglobulin expression and functional silencing of self- reactive B lymphocytes in transgenic mice. Nature 334:676-582.

Hayakawa, K., Asano, M., Shinton, S. A., Gui, M., Allman, D., Stewart, C. L., Silver, J., and Hardy, R. R. (1999). Positive selection of natural autoreactive B cells. Science 285:113-116.

Healy, J. I., Dolmetsch, R. E., Timmerman, L. A., Cyster, J. G., Thomas, M. L., Crabtree, G. R., Lewis, R. S., and Goodnow, C. C. (1997). Different nuclear signals are activated by the B cell receptor during positive versus negative signaling. Immunity 6:419-428.

Hodge, M. R., Ranger, A. M., Charles de la Brousse, F., Hoey, T., Grusby, M. J., and Glimcher, L. H. (1996). Hyperproliferation and dysregulation of IL-4 expression in NF-ATp- deficient mice. Immunity 4: 397-405.

Karras, J. G., Wang, Z., Huo, L., Howard, R. G., Frank, D. A., and Rothstein, T. L. (1997). Signal transducer and activator of transcription-3 (STAT3) is constitutively activated in normal, self-renewing B-1 cells but only inducibly expressed in conventional B lymphocytes [see comments]. J Exp Med 185:1035-1042.

Lam, K. P., and Rajewsky, K. (1999). B cell antigen receptor specificity and surface density together determine B-1 versus B-2 cell development. J Exp Med 190:471-477.

Lecomte, O., Bock, J. B., Birren, B. W., Vollrath, D., and Parnes, J. R. (1996). Molecular linkage of the mouse CD5 and CD6 genes. Immunogenetics 44:385-390.

Loh, C., Carew, J. A., Kim, J., Hogan, P. G., and Rao, A. (1996). T-cell receptor stimulation elicits an early phase of activation and a later phase of deactivation of the transcription factor NFAT1. Mol Cell Biol 16: 3945-3954.

Nemazee, D. A., and Burki, K. (1989). Clonal deletion of B lymphocytes in a transgenic mouse bearing anti-MHC class I antibody genes. Nature 337:562-566.

Okamoto, M., Murakami, M., Shimizu, A., Ozaki, S., Tsubata, T., Kumagai, S., and Honjo, T. (1992). A transgenic model of autoimmune hemolytic anemia. J Exp Med 175:71-79.

Oukka, M., Ho, I. C., de la Brousse, F. C., Hoey, T., Grusby, M. J., and Glimcher, L. H. (1998). The transcription factor NFAT4 is involved in the generation and survival of T cells. Immunity 9:295-304.

Ranger, A. M., Hodge, M. R., Gravallese, E. M., Oukka, M., Davidson, L., Alt, F. W., de la Brousse, F. C., Hoey, T., Grusby, M., and Glimcher, L. H. (1998). Delayed lymphoid repopulation with defects in IL-4-driven responses produced by inactivation of NF-ATc. Immunity 8:125-134.

Ranger, A. M., Oukka, M., Rengarajan, J., and Glimcher, L. H. (1998). Inhibitory function of two NFAT family members in lymphoid homeostasis and Th2 development. Immunity 9:627-635.

Rao, A., Luo, C., and Hogan, P. G. (1997). Transcription factors of the NFAT family: regulation and function. Annu Rev Immunol 15:707-747.

Rothstein, T. L., Kolber, D. L., Murphy, T. P., and Cohen, D. P. (1991). Induction of phorbol ester responsiveness in conventional B cells after activation via surface Ig. J Immunol 147:3728-3735.

Schuh, K., Avots, A., Tony, H. P., Serfling, E., and Kneitz, C. (1996). Nuclear NF-ATp is a hallmark of unstimulated B cells from B-CLL patients. Leuk Lymphoma 23:583-592.

Teutsch, M., Higer, M., Wang, D., and Wortis, H. W. (1995). Induction of CD5 on B and T cells is suppressed by cyclosporin A, FK- 520 and rapamycin. Int Immunol 7:381-392.

Venkataraman, L., Francis, D. A., Wang, Z., Liu, J., Rothstein, T. L., and Sen, R. (1994). Cyclosporin-A sensitive induction of NF-AT in murine B cells. Immunity 1:189-196.

Watanabe, N., Nisitani, S., Ikuta, K., Suzuki, M., Chiba, T., and Honjo, T. (1999). Expression levels of B cell surface immunoglobulin regulate efficiency of allelic exclusion and size of autoreactive B-1 cell compartment. J Exp Med 190:461-469.

Wortis, H. H., Teutsch, M., Higer, M., Zheng, J., and Parker, D. C. (1995). B-cell activation by crosslinking of surface IgM or ligation of CD40 involves alternative signal pathways and results in different B-cell phenotypes. Proc Natl Acad Sci U S A 92:3348-3352.

Role of CD5 in growth regulation of B-1 cells

S. Bondada, G. Bikah[*], D.A. Robertson, and G. Sen
Department of Microbiology and Immunology and the Sanders Brown Center on Aging.
University of Kentucky, Lexington, KY 40536-0230
[*] Dept. of Immunology, Duke University Medical Center, Durham, NC 27710

Abstract

CD5 is a membrane glycoprotein that is expressed on a subset of B lymphocytes called B-1 cells, thymocytes and T cells. The CD5+ B-1 cells are normally unresponsive to surface Ig receptor induced growth signals unless the CD5 gene is deleted or sequestered away. Here we show that CD5 mediated negative regulation is unique to B cell receptor (BCR) signaling. The CD5 molecule in normal B-1 cells is constitutively tyrosine phosphorylated and associates specifically with SHP-1, an SH2 domain containing protein tyrosine phosphatase. CD5 promotes a prolonged interaction between BCR and SHP-1, which may be inhibitory to BCR signaling. CD5 was shown to modulate the function of autoantibody producing B cells in transgenic mice expressing anti-DNA antibodies.

Introduction

CD5 is a 67 kDa differentiation antigen that is primarily expressed on T lymphocytes. It belongs to the scavenger receptor family of proteins[1]. Initially B-lymphocytes have been subdivided into B-1 and B-2 cells on the basis of CD5 and IgM/IgD expression. The B-1 cells arise early in ontogeny and are characterized by high IgM and low IgD expression [2]. All the B-1 cells in the spleen and a majority in the peritoneum express the T cell differentiation marker CD5. Only some B-1 cells (defined by high IgM and low IgD levels) in the peritoneum express CD5 but all of them express the macrophage differentiation antigen Mac-1 allowing their further subdivision into B-1a or B-1b subsets. B-1 cells tend to make antibodies that often cross-react with self molecules and are increased in autoimmune states such as rheumatoid arthritis as well as murine models of lupus. Positive selection appears to play an important role in the formation of B-1 cell compartment [3, 4]. Although B-1 cells in vivo appear to be in a state of activation based on their size and self-replenishment, B cell receptor (BCR) cross-linking fails to induce a growth response, just like in immature B cells [2, 5]. In vivo BCR cross-linking induces apoptosis of B-1 cells [6, 7].

Here we investigated the basis of this B-1 cell unresponsiveness. CD5 appeared to be a good candidate for a regulatory molecule as CD5 expression differed between the non responsive B-1 cells and the responsive B-2 cells and since CD5 has been shown to be associated with BCR in B-1 cells [8]. Furthermore, CD5 has been shown to negatively regulate TCR mediated activation of thymocytes and positive selection [9]. As shown previously, B-1 cells from CD5$^{-/-}$ mice but not wild type mice proliferated in response to BCR cross-linking in vitro [7] while B-1 cells from both sources responded equally well to stimulation with anti-CD40 antibody or lipopolysaccharide. Therefore, we hypothesized that CD5 is a negative regulator of BCR signaling in B-1 cells. In B cells, surface molecules with inhibitory properties like CD22 and Fc receptor, contain a motif known as ITIM (immune receptor tyrosine based inhibitory motif) and associate with protein tyrosine phosphatases (PTPases) [10, 11]. Since CD5 also contains an ITIM motif like these inhibitory receptors, we investigated the ability of CD5 to interact with PTPases SHP-1 and SHP-2 and the effect of such interaction on BCR signaling. We also measured the impact of CD5 regulation on autoantibody producing B cells.

Results

CD5 negatively regulates BCR but not CD72 specific B cell growth responses:
To determine if negative regulation by CD5 was specific to stimulation via BCR, we investigated the response of B-1 cells from wild type and CD5 -/- mice to cross-linking BCR and CD72, another B cell specific surface molecule. CD72 was chosen since it appears to stimulate the same subpopulations of B cells as anti-µ and uses some of the early signaling pathways employed by the BCR [12, 13]. As shown in Table 1, anti-µ stimulated proliferation in B-1 cells from CD5 -/- but not wild type mice, which is in agreement with our previous results [7]. Like anti-µ, CD72 induced proliferation of B-2 but not B-1 cells in the wild type mice. Flow cytometry did not detect any significant difference in expression of CD72 on B-1 and B-2 cells in the wild type mice. Unlike the anti-µ response, the anti-CD72 response of B-1 cells was not restored in CD5 -/- mice. This suggested that CD5 negatively regulates BCR but not CD72 signaling in B-1 cells. B-1 cells from both wild type and CD5 -/- mice and B-2 cells responded well to stimulation with anti-CD40 (Table 1) and LPS attesting to the competence of these B cell populations to proliferate in culture. The reason for the unresponsiveness of B-1 cells to anti-CD72 is unknown.

Table 1: Specificity of CD5 mediated negative regulation of B-1 cell growth responses[1]

Stimulus	B-2 (Wild type)	B-1 (Wild type)	B-1 (CD5 -/- mice)
Anti-µ	21,160	1,614	14,532
Anti-CD72	23,116	412	534
Anti-CD40	31,000	25,241	26,278

[1] Purified peritoneal B-1 and B-2 cells were cultured with 10 µg/ml of anti-µ, 50 µg/ml of anti-CD72 and 10 µg/ml of anti-CD40 antibodies for 48 hours. Proliferation was measured by pulsing the cultures with 1.0 µCi of [^{3}H]-thymidine for 4 hours. Cultures were harvested and counted on a Matrix 96 beta counter. Responses in the presence of medium alone were less than 1500 cpm for all the three B cell populations. Mean cpm responses of triplicate cultures are provided. The SE values were less than 10% of the mean.

CD5 specifically associates with SH2 domain containing protein tyrosine phosphatase (SHP-1): Next we investigated the biochemical basis of CD5 mediated negative regulation of BCR signaling. Previously we showed that CD5 modulated both BCR induced calcium mobilization and translocation of NF-κB into the nucleus, suggesting that CD5 might affect an early step in BCR signaling cascade [7]. We hypothesized that CD5 might associate with a PTPase and interfere with BCR induced protein tyrosine kinase activity. Since most PTPases contain an SH2 domain and bind to phosphotyrosine residues on proteins, we first investigated the tyrosine phosphorylation status of CD5 in B-1 cells. Interestingly, CD5 was constitutively tyrosine phosphorylated in normal B-1 cells and in BKS-2, a CD5$^+$ B cell lymphoma (data not shown). Probing of these blots with antibodies to the PTPases, SHP-1, SHP-2 and to the inositol phospatase, SHIP showed that CD5 immunoprecipitates contained SHP-1 (Fig. 1B) but not the other two phosphatases (data not shown). The interaction between CD5 and SHP-1 appeared to be specific since there was little or no association of SHP-1 with CD19 and lyn in B-1 cells, even though both these molecules associated with SHP-1 in B-2 cells (Fig. 1B) as reported previously [14]. Similarly there was very little association of SHP-1 with CD22 in B-1 cells (data not shown). Also, there was no SHP-1 in immunoprecipitates of either class II molecules from B-1 cells or of CD5 from wild type B-2 or CD5 $^{-/-}$ B-1 cells, populations that are deficient in CD5 (data not shown). During these studies, we also found that the amount of SHP-1 in B-1 cells is dramatically less (five fold) than in B-2 cells (Fig. 1A). The small amount of SHP-1 in B-1 cells appears to associate with CD5 more readily than CD22, CD19 or lyn further emphasizing the preferential association between CD5 and SHP-1.

CD5 modulates association of IgM and SHP-1 inB-1 cells: Just like in human B-CLL cells, we found that CD5 associated with IgM in mouse B-1 cells [15]

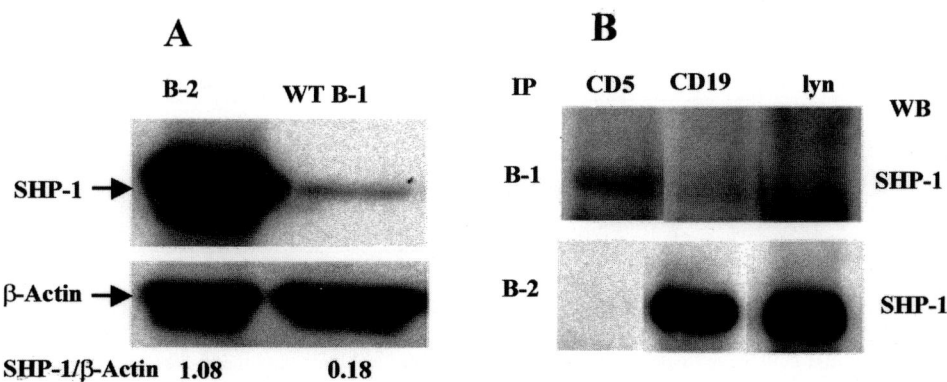

Fig. 1. Specific association of CD5 with SHP-1. Lysates of B-1 or B-2 cells from wild type mice were immunoprecipitated with antibodies to CD5, CD19 and lyn and were analyzed by Western blots which were probed with anti-SHP-1 antibodies. Western blots of B-1 and B-2 cell lysates were probed with anti-SHP-1 or anti-actin and the ratio of SHP-1 to actin is indicated.

CD5 modulates association of IgM and SHP-1 in B-1 cells: Just like in human Since IgM was also shown to transiently bind to SHP-1 in murine B-2 cells [16], we investigated the effect of CD5 on IgM-SHP-1 association in B-1 cells. As shown in Figure 2A, IgM from B-1 and B-2 cells as well as CD5 from wild type B-1 cells bound to SHP-1. In B-2 cells stimulated with anti-μ for 30 min. there was little SHP-1 in μ immunoprecipitates. But unlike in B-2 cells, IgM in B-1 cells remained associated with SHP-1 even after activation with anti-μ for 30 min. The IgM association with SHP-1 appeared to be specific since class I immunoprecipitates did not contain SHP-1. As noted above this is remarkable considering the small amount of SHP-1 available in B-1 cells. The persistent association between IgM and SHP-1 may interfere with the necessary PTK activation and down stream signaling contributing to B-1 cell unresponsiveness to BCR cross-linking. Accordingly, this IgM-SHP-1 association was reduced dramatically in anti-μ stimulated B-1 cells from CD5 $^{-/-}$ mice (Fig. 2A). Similarly prior sequestration of CD5 with botin-anti-CD5 and avidin to prevent its association with IgM reduced the levels of IgM bound SHP-1 in anti- μ treated wild type B-1 cells (Fig. 2B). Previously we showed that such a maneuver also restored anti-μ responsiveness in B-1 cells [7]. Also these results suggested that CD5 was in part responsible for the persistence of SHP-1 in IgM complex in B-1 cells

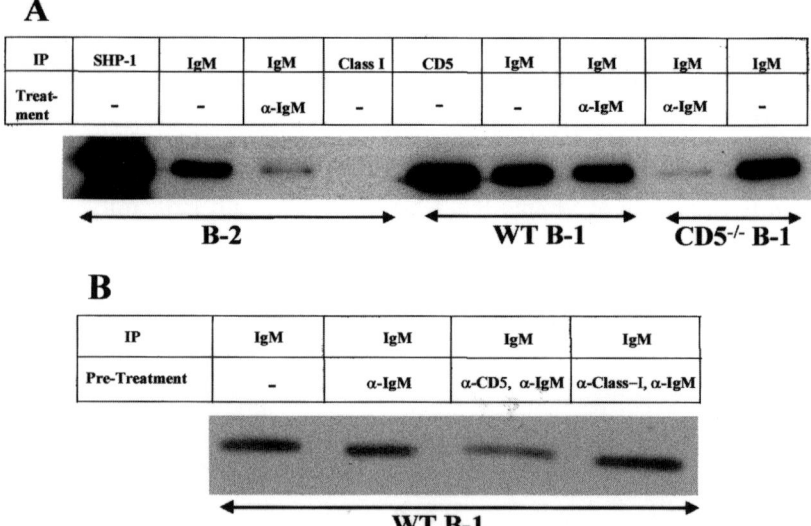

Fig. 2. Association of IgM with SHP-1 and its modulation by CD5. Panel A. Wild type B-2, B-1 and CD5$^{-/-}$ B-1 cells were treated with anti-IgM or medium for 30 min. and the lysates were prepared. Immunoprecipitates were prepared with anti-μ, anti-CD5 or anti-class I and were analyzed in Western blots by probing with anti-SHP-1 antibodies. Panel B: CD5 was pre-cross-linked by exposure of B-1 cells to biotin-anti-CD5 plus avidin before treatment with anti-μ and Western blot analysis as described for Panel A.

Since the surface IgM is capped and shed during the period of 30 min. activation with anti-μ, we measured the amount of μ heavy chain in these blots by stripping and reprobing with a μ-specific antibody. Even though the μ heavy chain was reduced after activation, the amount of reduction was comparable in B-1 and B-2 cells (data not shown). Expressing the amount of SHP-1 normalized to the μ content showed that the SHP-1 /μ ratio in anti-μ treated wild type B-1 cells was 256% of control while it was only 47% of control in similarly treated B-2 cells. Moreover, this SHP-1 /μ ratio decreased to 23 % of control in CD5 $^{-/-}$ B-1 cells stimulated with anti-μ. [15]. Thus the prolonged association of SHP-1 with IgM in B-1 cells appears to be highly specific.

CD5 regulates anti-DNA antibody production in 56R anti-DNA transgenic mice:

To test the impact of CD5 regulation on autoantibody producing B cells, we obtained transgenic mice in which heavy chain variable region of an anti-DNA antibody (3H9) from MRL/lpr mouse was introduced and crossed them to C57BL/6 or CD5 $^{-/-}$ mice. F2 offspring that express the transgenic VH gene and CD5$^{-/-}$ phenotype were identified by PCR analysis and flow cytometry respectively. The heavy chain variable region was mutated at position 56 to increase the affinity of the anti-DNA antibody [17]. This transgenic mouse was chosen for two reasons. First, the 56R heavy chain can overcome light chain restriction for binding to DNA. Secondly, the transgenic B cells in BALB/c

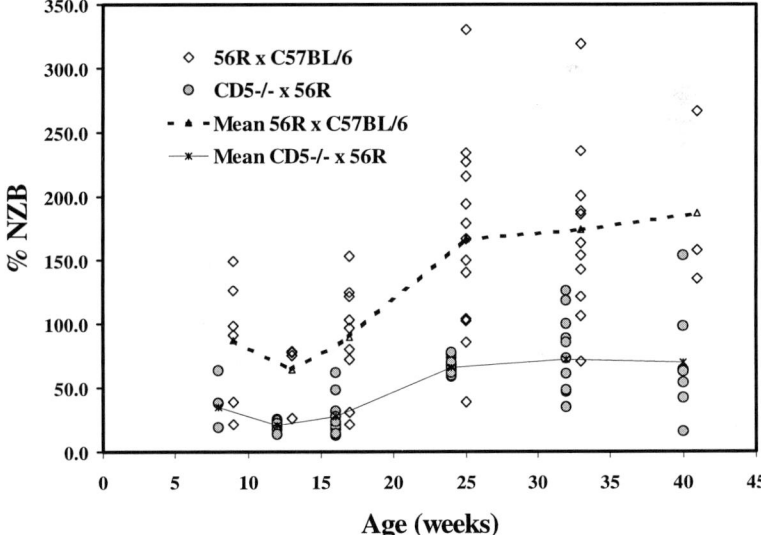

Fig. 3. Levels of single stranded (ss) anti-DNA antibodies in 56R mice crossed to wild type and CD5 $^{-/-}$ mice. Sera from mice of various ages were analyzed for the presence of anti-DNA antibodies using an ELISA assay with calf thymus DNA as antigen to coat microtiter wells. The levels of the anti-ss DNA antibody were expressed as % of such antibody in pooled NZB serum. Mean responses were indicated as lines. The differences between the two crosses were found to be statistically significant with a p<0.05.

background were regulated such that 56R mice had very few anti-DNA antibodies in their serum and there was a reduction in splenic B cells presumably due to B cell tolerance mechanisms. Interstingly, we found that in the C57BL/6 background, there was a spontaneous production of anti-DNA antibodies, which increased progressively with age (Fig. 3). Most surprisingly, the spontaneous production of anti-ss DNA antibody was significantly reduced in 56R mice bred to the CD5$^{-/-}$ mice. The decrease could be seen readily at later ages when the serum titer in the wild type background increased. More recently we measured the antibodies to the double stranded DNA which were also increased in the wild type but decreased in the CD5 $^{-/-}$ mice (data not shown). Thus CD5 appears to regulate anti-DNA antibody production but the cellular basis of this decrease in the mutant mice is not yet known.

Discussion

These studies have provided a biochemical basis for the negative regulation of BCR signaling by CD5 in B-1 cells. CD5 was shown to interact with SHP-1 specifically and modulated the association of SHP-1 with the BCR. The interaction between BCR and SHP-1 was transient in B-2 cells but was prolonged in B-1 cells. Elimination of CD5 either by gene deletion or by prior sequestration allowed BCR/SHP-1 complex to dissociate rapidly suggesting that CD5 was responsible for the increased interaction between these molecules. Consistent with the notion that SHP-1 inhibits BCR signaling in B-1 cells, pre-cross-linking of CD5 restored normal signaling pattern induced by BCR cross-linking. Thus the previously reported defect in BCR-induced calcium mobilization in wild type B-1 cells was no longer seen [15]. Also CD5

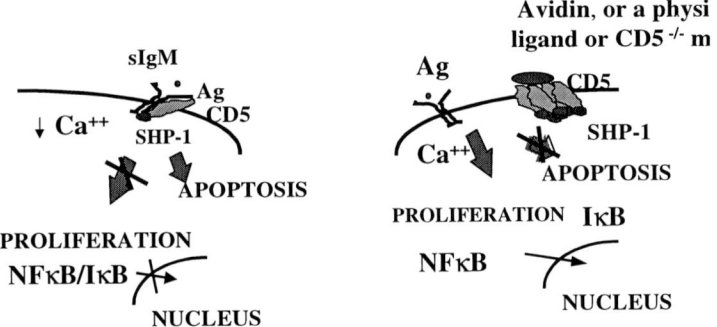

Fig. 4 A model to illustrate the role of SHP-1 in CD5 mediated negative regulation of BCR signaling in B-1 cells from wild type and CD5 $^{-/-}$ mice.

pre-cross-linking allowed anti-µ stimulation to induce nuclear translocation of NF-κB in wild type B-1 cells, a signaling event that fails to occur in the B-1 cells in which CD5 was not sequestered away [15]. Together with the finding that B-1 cells from CD5 $^{-/-}$ mice and CD5 pre-cross-linked wild type B-1 cells proliferate in response to anti-µ stimulation, these results suggest that CD5-SHP-1 association may be the basis of CD5 mediated negative regulation of BCR signaling in B-1 cells (Fig. 4).

A critical role for SHP-1 in regulation of BCR signaling was also inferred by the hyper reactivity of B cells from motheaten mice that have a mutation in SHP-1[16]. Almost all the B cells in these mice have been shown to be B-1 in nature by virtue of their CD5 expression. It has been suggested that the increased reactivity of B cells in these mice may be due to a lack of regulation by FcR or the CD22 co-receptors [10, 16]. However, it is conceivable that absence of CD5-mediated SHP-1 recruitment to the BCR plays an important role in the increased responses of CD5 positive B cells in these mice. In support of this idea we find that CD22 associated SHP-1 is much less in B-1 cells compared to B-2 cells, whereas SHP-1 is readily associated with CD5 in wild type B-1 cells. The concept that CD5 regulation utilizes SHP-1 is also supported by the recent observation by Perez-Villar et al. who showed that CD5 negatively regulates T cell receptor signaling in Jurkat T cells by associating with SHP-1 enzyme [18]. They identified Tyr-378 as critical for binding of SHP-1 to CD5, which is part of a sequence with homology to ITIM. We have also observed that CD5 was responsible for increased association of CD5 with T cell receptor in thymocytes (data not shown).

We hypothesize that CD5-mediated negative regulation is important to prevent accidental activation of B-1 cells by self-antigen that depends mainly on BCR cross-linking in the absence of T cell help. However, this negative regulation can be overcome with strong signals such as those provided by T cells via CD40L-CD40 interaction or by mitogenic moieties such as lipopolysaccharide present in bacteria allowing B-1 cells to participate in immune responses to foreign antigens.

This negative regulation of BCR signaling by CD5 is somewhat surprising in the context of the finding that B-1 cells are antigen selected and exhibit an activated phenotype [4, 19, 20]. One possible explanation of this paradox may lie in differences between newly emerging B-1 cells and the mature B-1 cells present in adults. Thus the antigen selected B-1 cells may have developed the novel means of regulation by CD5 to prevent their activation by self molecules while preserving their ability to respond to foreign antigens.

We tested the possible regulatory role of CD5 in autoantibody production by breeding transgenic mice containing an anti-DNA specific V gene. Mice expressing this transgene in BALB/c background were shown to undergo receptor editing to prevent production of anti-DNA antibodies[17]. Interestingly these 56R mice in the C56BL/6 background spontaneously produced large amounts of anti-DNA antibodies, which were significantly reduced in the absence of CD5. One possibility is that in the C57BL/6 background, there is a preferential expression of the transgenic V genes in the B-1 cell compartment accounting for increased levels of anti-DNA antibodies in these mice. Since the B-1 cells in the knockout mice behave more like normal B-2 cells in terms of their responses to BCR crosslinking, they may be under more stringent regulatory mechanisms like B-2 cells preventing their abnormal activation. Recently it had been shown that B-1 cells from wild type mice have increased recombinase activity [21] which may be further increased in the absence of CD5 resulting in removal of self-specificities by receptor editing mechanisms [22, 23]. We cannot also rule out the possibility that T cell regulation has a more important role in the decreased autoantibody production in 56RxCD5$^{-/-}$ mice. Our present work is directed at distinguishing these possibilities.

Acknowledgements

This work was supported in part by the NIH grants AI21340 and AG05731 to SB. Our thanks are due to Dr. Chelvarajan for a critical review of this manuscript.

References

1 Huang, HJ, Jones, NH, Strominger, JL and Herzenberg, LA (1987) Molecular cloning of Ly-1, a membrane glycoprotein of mouse T lymphocytes and a subset of B cells: molecular homology to its human counterpart Leu-1/T1 (CD5). *Proc.Natl.Acad.Sci.U.S.A.* . 84: 204-208.
2 Hardy, RR (1992) Variable gene usage, physiology and development of Ly-1+ (CD5+) B cells. *Curr.Opin.Immunol* . 4: 181-185.
3 Haughton, G, Arnold, LW, Whitmore, AC and Clarke, SH (1993) B-1 cells are made, not born. *Immunol. Today* . 14: 84-87.
4 Hayakawa, K, Asano, M, Shinton, SA, Gui, M, Allman, D, Stewart, CL, Silver, J and Hardy, RR (1999) Positive selection of natural autoreactive B cells. *Science* . 285: 113-116.
5 Morris, DL and Rothstein, TL (1994) Decreased surface IgM receptor-mediated activation of phospholipase C gamma 2 in B-1 lymphocytes. *Int. Immunol.* . 6: 1011-1016.
6 Murakami, M, Tsubata, T, Okamoto, M, Shimizu, A, Kumagai, S, Imura, H and Honjo, T (1992) Antigen-induced apoptotic death of Ly-1 B cells responsible for autoimmune disease in transgenic mice. *Nature* . 357: 77-80.
7 Bikah, G, Carey, J, Ciallella, JR, Tarakhovsky, A and Bondada, S (1996) CD5-mediated negative regulation of antigen receptor-induced growth signals in B-1 B cells. *Science* . 274: 1906-1909.
8 Lankester, AC, van Schijndel, GM, Cordell, JL, van Noesel, CJ and van Lier, RA (1994) CD5 is associated with the human B cell antigen receptor complex. *Eur. J. Immunol.* . 24: 812-816.
9 Tarakhovsky, A, Kanner, SB, Hombach, J, Ledbetter, JA, Muller, W, Killeen, N and Rajewsky, K (1995) A role for CD5 in TCR-mediated signal transduction and thymocyte selection. *Science* . 269: 535-537.
10 Cornall, RJ, Cyster, JG, Hibbs, ML, Dunn, AR, Otipoby, KL, Clark, EA and Goodnow, CC (1998) Polygenic autoimmune traits: Lyn, CD22, and SHP-1 are

limiting elements of a biochemical pathway regulating BCR signaling and selection. *Immunity* . 8: 497-508.
11. Ravetch, JV (1997) Fc receptors. *Curr.Opin.Immunol* . 9: 121-125.
12. Venkataraman, C, Muthusamy, N, Muthukkumar, S and Bondada, S (1998) Activation of lyn, blk and btk but not syk in CD72 stimulated B lymphocytes. *J. Immunol.* . 160: 3322-3329.
13. Venkataraman, C, Lu, P-J, Buhl, AM, Chen, C-S, Cambier, JC and Bondada, S (1998) CD72 mediated B cell activation involves recruitment of CD19 and activation of phosphatidylinositol 3-kinase. *Eur.J.Immunol.* . 28: 3003-3016.
14. O'Rourke, L, Tooze, R and Fearon, DT (1997) Co-receptors of B lymphocytes. *Curr Opin Immunol* . 9: 324-329.
15. Sen, G, Bikah, G, Venkataraman, C and Bondada, S (1999) Negative regulation of antigen receptor-mediated signaling by constitutive association of CD5 with the SHP-1 protein tyrosine phosphatase in B-1 B cells. *Eur J Immunol* . 29: 3319-3328.
16. Pani, G, Kozlowski, M, Cambier, JC, Mills, GB and Siminovitch, KA (1995) Identification of the tyrosine phosphatase PTP1C as a B cell antigen receptor-associated protein involved in the regulation of B cell signaling. *J. Exp. Med.* . 181: 2077-2084.
17. Chen, C, Nagy, Z, Radic, MZ, Hardy, RR, Huszar, D, Camper, SA and Weigert, M (1995) The site and stage of anti-DNA B-cell deletion. *Nature* . 373: 252-255.
18. Perez-Villar, JJ, Whitney, GS, Bowen, MA, Hewgill, DH, Aruffo, AA and Kanner, SB (1999) CD5 negatively regulates the T-cell antigen receptor signal transduction pathway: involvement of SH2-containing phosphotyrosine phosphatase SHP-1. *Mol. Cell. Biol.* . 19: 2903-2912.
19. Wortis, HH (1992) Surface markers, heavy chain sequences and B cell lineages. *Int Rev.Immunol* . 8: 235-246.
20. Clarke, SH and McCray, SK (1993) VH CDR3-dependent positive selection of murine VH12-expressing B cells in the neonate. *Eur.J.Immunol.* . 23: 3327-3334.
21. Qin, XF, Schwers, S, Yu, W, Papavasiliou, F, Suh, H, Nussenzweig, A, Rajewsky, K and Nussenzweig, MC (1999) Secondary V(D)J recombination in B-1 cells. *Nature* . 397: 355-359.
22. Chen, C, Nagy, Z, Prak, EL and Weigert, M (1995) Immunoglobulin heavy chain gene replacement: a mechanism of receptor editing. *Immunity* . 3: 747-755.
23. Kench, JA, Russell, DM and Nemazee, D (1998) Efficient peripheral clonal elimination of B lymphocytes in MRL/lpr mice bearing autoantibody transgenes. *J Exp Med* . 188: 909-917.

Life and death decisions in B1 lymphoma cells

Dubravka Donjerković, Gregory B. Carey, Carolyn M. Mueller, Sarah Liu, and David W. Scott§,

Department of Immunology, American Red Cross Holland Laboratory, Rockville, MD 20855

Department of Microbiology and Immunology, George Washington University School of Medicine, Washington, DC 20037

Abstract

Crosslinking of surface immunoglobulin (Ig) receptors with anti-IgM (anti-μ) but not anti-IgD (anti-δ) antibodies causes growth arrest and apoptosis in several extensively characterized B1-like lymphoma cell lines. While anti-μ stimulates a transient increase in *c-myc* mRNA and protein expression, followed by a rapid decline below the baseline level, anti-δ only causes a moderate increase in the expression of this oncogene, which returns to baseline levels within 24-48 hours. However, signals downstream from anti-δ can be converted into an apoptotic pathway by modulating PI3K activity, suggesting that PI3K is a critical rheostat controlling survival signals in B1 cell lines. Anti-μ-induced down-regulation of c-Myc is followed in time with an increase in the cyclin dependent kinase inhibitor, $p27^{Kip1}$, in all anti-μ sensitive lymphoma lines. This increase correlates with growth arrest and apoptosis. The anti-μ-mediated decrease in c-Myc, increase in $p27^{Kip1}$, growth arrest and apoptosis, can all be prevented *via* CD40/CD40L signaling. Inhibition of caspase activation, on the other hand, prevents anti-μ-induced apoptosis, but has no effect on c-Myc, $p27^{Kip1}$, and G1 arrest. Interestingly, we also found that steroids and retinoids can mimic anti-μ-mediated signaling and lead to a loss of c-Myc, an increase in $p27^{Kip1}$, G1 arrest, and apoptosis. Together, these data suggest that modulation of c-Myc and $p27^{Kip1}$ protein levels is crucial for the life *versus* death decisions in murine immature B1-like lymphoma cells lines.

Introduction

We have extensively studied a series of B1-like lymphoma cell lines as models of growth arrest and apoptosis. These include WEHI-231, CH12, CH31, CH33 and ECH408. WEHI-231, CH31, and CH33 are considered to be functionally immature murine B-cell lymphoma cells because anti-IgM treatment causes their growth arrest (Scott et al. 1986) and subsequent apoptosis (Benhamou et al. 1990; Hasbold and Klaus 1990). CH12 cells, in contrast, are insensitive to anti-IgM-mediated growth arrest and apoptosis. ECH408 cells, a derivative of CH33 cells transfected with a δ chain gene construct, display a $μ^{high}/δ^{high}$ surface Ig expression profile and are growth arrested by anti-μ but not anti-δ. We previously reported that growth arrest in these cells correlates with an increase in the ratio of the $p27^{Kip1}$ kinase inhibitor in Cdk2/cyclin A or Cdk2/cyclin E complexes (Ezhevsky et al. 1996) and that this is inversely correlated with the levels of c-Myc in these cells (Donjerkovic et al. 1999).

In this report, we show that anti-μ and anti-δ have differential effects on c-Myc and $p27^{Kip1}$ and we describe the critical controlling elements for IgD signaling, PI3K and $p70^{S6K}$, whose inactivation leads to a death signal *via* anti-δ. We also describe the effects of CD40/CD40L signaling on

NFκB, c-Myc, p27^{Kip1}, growth arrest and apoptosis induced by membrane IgM (mIgM) crosslinking. Additionally, we show that mIgM-mediated signaling can be mimicked by an insect steroid hormone.

Results and Discussion

Anti-IgM inactivates NFκB, down-regulates c-Myc and increases p27^{Kip1} protein expression

Previous data from our laboratory, as well as from Sonenshein and colleagues, suggested that anti-IgM induced growth arrest and apoptosis of WEHI-231 cells are regulated in part by c-Myc (for review see Scott et al. 1997; Wu et al. 1997). Down-regulation of c-Myc is necessary for anti-IgM-induced growth arrest and apoptosis since overexpression of exogenous c-Myc renders WEHI-231 cells resistant to anti-IgM (Wu et al. 1996a). Furthermore, *c-myc* transcription in WEHI-231 lymphoma cells is regulated by NFκB (Lee et al. 1995) and the inhibition of NFκB (leading to c-Myc down-regulation) induces apoptosis in WEHI-231 cells (Wu et al. 1996b). Anti-IgM treatment leads to an increase in *c-myc* mRNA and protein within one to two hours, a decrease to below the baseline level at four to eight hours (McCormack et al. 1984), and complete disappearance by 24 h in unsynchronized cells (Fischer et al. 1994). This dramatic down-regulation of c-Myc induced by anti-IgM treatment is preceded by the profound inactivation of NFκB as early as 1 hour after the mIgM crosslinking (Fig. 1). Furthermore, we previously established that immunoglobulin tyrosine activation motifs (ITAMs) which contain tyrosines at positions 23 and 34 in the cytoplasmic tail of Ig-α chain are critical for growth arrest and apoptosis in CH31 cells (Yao et al.1995). To determine if these ITAMs were also required for the modulation of c-Myc noted above, we examined a series of ITAM-mutated CD8-Igα transfectants of CH31 for c-Myc protein and message expression. The results in Table 1 indicate that intact ITAMs are also required for the increase and subsequent decrease in c-Myc expression in these cells lines.

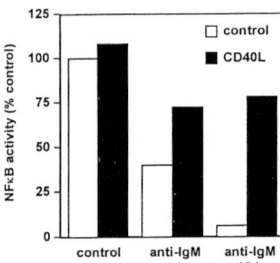

Fig. 1. Anti-IgM inactivates NFκB in WEHI-231 cells. Cells were cultured with anti-IgM (1 μg/ml) for indicated times, either in the absence or in the presence of CD40L (6 μg/ml). NFκB activity was determined by electrophoretic mobility shift assay (EMSA). Densitometric analysis was done using NIH Image software and the obtained values were analyzed by CA-Cricket Graph software

Table 1. Role of ITAM in signaling for modulation of c-Myc protein in CH31

Cell line	1 h	2 h	4 h	8 h	12 h	Result of anti-CD8 treatment
CD8:Igα	66¶	55	22	-35	-50	Growth arrest, apoptosis
M1*	7	10	16	20	14	No effect
M2	-4	-5	13	5	10	No effect
M3	1	14	13	6	12	No effect
M4	8	13	1	6	5	No effect
M5	59	103	51	16	-9	Growth arrest, apoptosis

¶Data reflect the percent change in c-Myc protein based on densitometric analysis of western blots at the indicated times after anti-CD8 treatment. Anti-IgM caused modulation of c-Myc, growth arrest and apoptosis in all clones.
*Igα cytoplasmic tail mutations (M) are as follows: M1, residues 23 and 34; M2, residues 17 and 34; M3, residue 34; M4, residue 23; M5, residue 17 (not ITAM).

Unlike crosslinking of mIgM, which leads to G1 arrest and apoptosis, crosslinking of IgD on the surface of ECH408 cells has no effect on cell cycle and apoptosis (Fig. 2A). Additionally, unlike anti-IgM treatment, which induces dramatic down-regulation of c-Myc, after the transient increase (McCormack et al. 1984; Tisch et al. 1988; Fischer et al. 1994), anti-δ causes a transient increase in *c-myc* mRNA and protein, but the levels of both mRNA and protein do not fall below the baseline (Tisch et al. 1988; Fig. 2B). Furthermore, only anti-IgM, but not anti-IgD, induced p27^{Kip1} accumulation (Fig. 2B). These data strongly suggest that the inability of IgD-mediated signa(s) to induce G$_1$ arrest and apoptosis correlates with maintenance of baseline c-Myc levels and low p27^{Kip1} levels in anti-µ sensitive murine B-lymphoma cell lines.

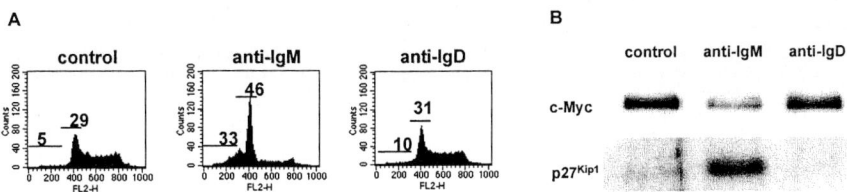

Fig. 2. Anti-IgM down-regulates c-Myc and increases p27^{Kip1}. **A)** ECH408 cells were incubated with 1 µg/ml anti-IgM or with 1 µg/ml anti-IgD Ab for 20 h. Cells were harvested, fixed and PI analysis done by flow cytometry. Data were analyzed by CellQuest software, and percentages of apoptotic and G1 phase cells are indicated. **B)** ECH408 cells were incubated as in Fig. 2A and the effects of anti-IgM and anti-IgD on c-Myc (upper panel) and on p27^{Kip1} (lower panel) protein levels determined by Western blot analysis of total cell lysates.

Anti-µ and anti-δ differentially affect PI3K and p70^{S6K} enzymatic activity

PI3K (for review see Fruman et al. 1998) has been implicated in BCR signaling (Campbell 1999). An 85 kDa regulatory subunit of this kinase contains SH3 and SH2 domains (for review see Pawson 1994), which enable it to associate with the BCR complex, thereby providing the p110^{PI3K} catalytic subunit access to inositol substrates at the plasma membrane. PI3K phosphorylates the third position of phosphatidylinositol and may alone, or in concert with PI4K and/or PI5K, generate phosphatidylinositol-3-phosphates such as PI (3)-phosphate, PI (3,4)-bisphosphate, and PI (3,4,5)-trisphosphate. The best characterized PI3K-dependent phosphatidylinositol products are PI (3,4,5)-trisphosphate, which mobilizes intracellular calcium, and PI (3,4)-bisphosphate, which is critical for the activation and/or translocation of phosphatidylinositol-dependent kinases (PDKs) such as PDK1 and PDK2, as well as protein kinase B (PKB, also called Akt). One downstream target of the PDK/PKB signaling pathway is p70^{S6K}, which is activated by multiple phosphorylation on serine and threonine residues. Consequent phosphorylation and activation of the ribosomal protein S6 by p70^{S6K} results in preferential translation of mRNAs which contain polypyrimidine rich 5', untranslated regions (Brown and Schreiber 1996). *C-myc* mRNA belongs to this class of mRNAs. Since anti-µ and anti-δ have differential effects on *c-myc* mRNA, it is possible that these two Ig isotypes differentially regulate the PI3K/p70^{S6K} signaling pathway. To test this, we examined if anti-µ and anti-δ differentially affect the phosphorylation state (*i.e.*, activity) of p70^{S6K} as well as levels of PIP$_3$ (*i.e.*, PI3K activity). As shown in Fig. 3, both anti-µ and anti-δ induce a decrease in PI3K (Fig. 3A) and p70^{S6K} (Fig. 3B) enzymatic activity at the 1 hour time point. However, similar to patterns of c-Myc protein expression, anti-µ-modulated p70^{S6K} and PI3K activity remains below baseline, while anti-δ-modulated activity of these two enzymes only transiently falls below baseline level, and then recovers by 8 h of treatment.

Together, these results suggest that anti-IgD and anti-IgM both stimulate a cellular signaling pathway, which includes inactivation of PI3K and consequently, p70^{S6K}. However, in the case of anti-δ signaling, PI3K and p70^{S6K} are only transiently inactivated and the final outcome is survival. Therefore, the inability of mIgD crosslinking to induce sustained c-Myc down-regulation, p27^{Kip1} accumulation, growth arrest and apoptosis correlates with the lack of sustained inactivation of

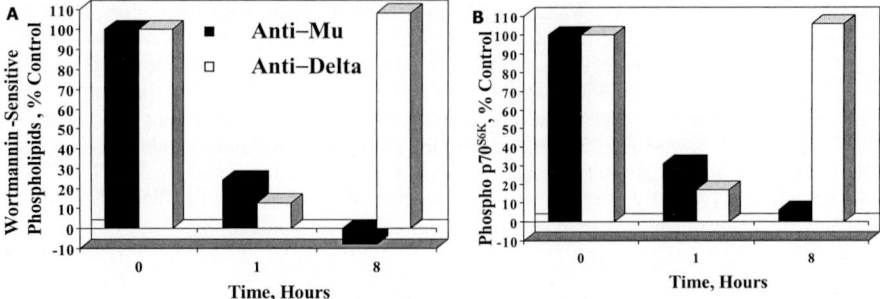

Fig. 3. Anti-μ and anti-δ differentially affect PI3K and p70^{S6K} enzymatic activity. **A)** Exponentially growing ECH408 cells were metabolically labeled for 48h with 2.5μCi/ml ^{32}[P]PO$_4$. Labeled cells were treated for 0, 1h, or 8h with monoclonal anti-μ or anti-δ (1μg/ml). To determine PI3K-dependent products, cells were incubated with 0.1 or 1μM wortmannin, followed by stimulation with appropriate antibody. Phospholipids were analyzed by the method of Whitman *et al.* (1987) in which labeled phospholipids were visualized by autoradiography. Densitometric analysis represented as percent control: 100 x (experimental intensity-wortmannin insensitive phospholipids)/(control intensity-wortmannin insensitive phospholipids). **B)** ECH408 cells were stimulated as above and phosphorylated (activated) p70^{S6K} kinase detected by Western blot analysis of the total cell lysates using the phospho-Thr389-specific p70^{S6K} antibody. Densitometric analysis of p70^{S6K} activation was calculated as: 100 x (experimental intensity - background intensity)/(control intensity - background intensity).

PI3K and p70^{S6K}. To test whether inactivation of PI3K can convert anti-δ into a death signal, we treated cells with the pharmacological PI3K inhibitor LY294002 (Vlahos et al. 1994) and anti-δ (Fig. 4). Indeed, anti-δ induces growth-arrest and apoptosis in the presence of LY294002 suggesting that PI3K activity is crucial for survival.

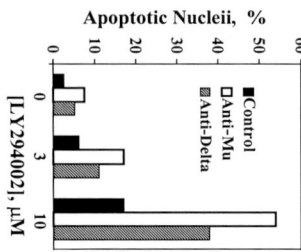

Fig. 4. Anti-IgD induces B-cell apoptosis in the presence of PI3K inhibitor, LY294002. ECH408 cells were stimulated with either anti-μ or anti-δ (1μg/ml) in the presence or absence of either 3μM or 10μM PI3K specific inhibitor 2-4-morpholinyl-8-phenyl-4H-1-benzopyran-4-one (LY294002) for 24hours. Cells were harvested, washed in PBS, fixed with 70 % ethanol, stained with PI and analyzed by flow cytometry. Data were analyzed by CellQuest and Cricket Graph software, and percentages of apoptotic cells are indicated.

CD40/CD40L signaling prevents IgM-mediated NFκB inactivation, c-Myc down-regulation, p27^{Kip1} accumulation, G1 arrest and apoptosis, while inhibition of caspases can only prevent IgM-mediated apoptosis

Engagement of CD40 on the surface of B cells with the CD40L on the surface of T cells provides a co-stimulatory signal, which is necessary for B cell activation, proliferation, and differentiation. Signaling through CD40 rescues WEHI-231 cells from anti-IgM-mediated apoptosis (Tsubata et al. 1993). Furthermore, CD40L treatment also prevents anti-IgM-induced inactivation of NFκB, down-regulation of c-Myc (Schauer et al. 1996; Schauer et al. 1998) and accumulation of p27^{Kip1} (Han et al. 1996). Here, we show that signaling *via* CD40L prevents anti-IgM-induced growth arrest in a dose-dependent manner as shown by thymidine incorporation assay (Fig. 5A). Furthermore, our results confirm that CD40 engagement maintains NFκB activity (Fig, 1), and high c-Myc levels, and also prevents p27^{Kip1} accumulation induced by mIgM crosslinking (Fig. 5B). Together, these data suggest that the survival signal mediated *via* CD40/CD40L is provided by preventing anti-IgM-induced NFκB inactivation, and by maintaining baseline c-Myc levels and low p27^{Kip1} levels in anti-μ sensitive murine B-lymphoma cells.

Caspases, a family of cysteine proteases that cleave after aspartic acid residues, are effectors of apoptotic cell death (for review see Thornberry and Lazebnik 1998). In WEHI-231 cells undergoing mIgM-mediated apoptosis, caspase 7 was reported to be active (Bras et al. 1999), while in CH31 cells treated with anti-IgM, caspase 3-like activity is reported, as shown by PARP cleavage (Andjelic and Liou 1998). We therefore examined if caspase activation was required for anti-Ig-induced growth arrest. Not surprisingly, synthetic peptide inhibitors of caspases prevent anti-IgM apoptosis in these cells. As shown in Fig. 6A by flow cytometric analysis of the cell cycle, at 24 h, Z-VAD-FMK prevents anti-IgM-induced apoptosis in WEHI-231 cells.

However, caspase inhibition has no effect on G1 arrest (Fig. 6A), nor on c-Myc and $p27^{Kip1}$ protein levels (Fig. 6B) as shown by the cell cycle and Western blot analyses, respectively. These results suggest that caspase activation is on a separate pathway(s) from, or downstream of, c-Myc down-regulation, $p27^{Kip1}$ accumulation and G1 arrest induced by mIgM crosslinking. Further studies are needed to distinguish between these two possibilities.

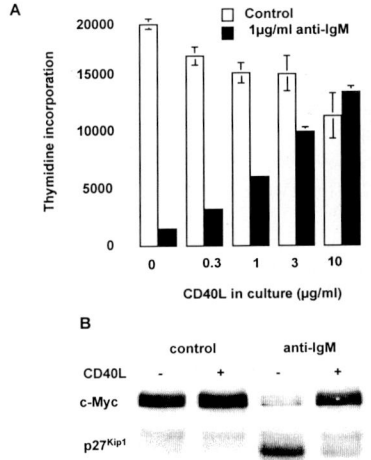

Fig. 5. CD40/CD40L signaling prevents c-Myc down-regulation, $p27^{Kip1}$ accumulation, G1 arrest, and apoptosis. A) Exponentially growing WEHI-231 cells were incubated with increasing concentrations of CD40L in the absence (open bars) or presence (black bars) of 1 µg/ml anti-IgM Ab for 24 hours, pulsed with radioactive deoxythymidine, harvested and thymidine uptake was quantitated using a Packard Matrix 9600 reader. Values are means ± standard deviation of at least three independent experiments. B) WEHI-231 cells were incubated as above ± 6 µg/ml CD40L and cell lysates analyzed for c-Myc (upper panel) and $p27^{Kip1}$ (lower panel) protein levels by Western blot analysis.

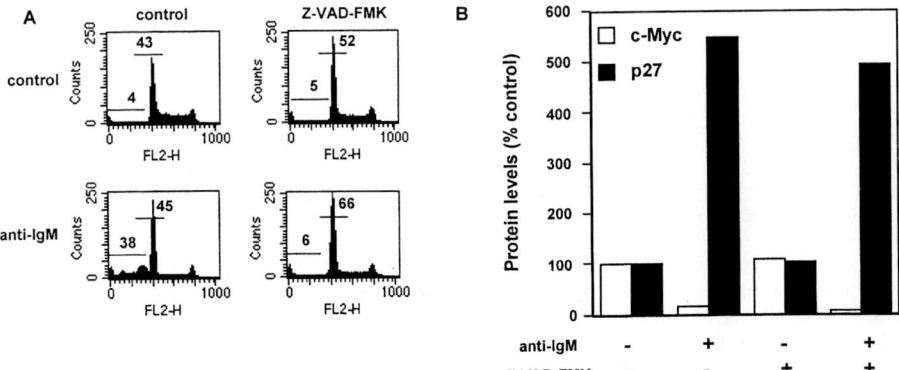

Fig. 6. Inhibition of caspases prevents IgM-mediated apoptosis but not growth arrest. A) WEHI-231 cells were cultured for 24 hours with anti-IgM (1 µg/ml), in the absence (left), or presence (right) of Z-VAD-FMK (50 µM). PI analysis was as in Figure 2A. Percentages of apoptotic and G1 cells are shown. B) C-Myc and $p27^{Kip1}$ protein levels were determined by Western blot analysis of the cells in Figure 6A.

Insect steroid hormone induces p27^{Kip1} accumulation, growth arrest, and apoptosis in WEHI-231 cells

Data from our laboratory (Ezhevsky et al. 1996; Donjerkovic et al. 1999) and from others (Han et al. 1996) show a very strong correlation between p27^{Kip1} accumulation and induction of G1 arrest and apoptosis by BCR crosslinking. However, it has not been directly proven that p27^{Kip1} is necessary for anti-IgM-induced growth arrest and apoptosis, nor that p27^{Kip1} up-regulation is sufficient for G1 arrest and possibly for apoptosis in WEHI-231 cells. To test this hypothesis, one has to overexpress p27^{Kip1} in either antisense or sense orientation. The prediction would then be that p27^{Kip1} deficient cells would be less sensitive (if not completely resistant) to mIgM-induced growth arrest and perhaps apoptosis. Conversely, p27^{Kip1} overexpression alone would induce G1 arrest and perhaps apoptosis.

The constitutive expression system can not be used because, based on our transfection studies, it is very likely that either complete down-regulation, or overexpression of p27^{Kip1} would prevent clonal expansion. Furthermore, transient transfection assays cannot be done readily because of the very low transfection efficiency of murine B-lymphoma cells. We therefore used the ecdysone-inducible system in which muristerone A (MA, a synthetic analogue of the insect steroid hormone, ecdysone) is an inducer. Surprisingly, while trying to overexpress p27^{Kip1}, we found that MA induces a dramatic up-regulation of p27^{Kip1} not only in cells transfected with p27^{Kip1} (Fig. 7A, lower panel), but also in control, parental, untransfected, WEHI-231 cells (Fig. 7A, upper panel). It also induces p27^{Kip1} in WEHI-231 cells that were stably transfected with RXR and modified ecdysone receptor, but not with p27^{Kip1} (WEHI-231/pVgRXR cells, Fig. 7A, middle panel). Furthermore, MA induces comparable levels of G1 arrest and apoptosis in parental, untransfected WEHI-231 cells, in WEHI-231/pVgRXR cells, and in cells that were stably transfected with pVgRXR and p27^{Kip1} containing pIND vector (Fig. 7B).

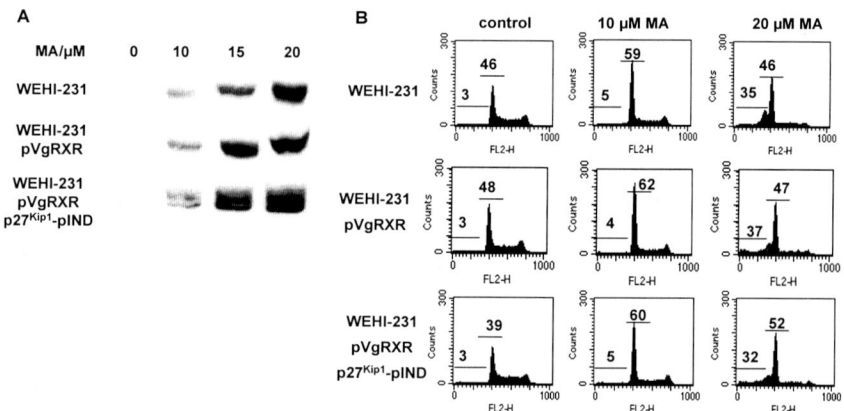

Fig. 7. Insect steroid hormone induces p27^{Kip1} accumulation, growth arrest, and apoptosis in WEHI-231.
A.) Exponentially growing WEHI-231, WEHI-231/pVgRXR 6.5, and WEHI-231/pVgRXR 6.5/p27^{Kip1} cells were cultured with MA (10 μM, 15 or 20μM) for 24 hours, harvested, and total cell lysates prepared for western analysis of p27Kip1. **B)** WEHI-231, WEHI-231/pVgRXR 6.5, and WEHI-231/pVgRXR 6.5/p27^{Kip1} cells were cultured as in Fig. 7A as above and PI analysis performed as in Figure 2A. Percentages of apoptotic and G1 cells are shown.

Additionally, we showed that not only muristerone A, but also other steroids (such as dexamethasone) and retinoids (such as *all trans-* and *9-cis-*retinoic acid) also induce p27^{Kip1} accumulation, G1 arrest, apoptosis, as well as c-Myc down-regulation (Donjerkovic *et al.*, submitted for publication). Furthermore, as is the case in anti-IgM signaling, all of the above effects of steroids and retinoids can be prevented by CD40/CD40L engagement. Inhibition of caspases (using pharmacological inhibitors such as Z-VAD-FMK) on the other hand, can prevent steroid/retinoid-mediated apoptosis, but has no effect on c-Myc, p27^{Kip1}, and G1 arrest, suggesting, again, that caspase activation is on a separate pathway(s) from, or downstream of, c-Myc down-regulation, p27^{Kip1} accumulation. and G1 arrest.

Based on our results, we postulate (Fig. 8) that the triggering of the mIgM or signaling *via* hormone-nuclear receptor each results in decreased expression of c-Myc, *via* inactivation of NFκB. Involvement of NFκB inactivation in down-regulation of c-Myc has already been established for anti-IgM-treated WEHI-231 cells by Sonenshein and colleagues (Schauer et al. 1996; Schauer et al. 1998), while we found that steroids can also inactivate NFκB in these cells (Donjerkovic *et al.* submitted for publication). The loss of c-Myc then leads to p27^{Kip1} accumulation and ultimately to G1 growth arrest and apoptosis. Signaling *via* CD40/CD40L prevents c-Myc down-regulation, p27^{Kip1} accumulation, growth arrest and apoptosis induced by anti-IgM or steroids/retinoids, by preventing NFκB inactivation. Inhibition of caspases, on the other hand, only prevents apoptosis induced by mIgM signaling and by steroids/retinoids. Finally, an anti-δ-mediated signal transduction pathway can lead to G1 arrest and apoptosis providing that PI3K is inactivated.

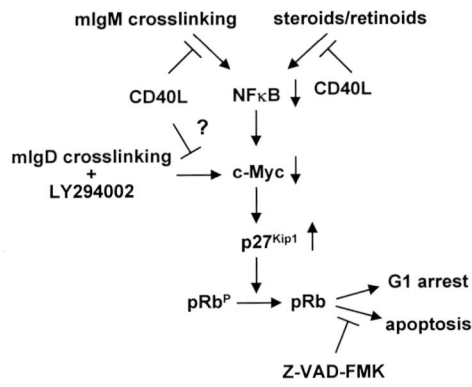

Fig. 8. Working model for steroid and BCR mediated growth arrest and apoptosis. Anti-IgM induces inactivation of NFκB and consequently c-Myc oncogene down-regulation, which leads to p27^{Kip1} accumulation, pRB hypophosphorylation, G1 arrest and cell death. Steroids and retinoids also inactivate NFκB, induce c-Myc down-regulation, up-regulate p27^{Kip1} and induce growth arrest and apoptosis. In both systems CD40/CD40L engagement rescues cells from G1 arrest and death by blocking NFκB inactivation, c-Myc down-regulation and p27^{Kip1} accumulation, while inhibition of caspases prevents only apoptosis induced by mIgM signaling and by steroids/retinoids. Anti-δ-mediated signal can also lead to G1 arrest and apoptosis providing that PI3K is inactivated. Hypothetical parts of this working model are indicated by question marks.

Murine B1-like lymphoma cell lines are a very attractive experimental system that have provided a great amount to our knowledge of the molecular mechanisms of B cell tolerance, as well as molecular mechanisms of cell cycle and cell death regulation in malignant cells. By studying the basic biochemical events that determine the molecular decision of the cell to die, growth arrest or proliferate, insights into the mechanisms by which immune tolerance is established and maintained, as well as insights into the mechanisms of neoplastic transformation can be obtained.

In this study, we show that modulation of transcription factors NFκB and c-Myc, as well as modulation of CKI, p27^{Kip1}, determines the proliferation *versus* apoptosis decision following BCR crosslinking. Additionally, our data strongly suggest the involvement of PI3K and p70^{S6K} in this decision-making process, providing a possible explanation for the differential effects of mIgM-*versus* mIgD-mediated signaling.

Acknowledgements

This work was supported by USPHS grant CA55644.

References

Andjelic S, Liou HC (1998) Antigen receptor-induced B lymphocyte apoptosis mediated via a protease of the caspase family. Eur J Immunol 28:570-581

Benhamou LE, Cazenave PA, Sarthou P (1990) Anti-immunoglobulins induce death by apoptosis in WEHI-231 B lymphoma cells. Eur J Immunol 20:1405-1407

Bras A, Ruiz-Vela A, Gonzalez de Buitrago G, Martinez AC (1999) Caspase activation by BCR cross-linking in immature B cells: differential effects on growth arrest and apoptosis. FASEB J 13: 931-944

Brown EJ, Schreiber SL (1996) A signaling pathway to translational control. Cell 86:517-520

Campbell, KS (1999) Signal transduction from the B cell antigen-receptor. Curr Opin Immunol 11:256-264

Donjerkovic D, Zhang L, Scott, DW (1999) Regulation of p27Kip1 accumulation in murine B-lymphoma cells: role of c-Myc and calcium. Cell Growth Differ 10:695-704

Ezhevsky SA, Toyoshima H, Hunter T, Scott, DW (1996) Role of cyclin A and p27 in anti-IgM induced G1 growth arrest of murine B-cell lymphomas. Mol Biol Cell 7:553-564

Fischer G, Kent SC, Joseph L, Green DR, Scott, DW (1994) Lymphoma models for B cell activation and tolerance. X. Anti-mu- mediated growth arrest and apoptosis of murine B cell lymphomas is prevented by the stabilization of myc. J Exp Med 179:221-228

Fruman DA, Meyers RE, Cantley LC (1998) Phosphoinositide kinases. Annu Rev Biochem 67:481-507

Han H, Nomura T, Honjo T, Tsubata T (1996) Differential modulation of cyclin-dependent kinase inhibitor p27Kip1 by negative signaling via the antigen receptor of B cells and positive signaling via CD40. Eur J Immunol 26:2425-2432

Hasbold J, Klaus GG (1990) Anti-immunoglobulin antibodies induce apoptosis in immature B cell lymphomas. Eur J Immunol 20:1685-1690

Lee H, Arsura M, Wu M, Duyao M, Buckler AJ, Sonenshein GE (1995) Role of Rel-related factors in control of c-myc gene transcription in receptor-mediated apoptosis of the murine B cell WEHI 231 line. J Exp Med 181:1169-1177

McCormack JE, Pepe VH, Kent RB, Dean M, Marshak-Rothstein A, Sonenshein GE (1984) Specific regulation of c-myc oncogene expression in a murine B-cell lymphoma. Proc Natl Acad Sci U S A 81:5546-5550

Pawson T (1994) SH2 and SH3 domains in signal transduction. Adv Cancer Res 64:87-110

Schauer SL, Bellas RE, Sonenshein GE (1998) Dominant signals leading to inhibitor kappaB protein degradation mediate CD40 ligand rescue of WEHI 231 immature B cells from receptor- mediated apoptosis. J Immunol 160:4398-4405

Schauer SL, Wang Z, Sonenshein GE, Rothstein TL (1996) Maintenance of nuclear factor-kappa B/Rel and c-myc expression during CD40 ligand rescue of WEHI 231 early B cells from receptor-mediated apoptosis through modulation of I kappa B proteins. J Immunol 157:81-86

Scott DW, Donjerkovic D, Maddox B, Ezhevsky S, Grdina, T (1997) Role of c-myc and p27 in anti-IgM induced B-lymphoma apoptosis. Curr Top Microbiol Immunol 224:103-112

Scott DW, Livnat D, Pennell CA, Keng P (1986) Lymphoma models for B cell activation and tolerance. III. Cell cycle dependence for negative signalling of WEHI-231 B lymphoma cells by anti- mu. J Exp Med 164:156-164

Thornberry NA, Lazebnik Y (1998) Caspases: enemies within. Science 281:1312-1316

Tisch R, Roifman CM Hozumi N (1988) Functional differences between immunoglobulins M and D expressed on the surface of an immature B-cell line. Proc Natl Acad Sci U S A 85:6914-6918

Tsubata T, Wu J, Honjo T (1993) B-cell apoptosis induced by antigen receptor crosslinking is blocked by a T-cell signal through CD40. Nature 364:645-648

Vlahos CJ, Matter WF, Hui KY, Brown RF (1994) A specific inhibitor of phosphatidylinositol 3-kinase, 2-(4- morpholinyl)-8-phenyl-4H-1-benzopyran-4-one (LY294002). J Biol Chem 269:5241-5248

Whitman M, Kaplan D, Roberts T, Cantley L (1987) Evidence for two distinct phosphatidylinositol kinases in fibroblasts. Implications for cellular regulation. Biochem J 247:165-174

Wu M, Arsura M, Bellas RE, FitzGerald MJ, Lee H, Schauer SL, Sherr DH, Sonenshein GE (1996a) Inhibition of c-myc expression induces apoptosis of WEHI 231 murine B cells. Mol Cell Biol 16:5015-5025

Wu M, Lee H, Bellas RE, Schauer SL, Arsura M, Katz D, FitzGerald MJ, Rothstein TL, Sherr DH, Sonenshein GE (1996b) Inhibition of NF-kappaB/Rel induces apoptosis of murine B cells. Embo J 15:4682-4690

Wu M, Yang W, Bellas RE, Schauer SL, FitzGerald MJ, Lee H, Sonenshein GE (1997) c-myc promotes survival of WEHI 231 B lymphoma cells from apoptosis. Curr Top Microbiol Immunol 224:91-101.

Yao XR, Flaswinkel H, Reth M, Scott DW (1995) Immunoreceptor tyrosine-based activation motif is required to signal pathways of receptor-mediated growth arrest and apoptosis in murine B lymphoma cells. J Immunol 155:652-661.

V

B-1 Cells in Inflammation

The Role of B-1 and B-2 Cells in Immune Protection from Influenza Virus Infection

N. Baumgarth[1], J. Chen[2], O. C. Herman[1], G. C. Jager[1] and L. A. Herzenberg[1]

[1]Dept. Genetics, Stanford University Medical School, Stanford, CA 94305, USA and [2]Center for Cancer Research and Dept. of Biology, Massachusetts Institute of Technology, Cambridge, Massachusetts 02139, USA

Introduction

Two distinct B cell populations, B-1 and B-2, which are present in the periphery of adult man and mice, differ considerably in development, surface phenotype, immunoglobulin-repertoire usage, and tissue distribution [1, 2]. Importantly, these types of B cells also differ in their responses to stimulation with antigen. Conventional B cells (B-2), which constitute most of the B cells in secondary lymphoid tissues of adults, undergo rapid proliferation when activated with anti-IgM antibody Fab_2 fragments. In contrast, (CD5$^+$) B-1 cells, which constitute the main peripheral B cell population in neonates and the main B cell population in the coelomic cavities of adults, undergo apoptosis rather then proliferation when stimulated through their B cell receptor [3-5]. On the other hand, B-1 cells, like their B-2 counterparts proliferate vigorously when stimulated with mitogens such as lipopolysaccharides [3, 5, 6], indicating that antigen recognition by B-1 and B-2 cells *in vivo* can, depending on the type of antigen encountered, induce different response patterns.

What the exact function B-1 cells is during immune responses to pathogens is still unresolved. B-1 cells are the producers of most of the circulating "natural" antibodies [7-9] and produce antibodies to a number of evolutionary conserved bacterial cell wall antigens, such as phosphoryl choline and phosphatidyl choline (PtC) (Reviewed in [10]). Recently, it was demonstrated that B-1 cell-derived natural IgM antibodies specific for PtC protect mice from rapid death following cecal ligation- and puncture-induced bacterial sepsis [11]. Thus, the study supports the long held view that B-1 cell-derived IgM antibodies constitute a first-line of defense against infection with bacteria [12]. The role of B-1 cells in immune protection from viral infections has not been studied. *In vitro* studies by Rott and colleagues [13] had shown, however, that certain strains of influenza virus act as mitogens for B-1 cells, inducing T-independent proliferation of these cells. We therefore compared the *in vivo* responses of B-1 and B-2 cells to infection with influenza virus and delineated their contribution to the protective immunity against this virus.

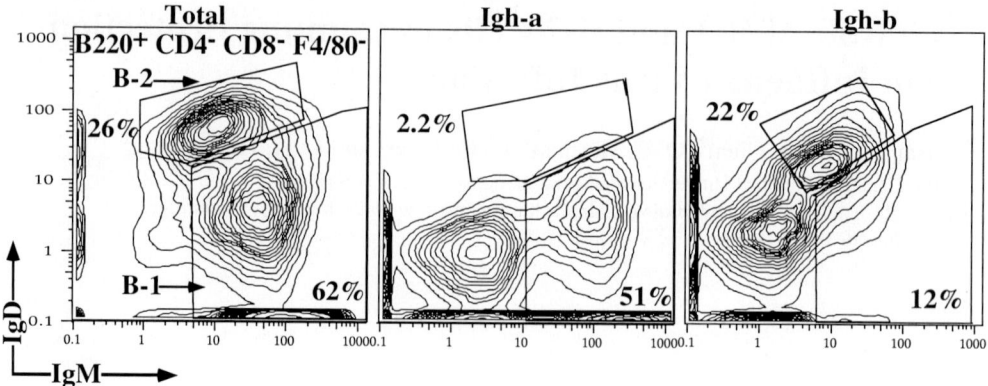

Fig. 1. Peritoneal cavity of B-1/B-2 allotype chimeras contain B-1 and B-2 cells of differing allotypes. Shown are 5% contour FACS plots of PerC from anti-IgM^b antibody-treated (Igh-b) CB.17 mice reconstituted with PerC from (Igh-a) BALB/c mice two months after end of antibody treatment. Cells were gated for expression of the B cell marker B220 and lack of expression of T cell and macrophage markers as indicated, and then stained with either total or allotype-specific anti-IgM and anti-IgD. Frequencies of B-1 and B-2 cells are indicated.

The Innate and Acquired Humoral Response to Influenza Virus is Provided by B-1 and B-2 Cells, Respectively

We created B-1/B-2 allotype chimeric mice in order to distinguish the cells and cell products *in vivo* using allotype-specific monoclonal antibodies against various classes of Ig [14]. Newborn Igh-b allotype-expressing CB.17 mice are treated from birth with anti-IgM^b antibodies to ablate host-derived B cell development. On day two after birth these mice receive peritoneal cavity wash out cells (PerC) as a source of B-1 cells, or FACS-sorted B-1 cells, from Igh-a allotype-expressing congenic BALB/c mice. The antibody treatment is continued for 6 weeks during which donor-derived B-1 cells expand. Within two months after the end of antibody treatment, host-derived B cell development leads to full reconstitution of the B-2 cell compartment. For at least 16 month after cell transfer, however, more then 80% of the B-1 cell compartment in peritoneal cavity and spleen of these chimeras consist of donor-derived B-1 cells. A maximum of 20% of the B-1 cell pool is derived from the host (Fig. 1). Thus, all Igh-a expressing B cells in these mice are B-1 cells and the vast majority of the Igh-b expressing B cells are B-2 cells. Sublethal infection of the chimeras was induced with the influenza A virus strain "Mem71" [15].

The results were clear-cut [8]. Whereas B-1 cells produce most (>80%) of the circulating total IgM, including the virus-binding natural IgM antibodies that are present prior to the infection, B-2 cells contribute only minimally, if at all, to the

Table 1. Total and virus-specific serum IgM and IgG2a levels in allotype-chimeras and controls

Mouse strain	Days after infection	Ig-allotype	Total IgM (μg/ml)*	Anti-Mem71 IgM (U/ml)*	Anti-Mem71 IgG2a (U/ml)*
Chimera	0	Igh-a	2,900	8	<0.001
	7	Igh-a	n.d.	9	<0.001
	0	Igh-b	400	3	<0.001
	7	Igh-b	n.d.	50	8,000
BALB/c	0	Igh-a	2,600	8	<0.001
	7	Igh-a	n.d	160	1,000
CB.17	0	Igh-b	2,900	20	<0.001
	7	Igh-b	n.d.	80	20,000

n.d., not determined; *mean levels

serum-IgM pool and make little natural anti-viral IgM (Table 1). In contrast, seven days after influenza virus infection the observed increases in the titers of virus-specific serum IgM were of the b-allotype, whereas the levels of Igh-a allotype virus-specific IgM were unaltered (Table 1). Similar results were obtained with a kinetic study in which antibody titers were measured daily for two weeks after the infection [8]. Thus, B-1 cells do not respond to the infection with increased IgM production. All increased IgM production is derived from B-2 cells.

In the chimeras, roughly similar amounts of total serum IgG2a is produced by B-1 and B-2 cells. Neither B-1 nor B-2 cell-derived virus-binding antibodies of the IgG2a isotype were detected prior to influenza virus infection, but they make the bulk of the virus-specific humoral response following infection [16, 17]. Similar to the IgM response, all of the virus-induced specific IgG2a in the chimera was secreted by the B-2 cells. Therefore, the natural IgM secreting B cells are not the precursors of the cells that respond to the infection with further IgM secretion, or the precursor cells that start the germinal center reaction for affinity maturation and isotype switching. B-1 cells function by providing steady-state levels of natural virus-binding IgM antibodies, unaffected by the presence of antigen. In contrast, B-2 cells function by rapidly inducing the secretion of virus-specific IgM and IgG antibodies, which are presumably of higher affinity then those secreted by the B-1 cells. Maintaining normal levels of poly-specific natural antibodies throughout an infection may be of importance for the maintenance of immune protection against infections with other pathogens, at a time when the adoptive immune system is engaged in the response to the primary pathogen. B-1 cell-derived IgM might therefore serve a function distinct from that of B-2 cell-derived IgM.

Non-Redundant Role for B-1 and B-2 Cells in influenza Virus Infection

We wanted to test the role of secreted IgM for immune protection from influenza virus infection. For this, we compared rates of survival in wild-type mice and mice deficient in the secreted, but not membrane-bound form of IgM (sIgM$^{-/-}$) [18]

following infection with influenza virus. Indeed, despite the fact that sIgM$^{-/-}$ mice have otherwise normal populations of B cells able to secrete all other classes of immunoglobulins, the lack of secreted IgM alone led to a significant increase in deaths from infection [19]. Roughly 50% of sIgM$^{-/-}$ succumbed to infection with a dose of influenza virus that does not cause deaths in the wild-type controls. These increased deaths in the sIgM$^{-/-}$ mice were associated with a significant increase in lung virus titers at days 5 and 7, and a significant reduction in the virus-specific serum IgG response during the first 3 weeks following infection, when compared to controls (Fig.2).

To delineate the contributions of the B-1 and B-2 cell-derived IgM antibodies for this strong immune protective effect, we created B-1/B-2 irradiation chimeras between sIgM$^{-/-}$ mice and wild-type controls. Groups of irradiated sIgM$^{-/-}$ mice were reconstituted with sIgM$^{-/-}$ PerC and wild-type bone marrow and vise versa to create mice that lack either secreted IgM from the B-1 cells or from the B-2 cells. Surprisingly, following influenza virus infection, both groups of chimeras had mortality rates similar to mice in which neither B-1 nor B-2 cells secreted IgM. Consistent with the increased deaths seen in both groups of chimeras, these mice had also significant lower IgG2a responses compared to controls that had received PerC and bone marrow from wild-type mice. Furthermore, reconstitution of sIgM$^{-/-}$ mice with purified serum IgM from normal mice did not alter the mortality rates, or the levels of virus-specific IgG2a. Hence, secretion of IgM by both B-1 and B-2 cells is necessary for maximal immune protection against an acute viral infection. The presence of only one type of IgM is insufficient to provide full immune protection.

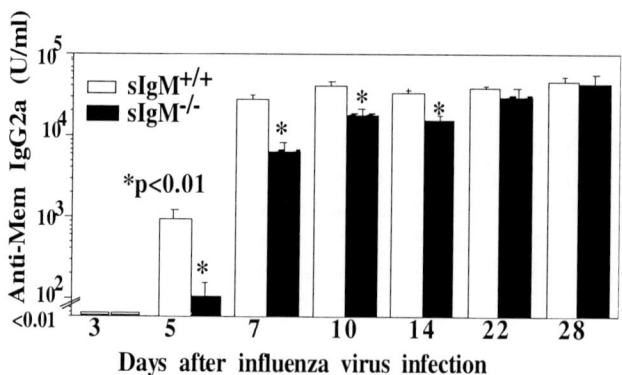

Fig. 2. Reduced anti-viral serum IgG2a titers in mice lacking secreted IgM. sIgM$^{-/-}$ mice and wild-type controls (sIgM$^{+/+}$) were infected with influenza virus Mem71 and serum levels of virus-specific IgG2a were determined by ELISA. Arbitrary units were determined by comparison to an allotype-matched hyperimmune serum.

How does secreted IgM provide immune protection against a viral infection? Various effects might operate simultaneously. IgM antibodies can neutralize the virus directly to prevent virus-attachment and internalization by host epithelial cells [20]. Moreover, our data show that secreted IgM affects virus neutralization in part indirectly by regulating the magnitude of the virus-specific (neutralizing) IgG response. Other effects such as opsonization of influenza virus for increased uptake by alveolar macrophages through binding of antigen-IgM-complement complexes to complement receptors might also play a role.

Secreted IgM Functions as Autocrine Regulator of the IgG Response

The requirement for secreted IgM in induction of optimal anti-viral IgG responses demonstrated here for influenza virus, is consistent with previous observations showing that secreted IgM is required for efficient IgG antibody responses to suboptimal doses of T-dependent antigens [18, 21]. This raises the question of how secreted IgM regulates the IgG response. Since complement and complement receptors are crucial for the induction of normal antibody responses [22, 23], previous studies using sIgM$^{-/-}$ mice proposed that natural IgM antibodies augment IgG responses by activating complement to form immune complexes [18, 21]. These complexes may then activate B cells by cross-linking the B cell receptors and/or by being trapped on follicular dendritic cells for activation of efficient germinal center reactions. Virus-induced B-2 cell-derived IgM is more effective in promoting the IgG response after influenza virus infection than B-1 cell-derived natural IgM. Chimeras that lacked the B-2 cell-derived IgM showed a more dramatic reduction in virus-specific IgG2a titers compared to chimeras that lacked B-1 cell-derived IgM [19]. Thus, focal secretion of virus-specific IgM immediately following B-2 cell activation seems crucial for directly activating the IgM-secreting B cells for IgG production.

As both virus-induced IgM and IgG are derived from B-2 cells, secreted IgM seems to act in an autocrine fashion. Since IgM-antigen complexes activate the classical complement cascade, antigen-IgM complexes present in the vicinity of the secreting antigen-specific B cells may bind to complement receptors expressed by these cells. These receptors are known to provide important costimulatory signals [22, 23], thus complement receptor engagement may further stimulate the antigen-activated IgM-secreting B cells. In addition, polymeric IgM may enhance antigen-mediated B cell receptor triggering by cross-linking surface Ig receptors on B cells that have bound the antigen.

T-independent secretion of IgM is induced in response to a number of antigens, including vesicular stomatitis virus [24]. It is therefore possible that the early activation of B cells in response to a viral infection is regulated independently of a helper T cell response. The early secretion of IgM by antigen-specific B cells might induce or enhance initial clonal expansion of these B cells in a manner similar to that of the early autocrine secretion of IL-2 by antigen-stimulated T cells which enhances clonal expansion of the T cells. Enhanced precursor frequencies of antigen-specific B cells increase the chances of T-B interaction, and therefore the speed and magnitude of T cell-dependent B cell differentiation events such as isotype switching and affinity maturation. Although speculative at this point, if these events are proven to occur *in vivo*, such an initial T cell-independent activation of B cells would bring into question the validity of the current two-signal model of T cell-dependent B cell activation.

Acknowledgements

The authors thank Dr. L. Brown for providing the virus material. This work was supported in part by NIH grant AI-34762-34 (to L.A.H) and AI41762 (to JC).

References

1. Stall AM, Wells SM, Lam K-P (1996) B-1 cells: unique origins and functions. Seminars in Immunology 8:45 - 59
2. Kantor AB, Herzenberg LA (1993) Origin of murine B cell lineages. Annu. Rev. Immunol. 11:501 - 538
3. Rothstein TL, Kolber DL (1988) Anti-Ig antibody inhibits the phorbol ester-induced stimulation of peritoneal B cells. J. Immunol. 141:4089 - 4093
4. Nawata Y, Stall AM, Herzenberg LA, Eugui EM, Allison AC (1990) Surface immunoglobulin ligands and cytokines differentially affect proliferation and antibody production by human CD5+ and CD5- B lymphocytes. Int. Immunol. 2:603 - 614
5. Bikah G, Carey J, Ciallella JR, Tarakhovsky A, Bondada S (1996) CD5-mediated negative regulation of antigen receptor-induced growth signals in B-1 B cells. Science 274:1906 - 1909
6. Morris DL, Rothstein TL (1993) Abnormal transcription factor induction through the surface immunoglobulin M receptor of B-1 lymphocytes. J. Exp. Med 177:857 - 861
7. Avrameas S (1991) Natural autoantibodies: from "horror autotoxicus" to "gnothi seauton". Immunol. Today
8. Baumgarth N, Herman OC, Jager GC, Herzenberg LA, Herzenberg LA (1999) Innate and acquired Immunities to influenza virus are provided by distinct B cells. Proc. Natl. Acad. Sci. USA 96:2250 - 2255
9. Coutinho A, Kazatchkine MD, Avrameas S (1995) Natural autoantibodies. Curr. Opin. Immunol. 7:812 - 818
10. Allison AC, Nawata Y (1992) Cytokines mediating the proliferation and differentiation of B-1 lymphocytes and their role in ontogeny and phylogeny. Annn. N. Y. Acad. Sci 651:200 - 219
11. Boes M, Prodeus AP, Schmidt T, Carroll MC, Chen J (1998) A critical role of natural immunoglobulin M in immediate defense against systemical bacterial infection. J. Exp. Med. 188:2381 - 2386
12. Herzenberg LA, Herzenberg LA (1989) Towards a layered immune system. Cell 59:953 - 954
13. Cash E, Charreire J, Rott O (1996) B-cell activation by superstimulatory influenza virus hemagglutinin: a pathogenesis for autoimmunity? Immunol. Rev. 152:67 - 88
14. Lalor PA, Stall AM, Adams S, Herzenberg LA (1989) Permanent alteration of the murine Ly-1 B repertoire due to selective depletion of Ly-1 B cells in neonatal animals. Eur. J. Immunol. 19:501 - 506

15. Baumgarth N, Brown L, Jackson D, Kelso A (1994) Novel features of the respiratory tract T-cell response to influenza virus infection: lung T cells increase expression of gamma interferon mRNA in vivo and maintain high levels of mRNA expression for interleukin-5 (IL-5) and IL-10. J. Virol. 68:7575 - 7581
16. Hocart MJ, Mackenzie MJS, Steward GA (1988) The IgG subclass responses induced by wild-type, cold-adapted and purified haemagglutinin from influenza virus A/Queensland/6/72 in CBA/CaH mice. J. General Virol. 69:1873 - 1882
17. Coutelier J-P, Logt JTMVd, Heessen FWA, Warnier G, Snick JV (1987) IgG2a restriction of murine antibodies elicited by viral infections. J. Exp. Med. 165:64 - 69
18. Boes M, Esau C, Fischer MB, Schmidt T, Carroll M, Chen J (1998) Enhanced B-1 cell development, but impaired IgG antibody responses in mice deficient in secreted igM. J. Immunol 160:4776 - 4787
19. Baumgarth N, Herman OC, Jager GC, Brown L, Boes M, Herzenberg LA, Chen J (1999) B-1 and B-2 cell-derived IgM antibodies are non-redundant components of the protective response to influenza virus infection, *submitted for publication.*
20. Taylor HP, Dimmock NJ (1985) Mechanisms of neutralization of influenza virus by IgM. J. Gen. Virol. 66:903 - 907
21. Ehrenstein MR, O'Keefe TL, Davies SL, Neuberger MS (1998) Targeted gene disruption reveals a role for natural secretory IgM in the maturation of the primary immune response. Proc. Natl. Acad. Sci USA 95:10089 - 1093
22. Tedder TF, Inaoki M, Sato S (1997) The CD19- CD21 complex regulates signal transduction thresholds governing humoral immunity and autoimmunity. Immunity 6:107 - 118
23. Carroll MC (1998) The role of complement and complement receptors in induction and regulation of immunity. Annu. Rev. Immunol. 16:545 - 568
24. Bachmann MF, Zinkernagel RM (1997) Neutralizing antiviral B cell responses. Annu. Rev. Immunol. 15:235 - 270

B-1 B Cell IgM Antibody Initiates T Cell Elicitation of Contact Sensitivity

P. W. Askenase[1], and R. F. Tsuji[2].

[1]Section of Allergy and Clinical Immunology, Dept. of Medicine, Yale University School of Medicine, New Haven, CT 06520 (supported by AI-43371); [2]Noda Institute for Scientific Research, Chiba, Japan.

Abstract. Although B-1 B cells have received considerable attention, their actual role in the normal functioning of the immune system is unclear. The hypothesized role of B-1 cell IgM in natural protective immunity is just being established. We have uncovered a separate and novel role for B-1 cell IgM in initiating the elicitation of acquired T cell-dependent contact sensitivity (CS), the prototype of *in vivo* T cell immunity, early after immunization (within 4 days). The recent recognition of a similarly unanticipated role of B cells in a variety of T cell responses, may indicate that B-1 cell IgM has a broader role in immunity than thought previously. We showed that 24 hr CS responses, and rises in local IFN-γ levels at 24 hrs later after antigen (Ag) challenge the ears, were absent in pan B cell and antibody deficient mice. The mechanism of B cell involvement in CS-initiation is via local C5a generation early (1-2 hrs) after antigen (Ag) challenge of the ears, in 4 day contact sensitized mice. C5a activates local mast cells to release serotonin (5-HT) and TNFα to induce endothelial ICAM-1 and VCAM-1, leading to T cell recruitment. We hypothesized that C5a was generated via complement activation due to antibodies forming local AgAb complexes, and that B-1 cell IgM was involved because isotype switching of B-2 cells to produce C-activating IgG isotypes, could not occur as early as day 4. Indeed, B-1 cell deficient CBA/N-*xid* mice lacked C5a in 2 hr ear extracts, and had impaired CS ear swelling and elaboration of IFN-γ at 24 hrs. Importantly, adoptive transfer of purified normal peritoneal B-1 cells, or just *i.v.* injection of Ag-specific IgM monoclonal antibodies in sensitized *xid*, restored deficient early C5a and late 24 hr ear swelling. These results suggest that early after Ag challenge, specific B-1 cell IgM, produced at distant sites by prior sensitization, forms AgAb complexes that trigger elaboration of C5a, to activate mast cell release of vasoactive TNFα and 5-HT to initiate CS, leading to T cell recruitment. We postulate that antibody of various isotypes possibly may lead to local vascular activation to aid in T cell recruitment in a variety of T cell responses, but that very early after immunization, Ag-specific IgM produced by B-1 cells, preferentially serves this important function.

A. Introduction. Contact sensitivity (CS) is a form of delayed-type hypersensitivity (DTH) that is used to investigate mechanisms of acquired T cell immunity *in vivo*. Challenge in the skin with hapten antigen (Ag) in sensitized animals, induces local recruitment of T cells, and their subsequent activation to produce inflammatory cytokines and 24-48 hr tissue swelling (1). How the T cells actually get recruited locally has received little attention. Herein we review recent data indicating that B-1 cell-derived immune IgM plays a crucial role in initiating this early T cell recruitment in CS. Previously, we described early local events, that peaked just 2 hours after hapten challenge in previously immunized mice. We called this "CS-initiation", because these early processes were *required* for T cell recruitment and subsequent function in CS. The components acting in CS-initiation were: mast cells (2), platelets (3,4), and unusual Ag-specific CS-initiating cells (5), which expressed CD5 and B220 but not CD3 (6), and led to early local release of vasoactive serotonin (5-HT) (2) and TNFα (7) (**Fig. 1**) from the mast cells and platelets. Recently, we also established a role for complement (C) in CS-initiation (8,9), and showed that CS was not elicited in B cell-deficient μMT mice (9), suggesting B cell involvement.

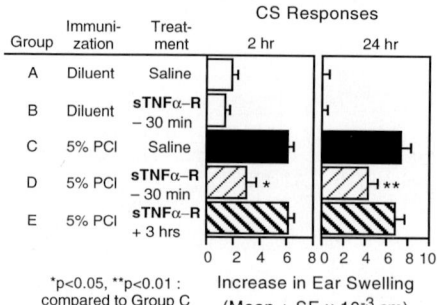

Fig. 1. Involvement of TNFα in the initiating phase of CS responses. Inhibition of both 2 and 24 hr CS ear swelling in soluble TNF-α Receptor treated PCl sensitized and challenged mice, that were treated prior to challenge (-30 min) (Gp D), but not when treated at +3 hrs (after CS-initiation) (Gp E), compared to positive controls treated with saline (Gp C).

B. Hypothesis For B-1 Cell Involvement in CS.

Finding involvement of complement and B cells, allowed us to organize the previously identified components of CS-initiation into a hypothesis concerning the function of these components. We hypothesize that skin applied sensitizing Ag is dispersed to distant sites like the peritoneal cavity, to activate B-1 cells to produce specific CS-initiating IgM antibodies, that are released systemically in the serum, to be available in the skin within 1 day following sensitization. B-2 cell-derived isotypes may similarly be involved, but later after immunization (10). At the time of skin challenge to elicit a secondary CS response on day 4, the locally applied Ag binds this IgM, producing AgAb complexes that activate C, to locally generate the active fragment C5a. Interaction of C5a with C5a receptors (C5aR) on local mast cells and platelets induces release of vasoactive 5-HT and TNFα that induce expression of endothelial adhesion molecules (ICAM-1 and VCAM-1), allowing recruitment of circulating CS-effector T cells into the local site. C5aR on activated T cells also may be involved (11) (**Fig. 2**). The potential power of DTH-initiating mechanisms is underscored by local cell DTH transfer experiments with limiting numbers of Ag-specific T cell clones, indicating that as few as *one* T cell could transfer DTH (12). Thus, recruitment of just a few T cells may lead to biologically relevant DTH responses.

C. DTH-Initiation Has Two Different "Early" Aspects.

These are: [1] early events after Ag immunization, and [2] early events after secondary Ag challenge. Quite early after immunization, i.e. within one day (4), B cells appear to become activated, and produce important antibodies, independent of T cells. Thus, the DTH-initiating cells are primed *and function* early after immunization. The other "early" is in the subsequent elicitation phase and refers to the effects of the generated DTH-initiating cell antibodies, early after secondary challenge. Thus, within only 1-4 hrs after challenge there is detectable local C-activation, edema, and endothelial activation (7). Thus, early-activated (1 day) B-1 cells produce IgM antibodies that function early after challenge (5,6), to mediate DTH-initiation for elicitation of a secondary response. The early-generated DTH-initiating cell, that acts early after challenge, is *required* for subsequent local T cell recruitment and activation to mediate DTH.

D. Complement Is Involved In CS and DTH.

The idea that B cells were involved in DTH-initiation arose following recognition of C involvement. Several C5aR antagonists inhibited CS and DTH, and four strains of C5-deficient mice had decreased CS elicitation (8). Further, correction of C5 levels with fresh serum reconstituted CS, and use of sera deficient in C components showed this was due to C5 (8) (**Fig. 3**). Injection of anti-C5 mAb into ears, to *locally* interfere with and deplete C5, inhibited both 2 and 24 hr CS responses, suggesting that local C activation was required to elicit CS. Timed adminstration of recombinant soluble human C receptor-1 showed that C was required <u>early</u> in CS and DTH elicitation (9).

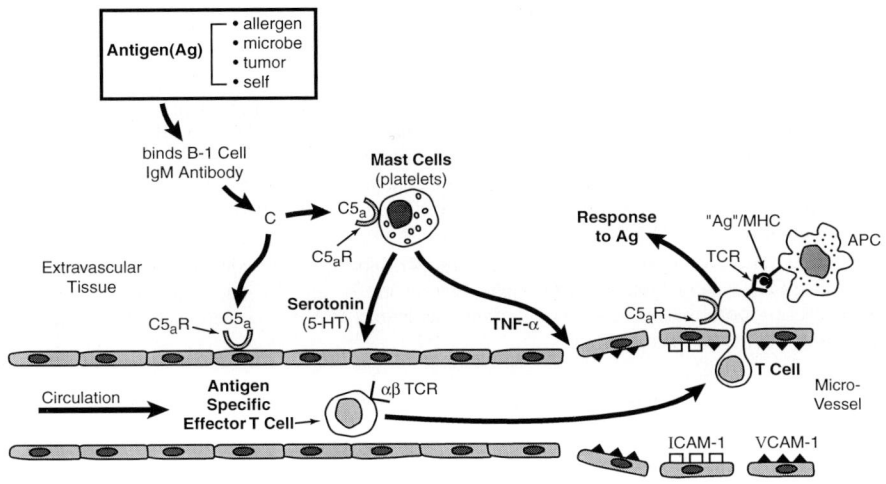

Fig. 2. Hypothesis for the delivery of T cells to Ag on APC in tissues in DTH and CS responses. Free eliciting Ag binds B-1 cell derived IgM, to activate C and generate C5a, to stimulate mast cells and platelets to release 5-HT and TNF-α, to induce endothelial ICAM-1 & VCAM-1, to extravasularly recruit T cells to bind Ag/MHC on APC and mediate response to Ag in tissues.

These findings were strengthened by showing absent CS in C5a-Receptor deficient mice (13), which led to assay for elaborated C5a in CS ear extracts for direct evidence of C-activation. At various times after hapten challenge, CS ear biopsies were extracted and assayed in vitro for C5a by macrophage chemotaxis. Increased C5a was detected 1 hr after challenge (13), confirming that C was activated early and locally. Using macrophages from C5a-R$^{-/-}$ mice, there was impaired chemotaxis to 2 hr but not 24 hr CS extracts, suggesting that C5a dominates early, while chemokines were found predominantly at 24 hrs (13). C5a induced 2 hr chemotactic activity was present in TCR$^{-/-}$ mice, showing it was T cell-independent (13) and importantly, was Ag-specific (13), leading us to postulate that B cell Ab were involved. To determine whether early C-activation was needed for later Th1 cell cytokine production, we measured IFN-γ in CS ear extracts by ELISA and RT-PCR. Local IFN-γ became detectable at 4 hrs, and was greatly increased at 24 hrs (13). Importantly, pretreatment with anti-C5 mAb decreased 24 hrs IFN-γ and cell infiltration (9), suggesting that early C5a led to local T cell recruitment and thus later production of IFN-γ.

Group	3% OX Sensitization	OX Elicitation	IV Transfer
A	Vehicle	0.2%	None
B	Vehicle	0.2%	NHS
C	+	0.2%	Saline
D	+	0.2%	NHS
E	+	0.2%	Heat-treated NHS
F	+	0.2%	C3(−)HS
G	+	0.2%	C6(−)HS
H	+	0.2%	C5(−)HS
I	Vehicle	0.8%	None
J	+	0.8%	None

Fig. 3. Effect of reconstitution with complement deficient human serum (NHS) in C5 deficient B10.D2/o mice, on 24 hr CS elicited with a suboptimal dose of 0.2% oxazolone (OX) Ag. In Gp D vs. C, NHS reconstituted impaired CS in these deficient mice. Gp E shows that 56°C x 30' heat treatment of NHS abolished the effect, while reconstitution with C-component deficient human sera (HS), showed that C3 [C3(−)] and C6-deficient sera reconstituted CS (Gps F & G), while C5-deficient HS did not (Gp H). Gp J shows the optimal CS response

E. B-1 Cell IgM is Involved in CS. We studied CS in B cell-deficient μMT mice that lack all B cells and serum immunoglobulins. Strikingly, 24 hr CS ear swellings were impaired in immunized and challenged μMT, compared to controls (9). However, μMT were on a B6 background poor for elicitation of CS. Thus, we examined $JH^{-/-}$ mice that have a different mutation resulting in total B cell deficiency (14), and their controls (C.B-17 Ig allotype congenics of BALB/c), are an excellent background for CS. Strong CS responses occurred in C.B-17 and minimal CS in $JH^{-/-}$, employing three different challenging doses (**Fig. 4**) definitively demonstrating involvement of B cells in CS ear swelling. We extracted ear samples and found large amounts of IFN-γ in 24 CS of positive control C.B-17, while ear extracts from sensitized and ear challenged $JH^{-/-}$ had no increase in IFN-γ (**Fig. 4**), biochemically confirming the absence of Th1 mediated CS. Taken together, these results, in a second B cell deficient strain, with 3 eliciting doses, measuring both macroscopic swelling and ear extract IFN-γ, showed that CS is virtually absent in B cell deficient mice.

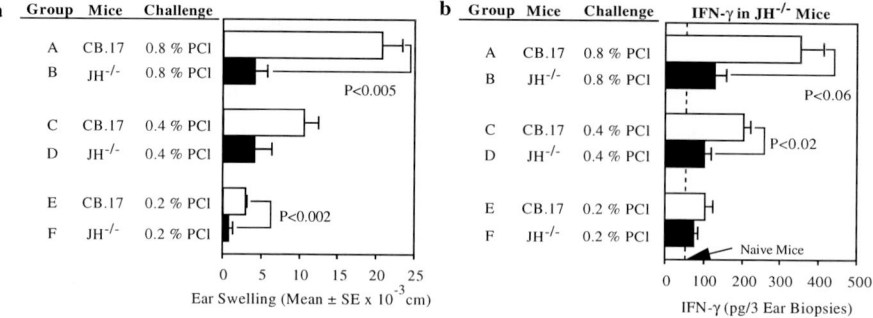

Fig. 4. Impaired elicitation of macroscopic CS (a.), and IFN-γ in ear extracts (b.) of B cell deficient $JH^{-/-}$ mice. $JH^{-/-}$ and control C.B-17 mice were contact sensitized with 5% PCl. Four days later, separate sensitized groups were ear challenged with either 0.8% or 0.4% or 0.2% PCl in acetone and olive oil (1:1), and 24 hr ear swelling responses were measured with a micrometer, and harvested ear extracts assayed for IFN-γ by ELISA.

1. B-1 cell Phenotype of DTH-Initiating Cells. Employing lymph node and spleen cells from 4 day or 1 day contact sensitized mice, we found that cells mediating CS-initiation had a unique putative phenotype of: $CD5^+$, $B220^+$ ($CD45RA^+$), Thy-1$^+$, $CD3^-$, αβ TCR$^-$, γδ TCR$^-$, $CD4^-$, and $CD8^-$ (6), later expanded to include: Mac1$^+$, $CD23^+$, IL-3R$^+$, IL-2R$^-$, Pgp-1$^+$ ($CD44^+$), J11d (HSA^+), and FcγRII$^+$ (15), and recently $CD19^+$ and $CD22^+$, consistent with B-1 B cells (16). These phenotypic characteristics of CS-initiating cells were determined by treatment of mixed immune cells with specific mAb + C, and then assay of their CS-initiating activity after i.v. cell transfer and challenge on the ears. This functional B-1 cell phenotype recently has been confirmed by transfer of FACS purified $CD5^+$, $CD19^+$ cells.

2. Defective CS-Initiation in CBA-N-*xid*. CBA/N-*xid* have a predominant B-1 cell deficiency (17), with some B-2 cell defects (18), and also partial mast cell (19), and macrophage (20) defects. Since C was involved in CS (8,9), and IgM was the only C-activating isotype made by B-1 cells, and no known knock outs of just B-2 cells exist, CBA/N-*xid* were chosen for the study. CS ear swelling responses at 24 hrs were decreased significantly in sensitized CBA/N-*xid* mice, compared to normal CBA/J (**Fig. 5**) and *xid* ear extracts had decreased 2 hr C5a and 24 hr IFN-γ. Absent CS-initiation in *xid* could be due to lack of the early C5a-mediated processes that were dependent on B-1 cell IgM antibody.

3. Reconstitution of CS-Initiation in CBA/N mice. CS-effector Th1 cells were not defective in CBA/N-*xid* mice, which could have explained their lack of CS, since Ag-specific in vitro T cell proliferation, and IFN-γ production were normal in contact sensitized *xid* mice (**Fig.**

6). Thus, despite the presence of the sensitized Th1 cells, *xid* were unable to elicit CS following challenge (**Fig. 5**). This suggested that the B cell defects had not impaired the *induction* of CS effector Th1 cells, as might have occurred if B cell APC function were required, and instead suggested that the B cell defect of *xid* mice affected T cell recruitment during elicitation of CS.

We undertook reconstitution of CBA/N-*xid* by transfer of purified normal B-1 cells harvested from the peritoneal cavity of CBA/J mice. Macrophages were removed by adherence at 37ºC, and then 98% B-1 cells (CD5$^+$ B220$^+$) were obtained by FACS sorting, were transferred i.p. to non-immune B-1 cell deficient *xid* mice, and 1 day later the *xid* recipients were contact sensitized with 5% PCl. Four days later they were tested on the ears with dilute PCl to attempt elicitation of CS.

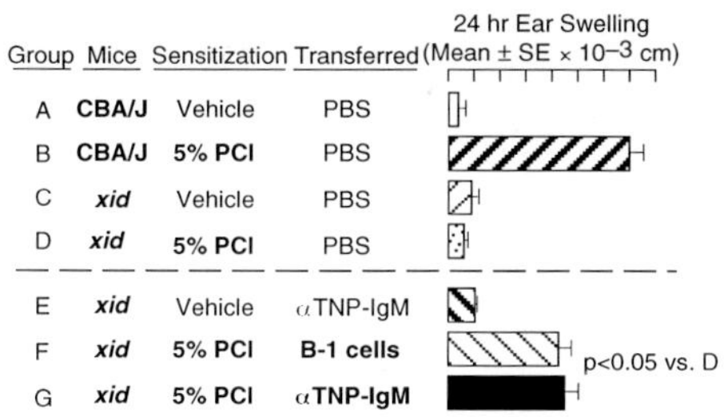

Group	Mice	Sensitization	Transferred	24 hr Ear Swelling (Mean ± SE × 10⁻³ cm)
A	CBA/J	Vehicle	PBS	
B	CBA/J	5% PCl	PBS	
C	*xid*	Vehicle	PBS	
D	*xid*	5% PCl	PBS	
E	*xid*	Vehicle	αTNP-IgM	
F	*xid*	5% PCl	B-1 cells	p<0.05 vs. D
G	*xid*	5% PCl	αTNP-IgM	

Fig. 5. B-1 cell reconstitution of CS in deficient CBA/N-*xid* mice. Shown are 24 hr ear swelling responses in positive control normal CBA/J mice (Gp B); compared to negative controls that were sensitized with vehicle alone (Gp A). Similar PCl sensitization and elicitation in *xid* mice, showed greatly decreased 24 hr responses (Gp D), that were partially and significantly reconstituted with enriched 98% B-1 cells (Gp F), given just prior to immunization, or with anti-TNP IgM given to PCl sensitized *xid* mice before challenge (Gp G), but not in non PCl sensitized *xid* recipients given anti-TNP-IgM (Gp E).

Fig. 5 Gp. B vs. Gp. A shows normal elicitation of 24 hr PCl CS in positive control sensitized CBA/J mice and that negligible CS was elicited in sensitized CBA/N-*xid* Gp. D vs. Gp. C. In contrast, transfer of normal B-1 cells into CBA/N recipients, that then were contact sensitized, allowed elicitation 4 days later of partial but significant 24 hr CS in these *xid* mice (Gp. F), while transfer of control peritoneal cells from B-1 cell deficient *xid* donors did not (data not shown). This reconstitution of CS by transfer of B-1 cells into *xid* mice (12), suggested that B-1 cells were responsible, at least partially, for CS-initiation early after immunization.

4. A Role For Specific IgM Antibodies in CS-Initiation (21). Since B-1 cells were involved in CS-initiation, their secretion of IgM might provide a mechanism. Reconstitution of CS-initiation was attempted with two different monoclonal anti-TNP IgM obtained from established B cell hybridomas called 13.4 and 32.17, that secrete anti-TNP/DNP IgM kappa. Purified 13.4 (**Fig. 5,** Gp. G), and 32.17 (not shown) anti-TNP IgM mAb, both partially reconstituted CS-initiation in previously sensitized CBA/N-*xid* mice. In contrast, control non-specific IgM had no activity (not shown), and 13.4 produced no 24 hr response in PCl challenged *xid* without prior PCl contact sensitization (**Fig. 5,** Gp. E).

5. Possible B Cell/Ig Initiation of Other T Cell Responses. Besides CS and DTH, there now are many examples of in vivo T cell responses that depend on B cells. We described a mouse model of asthma adoptively transferred by cells resembling B-1 cells (22), and previously found early responses in cellular immunity to intestinal Nematode (*Trichinella*) worms (23), and also in tumor immunity (24); as noted by others (25). Recent findings point to a role for B cells in T cell-dependent diabetes of NOD mice (26,27), and others employing μMT B cell deficient mice suggest a role for B cells in *initiation* of T cell-mediated collagen-induced arthritis (28,29), and in immunity to *M. tuberculosis* (30), *F. Tularensis* (31)*, Cryopotococcus* (32,33), and of *chlamydia* (31), which correlated with a lack of DTH to *chlamydia*. Moreover, studies in JH$^{-/-}$ B cell deficient mice indicate defective expression of T cell aspects of murine lupus (35,36), and other studies suggest that B cells facilitate T cell autoimmune diseases (37-40). Although it is possible that some of these examples of B cell involvement in T cell responses involve B cell APC functions in the afferent limb to induce effector T cells (41), in light of the possible involvement of B cell antibody-dependent initiating mechanisms to elicit T cell recruitment, as in CS and DTH, B cell antibodies in the efferent limb may initiate local Th1 T cell recruitment needed to elicit responses.

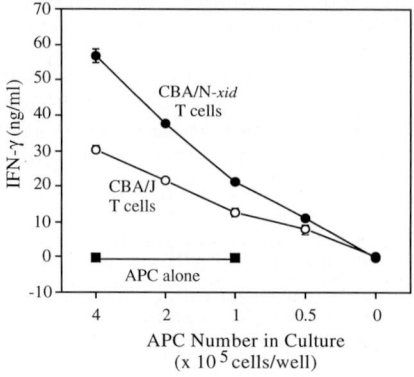

IFN-γ Production *in vitro* by Immune T Cells

Fig. 6. Intact *in* vitro **Th1 responsiveness in PCl sensitized *xid* mice.** TNP-APC induced *in* vitro IFN-γ production is not impaired in T cells of PCl contact sensitized B-1 cell deficient CBA/N-*xid* mice, compared to PCl sensitized normal CBA/J mice.

F. Summary. The absence of CS responses in pan B cell-deficient μMT and JH$^{-/-}$ mice indicated a role of B cells in these T cell responses. The similarity of phenotype between B-1 cells and previously described CS-initiating cells, suggested that B-1 cells are required for CS elicitation, at least early after sensitization. The absence of CS in B-1 cell-deficient CBA/N-*xid* mice, and partial reconstitution of 24 hr CS by transfer of B-1 cells, or by specific IgM Ab, indicate that B-1 cell IgM activation of C to elaborate C5a largely accounts for CS-initiation.

References

1. Askenase, PW. (1998)Effector and regulatory molecules and mechanisms in delayed-type hypersensitivity (DTH). In "Allergy: Principles and Practice," Fifth edition, E Middleton, Jr., CE Reed and EF Ellis, and NF Atkinson, JW Yunginger, and WW Busse. Eds. C.V. Mosby Co., St. Louis.
2. Askenase PW, Bursztajn S, Gershon MD, and Gershon RK. (1980) T cell dependent mast cell degranulation and release of serotonin in murine delayed-type hypersensitivity. J Exp Med. 152:1358-1374.
3. Geba GP, Ptak W, Anderson GA, Ratzlaff RE, Levin J, and Askenase PW. (1996) Delayed-type hypersensitivity in mast cell deficient mice: dependence on platelets for expression of contact sensitivity. J Immunol. 157:557-565.
4. Matsuda H, Ushio H, Geba GP, and Askenase PW. (1997) Human platelets can initiate T cell-dependent contact sensitivity through local serotonin release mediated by IgE antibody. J Immunol. 158:2891-2897.
5. Ptak W, Herzog WR, and Askenase PW. (1991) Delayed-type hypersensitivity initiation by early-acting cells that are antigen mismatched or MHC incompatible with late-acting, delayed-type hypersensitivity effector T cells. J Immunol. 146:469-475.
6. Herzog WR, Ferreri NR, Ptak W, and Askenase PW. (1989) The antigen-specific DTH-initiating Thy-1[+] cell is double negative (CD4[-], CD8[-]) and CD3 negative; and expresses IL-3 receptors, but no IL-2 receptors. J Immunol. 143:3125-3133.
7. McHale, JF, Harari OA, Marshall D, and Haskard DO. (1999) Vascular endothelial cell expression of intercellular adhesion molecule-1 and vascular cell adhesion molecule-1 at the onset of eliciting contact hypersensitivity in mice: Evidence for dominant role of TNFα. J Immunol 162:1648-1655.
8. Tsuji RF, Kikuchi M, and Askenase PW. (1996) Possible involvement of C5/C5a in the efferent phase of contact sensitivity. J Immunol. 156:4644-4650.
9. Tsuji RF, Geba GP, Wang Y, Kawamoto K, Matis LA, and Askenase PW. (1997) Required early complement activation in contact sensitivity with generation of C5-dependent chemotactic activity and late T cell IFN-γ: A possible initiating role of B cells. J Exp Med, 186:1015-1026.
10. Ptak W, Geba GP, and Askenase PW. (1991) Initiation of delayed-type hypersensitivity by low doses of monoclonal IgE antibody. Mediation by serotonin and inhibition by histamine. J Immunol. 146:3929-3936.
11. Naraf S, Davoust N, Ames RS, and Barnum SR 1999. Human T cells express C5a receptor and are chemoattracted to C5a. J. Immunol. 162: 4018-4023.
12. Marchal G, Seman M, Milon G, Truffa-Bachi P, Zilberfarb V. (1982) Local adoptive transfer of skin delayed-type hypersensitivity initiated by a single T lymphocyte. J Immunol. 129:954.
13. Tsuji RF, Kawikova I, Ramabhadran R, Taub D, Hugli TE, Gerard G, and Askenase PW. (2000) Early generation of C5a initiates the elicitation of contact sensitivity by leading to T cell recruitment. Submitted.
14. Chen J, Trounstine M, Alt FW, Young F, Kurahara C, Loring JF, and Huszar D. (1993) Immunoglobulin gene rearrangement in B cell deficient mice generated by targeted deletion of J_H locus. Intern Immunol 6:647-656.

15. Ishii N, Sugita Y, Nakajima H, Tanaka S, Askenase PW. (1995) Elicitation of nickel sulfate (NiSO4)-specific delayed-type hypersensitivity requires early-occurring and early-acting, NiSO4-specific DTH-initiating cells with an unusual mixed phenotype for an antigen-specific cell. Cell Immunol. 161:244-255.
16. Hardy RR, and Hayakawa K. (1994) CD5+ B cells, a fetal B cell lineage. Adv. Immunol. 55:297-339.
17. Rawlings DJ, Saffran DC, Tsukada S, Largaespada DA, Grimaldi JC, Cohen L, Mohr RN, Bazan JF, Howard M, Copeland NG, Jenkins NA, Witte ON. (1993) Mutation of unique region of Bruton's tyrosine kinase in immunodeficient *xid* mice. Science 261:358-361.
18. Quintans J. (1979) The immune response of CBA/N mice and their F_1 hybrids to 2,4,6-trinitrophenylated (TNP) antigens I. analysis of the response to TNP-coupled lipopolysaccharide in vivo and at the clonal level. Eur J Immunol 9:67-71.
19. Hata D, Kawakami Y, Inagaki N, Lantz CS, Kitamura T, Khan WN, Maeda-Yamamoto M, Miura T, Han W, Hartman SE, Yao L, Nagai H, Goldfeld AE, Alt FW, Galli SJ, Witte ON, and Kawakami T. (1998) Involvement of Bruton's tyrosine kinase in FcεRI-dependent mast cell degranulation and cytokine production. J Exp Med 187:1235-1247.
20. Mukhopadhyay S, George A, Bal V, Balachandran R, and Rath S. (1999) Bruton's tyrosine kinase deficiency in macrophages inhibits nitric oxide generation leading to enhancement of IL-12 induction. J Immunol 163:1786-1792.
21. Tsuji R, Kawikova I, Paliwal V, Akahira-Azuma M, and Askenase PW. (2000) A novel role for B-1 cells in initiating elicitation of contact sensitivity via IgM antibody. J Immunol Submitted.
22. Geba GP, Wegner CD, Wolyniec W, Yining L, and Askenase PW. (1997) In vivo airway hyperreactivity adoptively transferred to naive mice by THY-1+ and B220+ antigen-specific cells that lack surface expression of CD3. J Clin Invest. 100:629-638.
23. Identification and partial characterization of a T cell-derived antigen-binding factor from mice infected with the intestinal helminth *Trinchinella spiralis*. Int Arch Allergy and Appl Immunol 90:237-247.
24. Van Loveren H, Den Otter W, Meade R, Terheggen PMA, Askenase PW: (1985) A role for mast cells and the vasoactive amine serotonin in T cell dependent immunity to tumors. J Immunol. 134:1292-1299.
25. Trial J. (1988) Cooperation between early-acting delayed-type hypersensitivity T cells and cultured effector cells in tumor rejection. Cancer Res. 48:5922-5926.
26. Akashi T, Nagafuchi S, Anzai K, Kondo S, Kitamura D, Wakana S, Ono J, Kikuchi M, Niho Y, and Watanabe T. (1997) Direct evidence for the contribution of B cells to the progression of insulitis and the development of diabetes in non-obese diabetic mice. Int. Immunol. 9:1159-1164.
27. Serreze D, Chapman HD, Varnum DS, et al. B lymphocytes essential for the initiation of T cell mediated autoimmune diabetes: analysis of a new 'speed congenic' stock of NOD.Igµ^{null} mice. J Exp Med 184:2049-2053.
28. Taylor PC, Plater-Zyberk C, and Maini RN. (1995) The role of the B cells in the adoptive transfer of collagen-induced arthritis from DBA/1 (H-2^q) to SCID (H-2^d) mice. Eur J Immunol. 25:763-769.
29. Svensson L, Jirholt J, Holmdahl R, and Jansson L. (1998) B cell-deficient mice do not develop type II collagen-induced arthritis (CIA). Clin Exp Immunol 111:521-526.
30. Vordermeier HM, Venkataprasad N, Harris DP, and Ivany J. (1996) Increase of tuberculous infection in the organs of B cell-deficient mice. Clin Exp Immunol. 106:312-316.
31. Culkin SJ, Rhinehart-Jones T, and Elkins KL. (1997) A novel role for B cells in early protective immunity to an intracellular pathogen, *Francisella tularensis* strain LVS. J Immunol 158:3277-3284.
32. Yuan RR, Casadevall A, Oh J, and Scharff MD. (1997) T cell cooperate with passive antibody to modify *Cryptococcus neoformans* infection in mice. Proc Natl Acad Sci USA 94:2483-2488.
33. Feldmesser M, and Casadevall A. (1997) Effect of serum IgG1 to Cryptococcus neoformans glucuronoxylomannan on murine pulmonary infection. J Immunol 158:790-799.
34. Yang X, and Brunham RC. (1998) Gene knockout B cell-deficient mice demonstrate that B cells play an important role in the initiation of T cell responses to *Chlamydia trachomatis* (mouse pneumonitis) lung infection. J Immunol 161:1439-1446.
35. Chan OTM, Madaio MP, and Shlomchik MJ. (1999) B cells are required for lupus nephritis in the polygenic, fas-intact MRL model of systemic autoimmunity. J Immunol 163:3592-3596.
36. Shlomchik MJ, Madaio MP, Donghui N, Trounstein M, and Huszar D. (1994) The role of B cells in lpr/lpr-induced autoimmunity. J Exp Med 180: 1295-1306.
37. Genain CP, Nguyen MH, Letvin NL, Pearl R, Davis RL, Adelman M, Lees MB, Linington C, and Hauser SL. (1999) Antibody facilitation of multiple sclerosis-like lesions in nonhuman primate. J Clin Invest 96:2966-2974.
38. Kiely PDW, Pecht I, and Oliveira DBG. (1997) Mercuric chloride-induced vasculitis in the brown norway rat: αβ T cell-dependent and –independent phases. J Immunol 159:5100-5106.
39. Korganow AS, Ji H, Mangialaio S, Duchatelle V, Pelanda R, Martin T, Degott C, Kikutani H, Rajewsky K, Pasquali JL, Benoist C, and Mathis D. (1999) From systemic T cell self-reactivity to organ-specific autoimmune disease via immunoglobulins. Immunity 10:451-461.
40. Hayakawa K, Ishii R, Yamasaki K, Kishimoto T, Hardy RR. (1987) Isolation of high-affinity memory B cells: phycoerythrin as a probe for antigen-binding cells. Proc Natl Acad Sci USA 84:1379-1383.
41. Lyons JA, San M, Happ MP, and Cross AH. (1999) B cells are critical to induction of experimental allergic encephalomyelitis by protein but not by a short encephalitogenic peptide. Eur J Immunol 29:3432-3439.
42. Falcone M, Lee J, Patstone G, Yeung B, and Sarvetnick N. (1997) B lymphocytes are crucial antigen-presenting cells in the pathogenic autoimmunity response to GAD65 antigen in nonobese diabetic mice. J Immunol 161:1163-1168, 1998.

Role of B lymphocytes in host protection against the human filarial parasite, *Brugia malayi*.

T.V. Rajan and Natalia Paciorkowski.

Department of Pathology, University of Connecticut Health Center, Farmington CT 06030-3105

Introduction:

Lymphatic filariasis is a major public health problem in a number of tropical and sub-tropical countries. Over 2 billion people are at risk for contracting the infection and there are an estimated 120,000,000 active cases of the disease in 73 countries [1]. The causative agents of the disease are long, slender nematodes, *Wuchereria bancrofti* and *Brugia malayi*. L3 larvae are infective to certain mammals and are transmitted by mosquitoes. Mosquitoes deposit L3 larvae on the skin of the bitten mammalian host during a blood meal. The larvae burrow through the insect bite and migrate in the subcutaneous tissues to reach the lymphatics. Within the definitive mammalian host, they molt to the L4 stage 7-10 days after entry. Molting to the adult takes 4-6 weeks (*B. malayi*) or several months (*W. bancrofti*) later. Mature male and female parasites mate to produce microfilariae which are carried into circulation, from where they are picked-up by mosquitoes. Adults live for 8 to 10 years within the afferent lymphatics of lymph nodes. As long as the adults are alive, there are minimal but detectable changes in the flow of lymph in the inhabited lymphatics, but usually no blockage. Death of the adult worms results in a vigorous host inflammatory response, ending in granuloma formation, fibrosis and stricture of the affected lymphatics. These changes are followed by the lymphedema, which is characteristic of the disease (for a more complete review, please see [2]).

Host-Parasite Interactions in Filariasis:

The causative agents of lymphatic filariasis were discovered almost 100 years ago [3]. There is a voluminous literature on the epidemiology, clinical features, immunology, and more recently on the molecular biology of this disease. Nonetheless, critical aspects of the host-parasite interactions are obscure. Particularly unclear are the determinants of host immunity.

The mechanism(s) of host protection are difficult to define in human populations, where one can only obtain glimpses of the infectious process frozen in points of time. In the case of most infectious diseases, animal models have provided mechanistic insights that are not readily available from clinical studies. Unfortunately, there are no animal models for human filariasis that fully recapitulate the human infection. The commonly used animal model, the jird (*Meriones unguiculatus*) suffers from the lack of immunological and genetic tools. *Brugia pahangi* infection of cats approximates the human disease. While exquisitely detailed descriptions of *B. pahangi* infection of cats have been published by Denham and his collaborators over the last many years [4-8], these studies have also been limited by the lack of immunological and genetic tools for dissecting the details of the feline immune response. As a result, our ideas about host protection in lymphatic filariasis are, at best, sketchy.

While jirds are naturally permissive for infection, they can be rendered resistant by first immunizing them with X-irradiated L3 larvae [9]. Similar protection can also be seen in mice exposed to irradiated L3 larvae. This protection can be transferred to naive animals consistently by cell transfer, and less effectively by transfer of serum [10]. These data have been taken to mean that host protection is mediated by cellular rather than humoral immunity. Similarly, mice homozygous for $Hfh11^{nu}$ mutation (hereafter NUDE) can be protected from challenge infections by transfer of CD4+ T lymphocytes, but not by CD8+ T lymphocytes or serum from immunized animals [11]. These results therefore point to a conflicting and confusing set of data on the nature of host protection in filarial infection. It is however clear that host protection can be generated in suitably immunized mammalian hosts.

The mechanism by which irradiated larval immunizations might result in host protection has been addressed by Eisenbeiss [12]. He purified excretory/secretory products of *Acanthocheilonema viteae*, a natural pathogen of *Meriones unguiculatus* and immunized the natural host. He observed significant protection in this model, with most worms being killed precisely when they molt to the L4 stage. Eisenbeiss interpreted these data to mean that the worms are particularly vulnerable to host mediated attack while molting and are killed at this stage by host immune effectors.

It has been known for a long time that normal immunocompetent mice are non-permissive for infection by *Brugia malayi*, as well as the closely related filarial nematode, *Brugia pahangi*. In the early 1980s, Vincent, Vickery, Sodeman and co-workers in the US [13] and Suswillo and collaborators in the UK [14] showed that NUDE mice are permissive for *Brugia pahangi* infection, in contrast to their normal euthymic littermates. These results have led to the idea that the absence of T cells makes a mouse permissive for infection. Subsequent cell transfer experiments by Vickery and her collaborators [11] led to the further consolidation of this hypothesis. We have since shown that mice homozygous for the severe combined immunodeficiency (scid) mutation (hereafter SCID mice) are also permissive for infection [15].

Materials and Methods:

In using experimental animal models to study the biology of *Brugia malayi*, investigators have employed at least two different protocols. In a model that more closely resembles human infection, larvae are injected subcutaneously. In this approach, the larvae migrate through the subcutaneous tissue, enter lymphatics and take up residence within them in close proximity to lymph nodes. However, subcutaneous injection distributes the larvae widely over the entire body, making quantitative recovery from the carcass difficult (and not entirely reliable).

An alternative route is intra-peritoneal (i.p.) injection. This experimental model is clearly unnatural, in that *B. malayi* does not normally reside in the peritoneal cavity. Nonetheless, development of the injected L3 through the L4 larva to sexually mature productive adults takes place in the peritoneal cavity with the same kinetics and efficiency as the more natural route. Despite the unphysiological nature of the approach, the peritoneal model of filarial parasites has one major advantage – the larvae do not migrate and remain within this circumscribed space. Furthermore, filarial nematodes, in common with other multicellular organisms, do not undergo asexual reproduction and a larva gives rise to a single adult worm. These two factors permit precise quantitation of the yields of surviving parasites. All the experiments that will be discussed in this communication use this route of infection.

Results:

To examine the role of B cells in host protection against *B. malayi*, we injected 10 C57BL/6J +/+, C57BL/6J SCID and C57BL/6J $Igh6^{null}$ mice i.p. with 50 *B. malayi* L3 larvae. 5 mice in each group were necropsied at 2 weeks and 6 weeks post-infection. Since *B. malayi* L3 larvae develop into L4 larvae in 7-10 days and into adult worms in 4-6 weeks, we were able to study the role of B cells in both L4 larval and adult worm development. The data from one experiment are presented in Table 1 and indicate that C57BL/6J $Igh6^{null}$ mice are not significantly different from C57BL/6J SCID or $Rag1^{null}$ mice in worm burdens at both 2 and 6 weeks. They also exhibit significantly increased worm burdens compared to wild type C57BL/6J mice at 6 weeks ($p \ll 0.001$) indicating that B cells are necessary for elimination of *B. malayi* infection.

Table 1: *B. malayi* L4 and adult worm recoveries at two and six weeks respectively

Mouse Strain	2 wk Mean ± SD	6 wk Mean ± SD
C57BL/6J	23.6 ± 10	1.8 ± 4.3
C57BL/6J Rag-1^{null}	30.4 ± 10	20.8 ± 5.7
C57BL/6J SCID	33.2 ± 14.8	21.3 ± 8.4
C57BL/6J Igh6^{null}	48.8 ± 13	23 ± 10

To determine if this was a unique feature of the *Brugia malayi*/murine host interaction, we repeated the experiment using a closely related filarial organism, *Brugia pahangi*. *Brugia pahangi* is eliminated by immunocompetent mice with kinetics that approximate those of *B. malayi*. We injected cohorts of C57BL/6, C57BL/6 NUDE, and C57BL/6J *Igh6^{null}* mice with *Brugia pahangi* L3 and necropsied cohorts at different times after infection. Wild type C57BL/6 eliminate the *B. pahangi* quite rapidly, with essentially no detectable worm burdens at 2 and 3 weeks. C57BL/6 NUDE mice are significantly more permissive, containing as many as 20% of injected worms as late as 3 weeks in infection. In comparison to the course of infection in the T cell deficient C57BL/6 NUDE mice, worm burdens in C57BL/6J *Igh6^{null}* mice were higher at 1, 2 or 3 weeks (Paciorkowski and Rajan, *J. Exp. Med.*, in press). Thus, in contrast to what has been described in the literature, our data suggests that both for *B. malayi* and *B. pahangi*, C57BL/6J *Igh6^{null}* mice are significantly more permissive than the T cell deficient *Hfh11^{nu}* mice.

While we were obtaining these results using the C57BL/6J *Igh6^{null}* mice, two studies which corroborate our data appeared in the literature. The first showed that mice bearing the *Btk^{xid}* mutation are more permissive for *Litmosoides sigmodontis*, a natural pathogen of mice [16]. Similarly, Ravindran and collaborators also demonstrated an increased permissiveness of CBA/N in comparison to CBA/J mice to infection with *Brugia malayi* [17].

To determine if this is true in our hands as well, we injected cohorts of CBA/N and CBA/CaJ mice with *Brugia malayi*. As in other experiments, these mice were taken down in groups over the next several weeks and the data are published elsewhere (Paciorkowski and Rajan, , *J. Exp. Med*, in press). It was clear from these studies that CBA/CaJ mice behave like other immunocompetent mice, with yields declining over 4 or 5 weeks following infection. In contrast, CBA/N mice harbored significantly higher worm burdens at all time points. We have repeated these data with matched C57BL/6 and C57BL/6 *Btk^{xid}* mice, as well as matched BALB/c and C.CBAN mice. In all cases, mice bearing the *Btk^{xid}* mutation are significantly more permissive than wild type immunocompetent mice (data not shown). These results hold for both *B. malayi* and *B. pahangi*.

Since the *Btk^{xid}* mutation causes numerous defects in addition to the well-known problem with B1 B lymphocytes, we wanted to confirm that the defect in *Btk^{xid}* in mice was, in fact, solely mediated by the B cell deficit. In order to do so we have conducted cell transfer studies using both naïve and primed peritoneal exudate cells (PEC), which are regarded as a good source of B1 B lymphocytes. Data in literature suggest that transfer of peritoneal lymphocytes into recipient mice serves to repopulate primarily if not exclusively B1 B cells, since these are the only lymphocytes for which self-renewing stem cells are present in this population.

In order to examine the effect of naive B lymphocytes on the course of *B. malayi* infection, we harvested PEC from a cohort of CBA/CaJ adult males and injected them into age matched CBA/N male mice. The transfer protocol was used as described by Forster et al. [18]. Each recipient received 3×10^6 peritoneal lymphocytes. Six days following cell transfer, reconstituted animals (n = 9) and non-manipulated CBA/N controls (n = 10) were infected with 50 *Brugia pahangi* L3 larvae intraperitoneally. The worm recoveries were assessed two weeks post-infection. The data showed that re-population of CBA/N mice with naive peritoneal lymphocytes from an immunocompetent host resulted in a 50 % decrease of worm burdens at this early time point. FACS analysis of peritoneal lymphocytes of reconstituted animals confirmed the presence of CD5$^+$CD19$^+$ cells that were not seen in CBA/N controls. Moreover, reconstituted animals showed preferential expansion of B lymphocytes following infection (Paciorkowski and Rajan, , *J. Exp. Med*, in press).

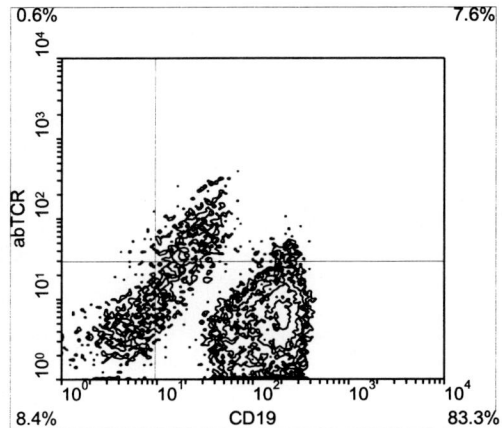

Fig. 1. The donor cell population that was used to reconstitute CBA/N mice. CBA/Ca males were infected with 50 *B. pahangi* L3 and necropsied 3 weeks post-infection. PEC were collected by peritoneal lavage, pooled, stained with anti-CD3-Biotin followed by streptavidin-magnetic beads, and passed through a magnetic column. Cells were stained for FACS analysis with anti-CD19-FITC and anti-$\alpha\beta$TCR-PE.

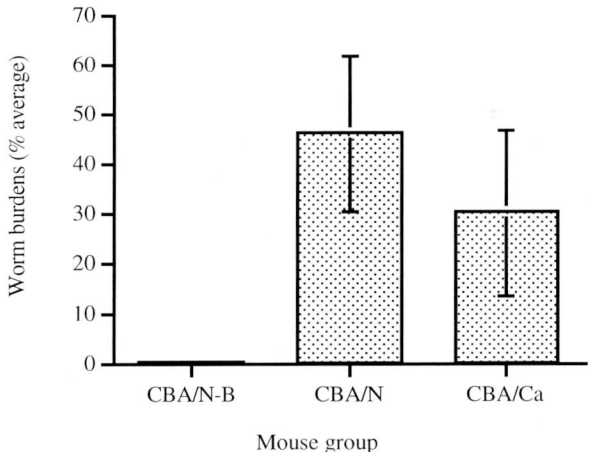

Fig. 2. *B. pahangi* recoveries in CBA/N mice reconstituted with T-depleted primed PEC from CBA/CaJ donors. T-depleted eluant was injected into CBA/N recipients (2 x 10^7 cells/mouse). 6 days later, reconstituted and control mice were challenged with *B. pahangi* L3. Worm burdens were assessed 2 weeks post-infection. CBA/N-B: reconstituted mice.

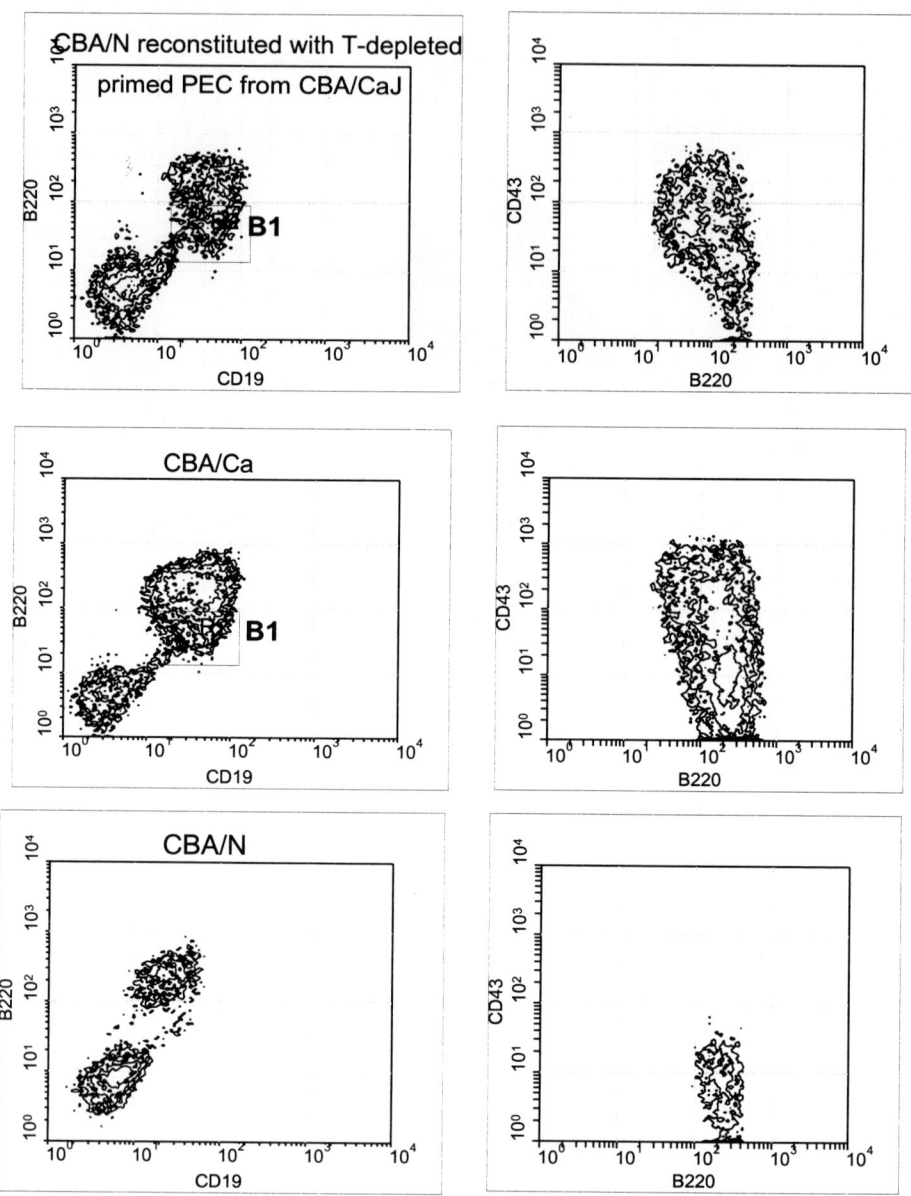

Fig. 3. Comparisons of peritoneal B lymphocyte populations in CBA/N mice (bottom row), CBA/CaJ mice (middle row), and CBA/N mice reconstituted with T-depleted PEC from primed CBA/Ca mice (top row). All FACS plots were gated on the lymphocyte window by forward and side scatter and analyzed for CD19 and B220 (left column) or CD43 and B220 (right column). B1 lymphocytes were identified as CD19+ B220 low.

To determine whether antigen-primed B cells were more effective in providing protection to a permissive host, we collected PEC from CBA/CaJ male mice that had been exposed to *Brugia pahangi* L3 once. The PEC were depleted of T lymphocytes using an anti-CD3-magnetic bead separation column. 2×10^7 T depleted cells were injected into naive CBA/N recipients (n = 10). Fig.1 shows that donor cells had less than 1% T lymphocytes by $\alpha\beta$TCR staining. Six days later, the reconstituted mice, non-manipulated CBA/N controls and naive CBA/CaJ mice were infected with *Brugia pahangi* L3. Worm yields were assessed two weeks post-infection. As Fig.2 shows, the transfer of primed B lymphocytes resulted in accelerated clearance of worms in CBA/N mice, with worm burdens that were significantly different not only from CBA/N controls but also from naive immunocompetent CBA/Ca mice. At the time of necropsy PEC were collected from all animals individually and analyzed for B lymphocyte populations by flow cytometry. B cells were identified as CD19+ B220+ and divided into B1 and B2 populations based on the surface expression of B220, CD43, CD23, and CD5. B1 cells were identified as B220 low CD43+ CD23- CD5+/-, and B2 lymphocytes as B220 high CD43- CD23+ CD5-. FACS analyses revealed a remarkable degree of homogeneity among the animals within a group; therefore, Fig.3 shows the data from one representative of each group. CBA/N mice reconstituted with primed T-depleted PEC from CBA/Ca had an expanded B1 cell population that was not observed in CBA/N controls (Fig.3, left column). The identity of these cells as B1 lymphocytes was confirmed by CD43 positive staining (right column), as well as with negative CD23 staining (data not shown). CD5 staining showed that both B1a and B1b lymphocytes are present at approximately equal levels (data not shown). Interestingly, reconstituted CBA/N tend to have higher percentage of B1 lymphocytes (of total lymphocyte population) than non-manipulated but immunocompetent CBA/Ca controls.

Discussion:

The mechanism by which the mammalian body protects itself against large extracellular, multicellular invertebrate parasites is unclear. As is true of most invertebrate organisms, nematode and cestode parasites are invested with a thick cuticle (exoskeleton). The cuticle is composed of highly cross-linked collagens, and is impermeable to large macromolecules, such as antibodies and/or complement. Thus, the mechanisms by which these organisms are killed by the host, and the cellular machinery that put these mechanisms in place are entirely unclear. Immunoparasitology has focused, at least in the last two decades, on the role played by the dichotomous T helper systems (Th1 vs Th2) in initiating and maintaining the appropriate cytokine milieu to mediate host protection. However, the precise mechanism(s) by which Th1 or Th2 effectors might actually mediate host defense remains unknown.

Given the impermeability of these organisms to antibody mediated mechanisms, B cells have not received much attention as a possible host protective mechanism. While normal immunocompetent mice are resistant to filarial organisms, it has been known for the last 20 years that NUDE mice, which congenitally lack T lymphocytes, support the development of infective larvae to the adult stage and the development of patent infection. We showed that B cell deficient $Igh6^{nul}$ mice on the segregating (C57BL/6 x 129/SvJ) background are not permissive to *B. malayi* infection. We subsequently determined that this background was non-permissive for filarial organisms, even in the complete lack of an adaptive immune response. This led us to re-examine the role of B and T cell immunity to filarial immunity. These findings, published earlier, strongly supported a view that B cells do indeed play a critical role in host protection [19].

The data presented here and elsewhere now support the view that B cells and particularly B1 cells may play an important role in host protection. Although the ability of B1 lymphocytes to develop immunological memory is controversial, our experiments suggest that in addition to "natural" antibodies responses in naïve mice, repeated immunization can boost this immunity resulting in complete protection. However, there are numerous factors that appear to suggest that the complete picture may be more complex. It is possible that while the primary response to the parasites utilizes B1 cells during the early phase of infection, accelerated secondary response is mediated by conventional B2 cells. Further experiments with transfers of purified naive and primed B1 lymphocytes will help to clarify their role in filariasis. Whether T cell help is required for the enhanced protection by primed B cells from multiply boosted mice remains unclear.

How might B cells mediate protection, given the non-permeability of the cuticle to antibodies and complement? B cells, in addition to making antibodies, clearly play other roles in the immune response, including the elaboration of cytokines and of antigen presentation to T lymphocytes. Our data do not rule out any of these three mechanisms; indeed, it is likely that all three mechanisms play some role. Our favorite model at the moment, however, is that antibodies play an important role by "neutralizing" excretory/secretory (E/S) molecules made by the parasite. We hypothesize that these parasite E/S products play a critical role in the host parasite interaction. For example, these molecules might be the parasite's tool for sensing its arrival in the mammalian system and for triggering appropriate developmental pathways. Data from a number of sources that suggest that immunization with excretory/secretory molecules can provide host protection is consistent with this model [12]. Given the fact that the mammalian system may lack appropriate mechanisms for directly killing of these organisms, the possibility that antibodies might block developmental pathways by interfering with host parasite interaction is an attractive one. However, we do not have sufficient data to make this the only viable model. Further work is in progress in our laboratory to document this model in greater detail.

References

1. Ottesen, E.A., B.O. Duke, M. Karam, and K. Behbehani. (1997). Strategies and tools for the control/elimination of lymphatic filariasis. Bull World Health Organ. 75: 491.

2. Rajan, T.V., and A.V. Gundlapalli. (1997). Lymphatic filariasis. Chem Immunol. 66: 125.

3. Lewis, T.R., (1872). On a hematozoan inhabiting human blood, its relation to chyluria and other diseases, . Government of India.

4. Denham, D.A., T. Ponnudurai, G.S. Nelson, R. Rogers, and F. Guy. (1972). Studies with Brugia pahangi. II. The effect of repeated infection on parasite levels in cats. Int J Parasitol. 2: 401.

5. Denham, D.A., T. Ponnudurai, and G.S. Nelson. (1972). Effect of continual reinfection on B. pahangi infections in cats. Trans R Soc Trop Med Hyg. 66: 20.

6. Denham, D.A., and J.A. Turton. (1972). The effect of levamisole on the development of Ostertagia circumcincta in lambs. J Helminthol. 46: 143.

7. Denham, D.A., T. Ponnudurai, G.S. Nelson, F. Guy, and R. Rogers. (1972). Studies with Brugia pahangi. I. Parasitological observations on primary infections of cats (Felis catus). Int J Parasitol. 2: 239.

8. Rogers, R., and D.A. Denham. (1974). Studies with Brugia pahangi. 7. Changes in lymphatics of injected cats. J Helminthol. 48: 213.

9. Yates, J.A., and G.I. Higashi. (1985). Brugia malayi: vaccination of jirds with 60cobalt-attenuated infective stage larvae protects against homologous challenge. Am J Trop Med Hyg. 34: 1132.

10. Hayashi, Y., K. Nakagaki, S. Nogami, B. Hammerberg, and H. Tanaka. (1989). Protective immunity against Brugia malayi infective larvae in mice. I. Parameters of active and passive immunity. Am J Trop Med Hyg. 41: 650.

11. Vickery, A.C., A.L. Vincent, and W.A. Sodeman, Jr. (1983). Effect of immune reconstitution on resistance to Brugia pahangi in congenitally athymic nude mice. Journal of Parasitology. 69: 478.

12. Eisenbeiss, W.F., H. Apfel, and T.F. Meyer. (1994). Protective immunity linked with a distinct developmental stage of a filarial parasite. J Immunol. 152: 735.

13. Vincent, A.L., W.A. Sodeman, and A. Winters. (1980). Development of Brugia pahangi in nude (athymic) and thymic mice, C3H/HeN. Journal of Parasitology. 66: 448.

14. Suswillo, R.R., D.G. Owen, and D.A. Denham. (1980). Infections of Brugia pahangi in conventional and nude (athymic) mice. Acta Trop. 37: 327.

15. Nelson, F.K., D.L. Greiner, L.D. Shultz, and T.V. Rajan. (1991). The immunodeficient scid mouse as a model for human lymphatic filariasis. J Exp Med. 173: 659.

16. Al-Qaoud, K.M., B. Fleischer, and A. Hoerauf. (1998). The Xid defect imparts susceptibility to experimental murine filariosis- -association with a lack of antibody and IL-10 production by B cells in response to phosphorylcholine. Int Immunol. 10: 17.

17. Ravindran, B., P.K. Sahoo, M. Mohanty, S. Mukhopadhyay, and A.P. Dash. (1999). Increased susceptibility of mice with XID mutation to Brugia malayi infection. Medical Science Research. 27: 135.

18. Forster, I., and K. Rajewsky. (1987). Expansion and functional activity of Ly-1+ B cells upon transfer of peritoneal cells into allotype-congenic, newborn mice. Eur J Immunol. 17: 521.

19. Babu, S., L.D. Shultz, T.R. Klei, and T.V. Rajan. (1999). Immunity in experimental murine filariasis: roles of T and B cells revisited [In Process Citation]. Infect Immun. 67: 3166.

Neo-self Antigens and the Expansion of B-1 Cells: Lessons from Atherosclerosis-prone Mice

G. J. Silverman[1], P.X. Shaw[2], L. Luo[1], D. Dwyer[1], M. Chang[2], S. Horkko[2], W. Palinski[2], A. Stall[3] and J. L. Witztum[2]

[1]Division of Rheumatology; [2]Division of Endocrinology, Department of Medicine, University of California, San Diego, La Jolla, California 92093-0663; [3]BD PharMingen, 10975 Torreyana Road, San Diego, CA 92121/ USA

Introduction

In the mouse, B-1 cells are a phenotypically distinct set of mature B lymphocytes that predominate in the peritoneal and pleural cavities and in the lamina propria in gut mucosa-associated lymphoid tissue. During fetal and neonatal periods, B-1 cells differentiate according to a developmentally regulated schedule, and later dominate certain immune functions throughout life. These B cells produce at least half of circulating IgM in naïve hosts that arise without specific immunization, also termed natural antibodies, and they are also a major source of secreted IgA in the gut [1]. While the antibodies from B-1 cells are often described as multireactive and of low binding affinity, in certain cases these antibodies have been shown to play important roles in immune defense from infections [2-4]. The characterization of the binding specificities, antibody gene usage, and in vivo prevalence of certain B-1 clones have led to the notion that they convey evolutionarily selected activities. B-1 antibodies also commonly display self-reactivities that have been postulated to be involved in currently poorly defined "housekeeping" functions.

The immunobiologic origins of B-1 cells has long been a topic of speculation. Initially, these lymphocytes were hypothesized to represent a separate lineage which derive from committed precursors that cannot give rise to conventional (i.e. B- 2) lymphocytes [5;6]. In an alternative view, sometimes called the induced differentiation hypothesis, B-1 cells are posited to derive from cellular triggering by polyvalent T-independent type 2 (TI-2) antigens in an environment devoid of cognate T-cell help [7-10]. B-1 cells display many of the features of activated cells. In contrast to conventional B cells, which are generally small cells believed to be resting and naive, peritoneal B-1 cells appear activated as they are larger and more granular [11]. These B-1 cells also constitutively express activated signal transducer and activator of transcription 3 (termed STAT3), while conventional splenic B cells only transiently express STAT3 upon B-cell antigen-receptor (BcR) triggering [12]. Mounting evidence indicates that B-1 cell development is affected by cell surface receptors and signaling molecules that modulate BcR-mediated interactions, which presumably contribute to the altered activation state of these B cells and influence their expansion. For example, mutant mice deficient in CD19, CD81/TAPA-1 and CD21/35, membrane-associated co-receptors that amplify BcR-mediated signals [13-15], or mice deficient in the intracellular molecules, Btk and Vav, which are involved in BcR signal transduction, have reduced levels of B-1 cells [16;17], whereas mice deficient in the inhibitory BcR co-receptor, CD22, or the inhibitory signaling molecule, SHP-1, have expanded B-1 populations [18;19]. Furthermore, transgenic mice that over express molecules enhancing BcR signaling (i.e. CD19) have elevated levels of B-1 cells [20-23]. These findings have contributed to the notion that lymphocytes in general, and B-1 cells in particular, require some constant level of receptor stimulation, or "tone", to sustain their survival in the periphery [24;25]. The self-reactivity of B-1 cells has been postulated to contribute to their clonal maintainence, but the biologic significance of these non-pathologic self-reactive binding activities is poorly understood.

Immune responses associated with atherosclerosis

It is valuable to consider this background when interpreting recent unexpected findings from investigations of animal models of atherosclerosis, the most common cause of morbidity and mortality in western nations. In atherosclerosis, the early vascular lesion results from the migration of monocytes into the arterial wall and differentiation into macrophages, where they take up modified lipoprotein such as oxidized low density lipoprotein (oxLDL), leading to "foam-cell" formation. As a result of hypercholesterolemia and/or a local prevalence of pro-oxidant factors and conditions, increased formation of oxLDL may occur in the artery wall. OxLDL is rapidly taken up by macrophages through a family of specific receptors, which promotes lesion growth (reviewed in [26]).

Responses to oxLDL within the immune system appear to modulate the pathogenesis of atherosclerosis in ways that are incompletely understood. In recent studies, T cell-dependent responses, mediated by CD40-CD40L interactions, have been shown to contribute to advanced disease [27], but other studies suggest that certain immune responses can also ameliorate the atherosclerotic process [28]. Importantly, while native LDL is generally a poor immunogen, even subtle modifications of autologous LDL, that include non-enzymatic glycation, methylation, ethylation, or carbamoylation, create moieties that are highly immunogenic. In particular, the oxidative modification of LDL produces a variety of "neo-self" determinants that can elicit strong B-cell and T-cell responses.

The generation of atherosclerosis-prone mice, such as apolipoprotein E-deficient ($ApoE^{-/-}$) mice and LDL receptor-deficient ($LDLR^{-/-}$) mice, has provided invaluable animal models for the investigation of the pathogenesis of the disease. Reiterating many features of clinical disease in man, even without a special diet, $ApoE^{-/-}$ mice spontaneously develop hypercholesterolemia and show accelerated atheromatous plaque formation. In addition, B-cell and T-cell immune responses specific for oxidative neoepitopes present on oxLDL, and not native LDL, are prominent in these mice, especially in aging males. Moreover, while ApoE$^{-/-}$ and LDLR$^{-/-}$ mice have no known shared intrinsic immunologic abnormalities or defects, they both spontaneously develop prominent IgM and IgG responses specific for the characteristic "oxidation-specific" epitopes in atherosclerotic lesions.

The B-cell response in atherosclerosis-prone mice

To investigate the B-cell response that arises in ApoE$^{-/-}$ mice, a panel of specific B-cell hybridomas was created, as previously reported [29]. Spleens from two male ApoE$^{-/-}$ mice, that never received exogenous immunization, were used in cell fusions without in vitro stimulation. Thereby, a large number of hybridomas were created that were subsequently selected based on reactivity for copper-oxidized LDL or malonaldehyde (MDA)-substituted LDL, both highly characteristic components of atheromatous lesions that elicit strong immune responses in diseased hosts [30;31]. From 768 pooled wells, 194 reactive wells were identified based on these binding reactivities. As previously reported [29], these clones were found to express IgM antibodies that generally displayed specific reactivity with either one or the other of these "oxidation-specific" modifications of LDL. In fine specificity analyses, it was shown that these antibodies bound oxidized phospholipids, but not native phospholipids, with specificity for 1-palmitoyl-2-oxovaleryl-3-phospatidylcholine (POVPC), an oxidized product of 1-palmitoyl-2-arachidonyl-3-phospatidylcholine (PAPC), a common phospholipid of LDL. Significantly, these oxLDL/POVPC-specific (auto)antibodies were shown to inhibit the uptake and degradation of oxLDL by macrophages, suggesting a potential mechanism by which these IgM antibodies might affect atherogenesis [31].

Based on the finding that oxLDL, but not native LDL, can inhibit macrophage uptake of apoptotic cells [32], it was hypothesized that oxLDL and apoptotic cells share certain oxidation-specific epitopes. To consider this hypothesis, the in vitro reactivities of these POVPC-reactive antibodies were evaluated for endothelial cells under conditions of serum starvation that cause the characteristic changes of apoptosis. In these studies, the POVPC/oxLDL-reactive antibodies were found to specifically react with apoptotic endothelia but not with healthy cultured cells. These findings suggest that antibodies that recognize antigenic determinant(s) on oxidatively modified adducts on LDL also specifically interact with neo-determinants on cell membranes that only become accessible during cell turnover.

Restricted genetic origins of oxLDL-specific antibodies from ApoE$^{-/-}$ mice

To investigate the genetic origins of these atherosclerosis-associated antibodies, we cloned the encoding genes. In these RT-PCR experiments, we were surprised to discover that several of these oxLDL specific antibodies utilized the same unmutated V_H gene representing S107.1-DFL16.1-J_H1 rearrangement paired with a V_κ22 light chain rearrangement. In fact, the genes expressed in each of four different oxLDL/POVPC-specific hybridomas represented the exact same canonical V_H and V_L genes that lack non-templated nucleotide (N) insertions at gene splice sites. These findings are consistent with an origin from fetal liver lymphocytes, which are devoid of terminal deoxy-transferase activity and cannot create N insertions.

Even more unexpected was the discovery that the genes encoding the oxLDL/POVPC-specific antibodies, with the exact same V_H-V_L pairing, are identical to antibody genes previously characterized by several other laboratories in cell lines isolated under very different conditions. More than three decades ago, in early studies of IgA plasma cell tumors that arise in mineral oil-treated mice, Cohn and coworkers characterized the S63 line [33] and Potter and coworkers described the T(EPC)15 cell line [34], which were later found to be the same. In these seminal studies, these B-cell lines were found to express diagnostic T15 idiotypic determinants (reviewed in [35]), which later studies confirmed was due to expression of the same V_H and V_K regions. Over the intervening years, the T15 B-cell clone was found to be the source of an important type of natural antibody to phosphorylcholine [36;37], the choline-phosphate head group in the phosphatidylcholine phospholipid. Phosphorylcholine is also prevalent in the cell wall polysaccharide of pneumococci, and it is also a common component in a variety of bacterial nematode pathogens [38]. Notably, in many inbred mouse strains, B cells expressing the canonical antibody genes of the T15 clone arise during the first week of life. Moreover, by day 10 they are represented at high frequency, even without prior immune exposure, and thereafter T15 is the dominant clone in anti-phosphorylcholine responses [39] even in germ-free mice [40]. These antibodies also provide optimal antibody-mediated defense from systemic infection by virulent pneumococci [2]. Moreover, rigorous adoptive transfer experiments have demonstrated that T15 B cells exist predominantly, if not solely, within the B-1 pool [41].

Confirming the clonal origins of the oxLDL specific antibodies cloned from ApoE$^{-/-}$ mice, we evaluated the Ig products of the ApoE B-cell lines using several serologic agents specific for T15-associated idiotypic determinants. In these studies, each of the oxLDL reactive antibodies, termed "EO" antibodies, such as E06, was recognized by idiotypic serologic reagents specific for the T15V_H region, the T15V_L region, or specific for a determinant requiring co-expression of both the T15 V_H and V_L regions. Moreover, pre-incubation of E06 an anti-idiotypic antibody inhibited E06 binding to oxLDL. Competition studies also documented that either purified phosphorylcholine salt, or phosphorylcholine-containing pneumococcal cell wall polysaccharide, inhibited binding of these antibodies to oxLDL, indicating the T15 idiotope and phosphorylcholine binding sites are functionally equivalent to the oxLDL binding site on these antibodies. Furthermore, purified T15 antibody, or the E06 antibody, could also inhibit the phagocytosis of oxLDL by macrophages [32;42].

Direct examination of the vascular lesions in ApoE$^{-/-}$ mice provided further evidence that these hybridomas are representative of the in vivo atherosclerosis-associated immune response. During the development of atherosclerosis, Ig have been shown to be deposited in atherosclerotic plaques, in part as immune complexes with oxLDL. However, in most studies these lesions have not been associated with a local infiltration of B cells. In carefully controlled immunohistochemical studies, Ig in the atherosclerotic lesions from ApoE$^{-/-}$ and LDLR$^{-/-}$ mice were shown to specifically stain with T15-specific idiotypic markers. These T15 Ig were deposited in the same patterns seen for the total Ig deposits in adjacent sections from these same lesions. These studies clearly implicate increased in vivo production and local lesional deposition of T15-related Ig during atherosclerosis.

Reconsidering the origins of T15 B cells

Our investigations of the immune response in atherosclerosis-prone mice have provided unexpected findings of preferential expression of the T15 B-cell clone, which was originally isolated from mineral oil induced plasma cell tumors, and later found to be the dominant source of protective anti-phosphorylcholine antibodies that derive from the B-1 pool. In more recent studies, we have confirmed that circulating T15 and anti-phosphorylcholine antibody levels progressively increase with aging and progression of atherosclerosis in these animals (manuscript in preparation), and that these levels directly parallel the spontaneous development of the anti-oxLDL response. While these natural antibodies were previously believed to be committed to the immune recognition and defense from common microbial pathogens, our findings provide unexpected insights into the full range of immune reactivity of this clone. Central to these findings is evidence that the archetypic IgA antibody, T15, or the IgM analogues from ApoE$^{-/-}$ derived B-cell hybridomas, such as E06, recognize determinants on apoptotic endothelial cells, which are also expressed on oxLDL. Furthermore, E06/T15 antibodies can block the in vitro uptake of oxLDL by elicited macrophages, which may suggest that this same, or a closely related, determinant is involved in the lipoprotein clearance processes associated with atherosclerosis. Therefore, in addition to the earlier known role of T15 antibodies in providing protection from bacterial sepsis, our findings characterize possible roles in inflammation and in the maintainence of cellular homeostasis that, to the best of our knowledge, have not previously been suspected.

Our evidence that the regulation of the T15 B-cell clone is altered in atherosclerosis-prone mice has led us to reconsider the in vivo roles of B-1 cells. Amongst the best explored, is the system of dominant B-1 clones which were originally characterized based on their recognition of bromelain-treated red blood cells. These B cells were subsequently shown to be specific for liposomes composed of phosphatidylcholine, which is distinct from T15 antibodies that recognize only the phosphorylcholine head group. Akin to the patterns seen for the T15 clone, anti-phosphatidylcholine B-1 cells are rare at birth, but later rapidly increase, to represent at least 2% of peritoneal B-1 cells by ~3 weeks of age [43]. Peritoneal anti-phosphatidylcholine B-1 cells display phenotypic changes consistent with chronic in vivo antigen receptor occupancy, which would be consistent with a role for endogenous ligand in the maintenance of these B-1 clones. Paralleling our own findings, an anti-phosphatidylcholine antibody was shown to provide protection in a murine model of bacterial peritonitis [44;45]. Importantly, Clarke and associates have reported elegant studies documenting in vivo clonal expansion, based on preferential V_H gene expression and H-L chain pairing, in patterns which may reflect selection by an undefined endogenous ligand. However, the red blood cell and anti-phosphatidylcholine binding specificities are clearly not functionally equivalent, the identity of the true natural ligand for these prevalent clones is poorly defined, and the physiologic roles of these anti-phosphatidylcholine B-1 antibodies have not been established.

Based on the emerging findings, we wonder whether the immune response of atherosclerosis-prone mice in part reflects a homeostatic response to in vivo elevated levels of natural neo-self ligands for the T15 clone, and possibly other dominant members of the B-1 pool. Whereas Carroll and co-workers have demonstrated that natural antibodies can specifically interact with endothelial neo-determinants created following post-ischemic injury [46], we speculate that in health certain B-1 clones may in fact be selected in vivo, and later expanded in adult life, by oxidation and/or apoptosis-related neo-self epitopes on phospholipid moieties. This process may be mediated by increased levels of circulating oxLDL, which have been documented in patients with advanced atherosclerosis, and we speculate that these levels of circulating ligands are responsible for the induction of T15 antibodies. Our current understanding of the trafficking patterns of B-1 cells is limited, but it is clear that they differ significantly from conventional B cells that are recruited into germinal center responses where they compete based on ligand binding affinities in reactions often assisted by specific T-cell responses. In contrast, B-1 cell are considered extra-follicular B cells that generally reside at sites distant from lymph nodes, and even in the spleen B-1 cells or their differentiated products do not enter germinal centers. How specific ligands may recruit B-1 cells into responses remains to be defined, however there is evidence that levels of circulating IgM antibodies may feedback to influence B-1 expansion [47]. Hence, for this primordial tier of B lymphocytes, immunologic homeostasis may seek to maintain a balance between the levels of circulating natural B-1 antibodies and their neo-self ligands. In part, these specific stimuli may contribute to specific B-1 clonal proliferation, trafficking and differentiation into Ig-secreting cells.

Our hypothesis appears to fit well with findings by Hyakawa et al. regarding the in vivo behavior of an anti-Thy-1 (CD90) B-1 cell clone [48]. Earlier, these investigators characterized an autoreactive T-cell specific antibody, encoded by a specific V_H3609-DQ52-J_H2 rearrangement paired with a specific $V_\kappa 21C$-$J_\kappa 2$ gene [49]. Recently, they reported that in the presence of the Thy-1 self antigen, which is arrayed on all T lymphocytes, there was dramatic expansion of V_H3609 transgene-expressing anti-Thy-1 B-cells paired with endogenous V_L regions encoded by the exact canonical natural V_κ rearrangement. In contrast, selection did not occur in

Thy-1$^{-/-}$ mice. Relevant to our hypothesis, this anti-Thy-1 clone in fact interacts with a glycodeterminant. Although little has been reported on the molecular basis for this Thy-1 determinant, it is not expressed during early fetal life (K. Hayakawa, personal communication), and it is likely to be later created through the activities of yet undefined developmentally regulated glyco-transferases. Hence, in part because this same glycodeterminant is not expressed during early phases of development, strict immunologic tolerance may not be established.

Summary

The pathogenesis of atherosclerosis involves an inflammatory process that is modulated by the immune system, and within these complex responses we have discerned a possible role for an archetypic B-1 clone. We speculate that due to their immunogenicity and in vivo distribution the "neo"-self determinants created in oxidatively modified LDL are highly stimulatory for certain B-1 cell clones. These neo-self determinants, which can be created chemically, by somatic processes, may in fact represent the molecular analogues of somatic maturation, or even aging. These changes, including those on non-protein antigens induced by oxidative metabolism, amongst others, create neo-determinants against which the host no doubt can not develop rigorous B-cell tolerance. The onset of expression of these oxidative neo-determinants relatively late in development may well serve a useful function for the highly evolved mammalian immune system, as targeting by evolutionarily selected B-1 clones may facilitate the amplification of other useful antibody-mediated physiologic functions. As in the case of the T15 clone, these antibodies may aid in protection against common microbial pathogens. Hence we postulate that during the evolution of the adaptive immune system the neo-self antigenic milieu may have been exploited for the natural selection of primordial clonal specificities. The T15 B-1 clone may then illustrate a common paradigm in which there has been natural selection based on utility for the defense of the individual from environmental threats, as well as for possible "housekeeping" role(s) and the maintenance of cellular homeostasis.

References

1. Kroese, F. G., R. de Waard, and N. A. Bos. 1996. B-1 cells and their reactivity with the murine intestinal microflora. Semin.Immunol. 8: 11-18.

2. Briles, D. E., C. Forman, S. Hudak, and J. L. Claflin. 1982. Antiphosphorylcholine antibodies of the T15 idiotype are optimally protective against Streptococcus pneumoniae. J.Exp.Med. 156: 1177-1185.

3. Boes, M., A. P. Prodeus, T. Schmidt, M. C. Carroll, and J. Chen. 1998. A critical role of natural immunoglobulin M in immediate defense against systemic bacterial infection. J Exp Med 188: 2381-2386.

4. Ochsenbein, A,F., T. Fehr, C. Lutz, M. Suter, and R. M. Zinkernagel. 1999. Control of early viral and bacterial distribution and disease by natural antibodies. Science 286: 2156-2159.

5. Hardy, R. R. and K. Hayakawa. 1991. A developmental switch in B lymphopoiesis. Proc.Natl.Acad.Sci.U.S.A. 88: 11550-11554.

6. Herzenberg, L. A. and A. B. Kantor. 1993. B-cell lineages exist in the mouse. Immunol.Today. 14: 79-83.

7. Wortis, H. H., M. Teutsch, M. Higer, J. Zheng, and D. C. Parker. 1995. B-cell activation by cross-linking of surface IgM or ligation of CD40 involves alternative signal pathways and results in different B-cell phenotypes. Proc.Natl.Acad.Sci.U.S.A. 92: 3348-3352.

8. Cong, Y. Z., E. Rabin, and H. H. Wortis. 1991. Treatment of murine CD5- B cells with anti-Ig, but not LPS, induces surface CD5: two B-cell activation pathways. Int.Immunol. 3: 467-476.

9. Haughton, G., L. W. Arnold, A. C. Whitmore, and S. H. Clarke. 1993. B-1 cells are made, not born. Immunol.Today. 14: 84-87.

10. Clarke, S. H. and L. W. Arnold. 1998. B-1 cell development: evidence for an uncommitted immunoglobulin (Ig)M+ B cell precursor in B-1 cell differentiation. J Exp Med 187: 1325-34.

11. Hayakawa, K., R. R. Hardy, and L. A. Herzenberg. 1986. Peritoneal Ly-1 B cells: genetic control, autoantibody production, increased lambda light chain expression. Eur.J.Immunol. 16: 450-456.

12. Karras, J. G., Z. Wang, L. Huo, R. G. Howard, D. A. Frank, and T. L. Rothstein. 1997. Signal transducer and activator of transcription-3 (STAT3) is constitutively activated in normal, self-renewing B-1 cells but only inducibly expressed in conventional B lymphocytes. J Exp Med 185: 1035-1042.

13. Rickert, R. C., K. Rajewsky, and J. Roes. 1995. Impairment of T-cell-dependent B-cell responses and B-1 cell development in CD19-deficient mice. Nature 376: 352-355.

14. Sato, S., D. A. Steeber, and T. F. Tedder. 1995. The CD19 signal transduction molecule is a response regulator of B-lymphocyte differentiation. Proc.Natl.Acad.Sci.U.S.A. 92: 11558-11562.

15. Miyazaki, T., U. Muller, and K. S. Campbell. 1997. Normal development but differentially altered proliferative responses of lymphocytes in mice lacking CD81. EMBO J 16: 4217-25.

16. Khan, W. N., A. Nilsson, E. Mizoguchi, E. Castigli, J. Forsell, A. K. Bhan, R. Geha, P. Sideras, and F. W. Alt. 1997. Impaired B cell maturation in mice lacking Bruton's tyrosine kinase (Btk) and CD40. Int Immunol 9: 395-405.

17. Zhang, R., F. W. Alt, L. Davidson, S. H. Orkin, and W. Swat. 1995. Defective signalling through the T- and B-cell antigen receptors in lymphoid cells lacking the vav proto-oncogene. Nature 374: 470-3.

18. O'Keefe, T. L., G. T. Williams, S. L. Davies, and M. S. Neuberger. 1996. Hyperresponsive B cells in CD22-deficient mice. Science 274: 798-801.

19. Cyster, J. G. and C. C. Goodnow. 1995. Protein tyrosine phosphatase 1C negatively regulates antigen receptor signaling in B lymphocytes and determines thresholds for negative selection. Immunity. 2: 13-24.

20. Sidman, C. L., L. D. Shultz, R. R. Hardy, K. Hayakawa, and L. A. Herzenberg. 1986. Production of immunoglobulin isotypes by Ly-1+ B cells in viable motheaten and normal mice. Science 232: 1423-1425.

21. O'Keefe, T. L., G. T. Williams, S. L. Davies, and M. S. Neuberger. 1996. Hyperresponsive B cells in CD22-deficient mice. Science 274: 798-801.

22. Sato, S., A. S. Miller, M. C. Howard, and T. F. Tedder. 1997. Regulation of B lymphocyte development and activation by the CD19/CD21/CD81/Leu 13 complex requires the cytoplasmic domain of CD19. J Immunol 159: 3278-87.

23. Shultz, L. D., P. A. Schweitzer, T. V. Rajan, T. Yi, J. N. Ihle, R. J. Matthews, M. L. Thomas, and D. R. Beier. 1993. Mutations at the murine motheaten locus are within the hematopoietic cell protein-tyrosine phosphatase (Hcph) gene. Cell 73: 1445-54.

24. Neuberger, M. S. 1997. Antigen receptor signaling gives lymphocytes a long life . Cell 90: 971-973.

25. Torres, R. M., H. Flaswinkel, M. Reth, and K. Rajewsky. 1996. Aberrant B cell development and immune response in mice with a compromised BCR complex. Science 272: 1804-8.

26. Steinberg, D. and J. L. Witztum. 1999. Lipoproteins, lipoporatein oxidation and atherogenesis. In Molecular Basis of Atherosclerosis. W.B. Saunders, Philadelphia, PA, pp. 458-475.

27. Mach, F., U. Schonbeck, G. K. Sukhova, E. Atkinson, and P. Libby. 1998. Reduction of atherosclerosis in mice by inhibition of CD40 signalling. Nature 394: 200-3.

28. Freigang, S., S. Horkko, E. Miller, J. L. Witztum, and W. Palinski. 1998. Immunization of LDL receptor-deficient mice with homologous malondialdehyde-modified and native LDL reduces progression of atherosclerosis by mechanisms other than induction of high titers of antibodies to oxidative neoepitopes. Arterioscler Thromb Vasc Biol 18: 1972-82.

29. Palinski, W., S. Horkko, E. Miller, U. P. Steinbrecher, H. C. Powell, L. K. Curtiss, and J. L. Witztum. 1996. Cloning of monoclonal autoantibodies to epitopes of oxidized lipoproteins from apolipoprotein E-deficient mice. Demonstration of epitopes of oxidized low density lipoprotein in human plasma. J.Clin.Invest. 98: 800-814.

30. Palinski, W., M.E. Rosenfeld, S. Yla-Herttuala , G. C. Gurtner, S. S. Socher, S. W. Butler, S. Parthasarathy, T. E. Carew, D. Steinberg and J. L. Witztum . 1989. Low density lipoprotein undergoes oxidative modification in vivo. Proc Natl Acad Sci U S A 86:1372-6.

31. Wu, R., U. de Faire, C. Lemne, J. L. Witztum, and J. Frostegard. 1999. Autoantibodies to OxLDL are decreased in individuals with borderline hypertension. Hypertension 33: 53-59.

32. Chang, M. K., C. Bergmark, A. Laurila, S. Horkko, K. H. Han, P. Friedman, E. A. Dennis, and J. L. Witztum. 1999. Monoclonal antibodies against oxidized low-density lipoprotein bind to apoptotic cells and inhibit their phagocytosis by elicited macrophages: evidence that oxidation-specific epitopes mediate macrophage recognition. Proc Natl Acad Sci U S A 96: 6353-6358.

33. Cohn, M. Natural history of the myeloma. 1967. Cold Spring Harbor Symp. Quant. Biol. 32: 211-221.

34. Potter, M. and M. A. Leon. 1968. Three IgA myeloma immunoglobulins from the BALB/c mouse: precipitation and pneumococcal C polysaccharide. Science: 369-371.

35. Potter, M. Antigen-binding myeloma proteins of mice. 1977. Adv Immunol 25, 141-211.

36. Sher, A. and M. Cohn. 1972. Inheritance of an idiotype associated with the immune response of inbred mice to phosphorylcholine. Eur J Immunol 2: 319-326.

37. Sigal, N. H., A. R. Pickard, E. S. Metcalf, P. J. Gearhart, and N. R. Klinman. 1977. Expression of phosphorylcholine-specific B cells during murine development. J.Exp.Med. 146: 933-948.

38. Harnett, W. and M. M. Harnett. 1999. Phosphorylcholine: friend or foe of the immune system. Immunology Today 20: 125-129.

39. Gearhart, P. J., N. H. Sigal, and N. R. Klinman. 1977. The monoclonal anti-phosphorylcholine antibody response in several murine strains: genetic implications of a diverse repertoire. J.Exp.Med. 145: 876-879.

40. Sigal, N. H., P. J. Gearhart, and N. R. Klinman. 1975. The frequency of phosphorylcholine-specific B cells in conventional and germfree BALB/C mice. J.Immunol. 114: 1354-1358.

41. Masmoudi, H., T. Mota-Santos, F. Huetz, A. Coutinho, and P. A. Cazenave. 1990. All T15 Id-positive antibodies (but not the majority of VHT15+ antibodies) are produced by peritoneal CD5+ B lymphocytes. Int.Immunol. 2: 515-520.

42. Horkko, S., D. A. Bird, E. Miller, H. Itabe, N. Leitinger, G. Subbanagounder, J. A. Berliner, P. Friedman, E. A. Dennis, L. K. Curtiss, W. Palinski, and J. L. Witztum. 1999. Monoclonal autoantibodies specific for oxidized phospholipids or oxidized phospholipid-protein adducts inhibit macrophage uptake of oxidized low-density lipoproteins. J Clin Invest 103: 117-128.

43. Mercolino, T. J., L. W. Arnold, L. A. Hawkins, and G. Haughton. 1988. Normal mouse peritoneum contains a large population of Ly-1+ (CD5) B cells that recognize phosphatidyl choline. Relationship to cells that secrete hemolytic antibody specific for autologous erythrocytes. J Exp Med 168: 687-698.

44. Reid, R. R., A. P. Prodeus, W. Khan, T. Hsu, F S. Rosen, and M. C. Carroll. 1997. Endotoxin shock in antibody-deficient mice: unraveling the role of natural antibody and complement in the clearance of lipopolysaccharide. J Immunol 159: 970-975.

45. Williams, J. P., T. T. Pechet, M. R. Weiser, R. Reid, L. Kobzik, F. D. Moore Jr, M. C. Carroll and H. B. Hechtman. 1999. Intestinal reperfusion injury is mediated by IgM and complement. J Appl Physiol 86: 938-942.

46. Weiser, M. R., J. P. Williams, F. D. Moore, Jr, L. Kobzik, M. Ma, H. B. Hechtman and M. C. Carroll. 1996. Reperfusion injury of ischemic skeletal muscle is mediated by natural antibody and complement. J Exp Med 183: 2343-2348.

47. Boes, M., C. Esau, M. B. Fischer, T. Schmidt, M. Carroll and J. Chen. 1998. Enhanced B-1 cell development, but impaired IgG antibody responses in mice deficient in secreted IgM. J Immunol 160: 4776-4787.

48. Hayakawa, K., M. Asano, S. A. Shinton, M. Gui, D. Allman, C. L. Stewart, J. Silver, and R. R. Hardy. 1999. Positive selection of natural autoreactive B cells. Science 285:113-116.

49. Hayakawa, K., C. E. Carmack, R. Hyman and R. R. Hardy. 1990. Natural autoantibodies to thymocytes: origin, VH genes, fine specificities, and the role of Thy-1 glycoprotein. J Exp Med 172: 869-788.

Note in proof: Greater detail on our investigations will appear in,
Shaw, P.X., S. Horkko, M. Chang, L.K. Curtiss, W. Palinski, G.J. Silverman, J.L. Witztum. 2000. Natural antibodies with the T15 idiotype have functions in atherosclerosis, apoptotic clearance and protective immunity. J Clin invest. In press.

B Cell Subsets in Pristane-induced Autoimmunity

H. B. Richards, E. A. Reap[*], M. Shaw[*], M. Satoh, H. Yoshida, and W. H. Reeves[1]
Division of Rheumatology and Clinical Immunology, University of Florida, Gainesville, FL 32610; [*]Thurston Arthritis Research Center, University of North Carolina, Chapel Hill, NC 27599

B-1 cells are a self-replenishing B cell lineage with an IgM^{high}, $CD5^{intermediate}$ IgD^{low} phenotype (1). In the adult, they reside predominantly in the peritoneal and pleural cavities and to a lesser extent the spleen (1). In mice, their numbers vary from strain to strain and are down-regulated by the Th1 cytokines IFNγ and IL-12 and up-regulated by Th2 cytokines (IL-4, IL-5 and IL-10) (2). Maintenance of this subset depends on CD19 mediated signals (3,4) and also appears to require CR2 (CD21) (5). B-1 cells are involved in the production of autoantibodies of the IgM subclass (6) and may play a critical role for the development of the lupus-like syndrome of NZB/W mice (7,8,9). In contrast, conventional B cells appear responsible for autoantibody production in the Fas deficient mice as intraperitoneal B-1 cell counts progressively decrease with age in B6/*lpr* mice (10).

Mice with high numbers of B-1 cells (e.g. NZB and BALB/cAn) are susceptible to the induction of plasmacytomas following intraperitoneal pristane injection (11). In contrast, strains with low or absent B-1 cells, such as mice carrying the *xid* mutation, do not develop these B cell neoplasms (12). Pristane also induces a lupus-like syndrome in most normal strains of mice (13,14). Plasmacytomagenesis (15) and autoantibody induction (16) both are strongly influenced by IL-6. Moreover, the frequency of plasmacytomas and autoantibodies is reduced if the mice are housed in a specific pathogen free environment (17,18). However, pristane can induce autoantibody formation in strains that are resistant to the induction of plasmacytomas (14,19), suggesting that the two processes are at least partially independent. In the present study we examined the effect of pristane on B-1 cells and its relationship to the development of the lupus-like syndrome.

Materials and Methods

Animals
The following strains of mice were obtained from Jackson Laboratory (Bar Harbor, ME): female BALB/c ByJ (age 8-10 months, conventionally housed, n = 8); BALB/c J (+/+) (age 3 months, specific pathogen free, n = 8) and sex-matched

---

[1] This work was supported by research grant AR44731 from the United States Public Health Service. Dr. Richards is an Arthritis Foundation Postdoctoral Fellow.

BALB/c J IFNγ deficient (IFNγ-/-) mice (age 3 months, specific pathogen free, n = 8). The mice received a single 0.5 ml i.p. injection of either pristane (Sigma) or PBS and were euthanized 2 weeks later.

Cell Staining and Analysis

Flow cytometry was performed as described (10). Briefly, peritoneal cells were obtained by lavage with 5 ml HBSS / 10% newborn calf serum (NCS) and single cell suspensions of spleen cells were prepared. Erythrocytes were lysed and the remaining cells were washed and suspended in HBSS 10%NCS at 10^7 cells/ml. Cells were either left unstained (to adjust for autofluorescence) or stained with biotinylated rat anti-mouse CD5 (53-7.3), biotinylated rat anti-mouse IgD (AMS 9.1) and FITC conjugated rat anti-mouse IgM (R6-60.2) (PharMingen) for 30 min at 4° C. Cells stained with biotinylated mAb were incubated subsequently for 30 min with streptavidin-phycoerythrin (PE). Cells were fixed with 1% paraformaldehyde/PBS after staining and analyzed by flow cytometry (Facscan, Becton Dickenson). Only cells within a defined lymphocyte gate were evaluated.

Results

Pristane alters the composition of intraperitoneal cell populations in mice by recruiting macrophages/monocytes and CD4+ T cells into the peritoneal cavity (20). It appears that the total number of peritoneal B cells is not affected significantly by pristane, but careful studies of B cell subsets have not been carried out. B-1 cells have been implicated in pristane-induced plasmacytomagenesis (11) and may play an essential role for the development of autoimmunity (6,8,21). It was therefore of interest to study the effect of pristane on this subset of B cells.

Pristane Depletes Peritoneal B-1 Cells

Initial experiments were performed with conventionally housed, aged BALB/c ByJ mice, which have large numbers of peritoneal B-1 cells (22). Four mice were treated with pristane and 4 with PBS. Splenocytes and peritoneal cells were analyzed two weeks later by flow cytometry, gating on lymphocytes. There was a striking decrease in the number of peritoneal cells with a B-1a phenotype (IgM^{high}, $CD5^+$, IgD^{low}) in pristane-treated compared with PBS-treated mice (Fig. 1). B-1b (IgM^{high}, $CD5^-$, IgD^{low}) cells appeared to be depleted as well.

On average IgM^+, $CD5^+$ lymphocytes were 51% in controls vs. 5% in pristane-treated mice. Similarly, IgM^+, IgD^+ cells were reduced from 77% to 12% (Fig. 2). Conventional B cells (IgM^+, $CD5^-$) also decreased in pristane vs. PBS-treated mice (30% vs. 15%), but this was less marked than the reduction of B-1 cells. The lower percentage of conventional B cells in the pristane-treated group may reflect the increased numbers T cells (CD5$^+$, IgM$^-$, IgD$^-$) present within the peritoneal cavity due to pristane treatment. The fraction of T cells (CD5$^+$, IgM$^-$) increased from 10% in PBS-treated to 66% in pristane-treated mice. This was accompanied by a less dramatic increase in CD5$^-$, IgM$^-$ cells (Fig. 2).

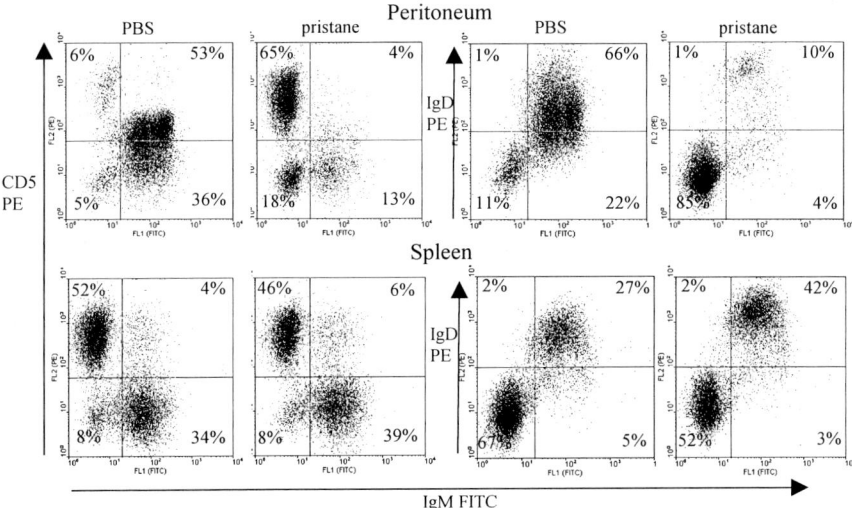

Fig. 1. Peritoneal cells (top) and splenocytes (bottom) from BALB/c ByJ mice 2 weeks after treatment with either pristane or PBS were stained with FITC-conjugated anti-IgM plus PE-conjugated anti-CD5 or anti-IgD antibodies and analyzed by FACS. Diagrams from representative mice show percentages of cells within a lymphocyte gate. Percentages in each quadrant are based on 20,000 events collected per sample.

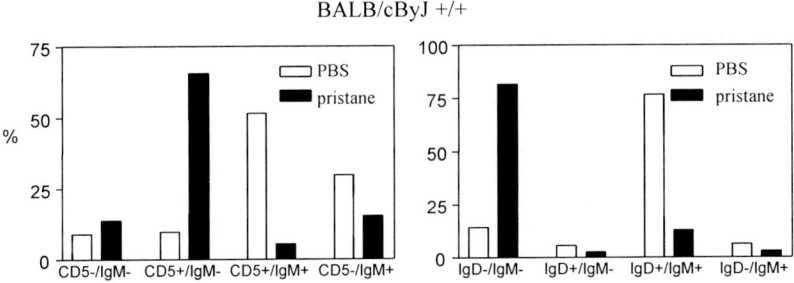

Fig. 2. Peritoneal cells from BALB/c ByJ mice 2 weeks after treatment with either PBS (n = 4) or pristane (n = 4) were stained with FITC-conjugated anti-IgM and PE-conjugated anti-CD5 or anti-IgD and analyzed by FACS. Only cells within a lymphocyte gate were analyzed. Mean percentages of cells based on 20,000 events collected per sample are shown.

In contrast, pristane had little effect on the percentage of splenic B-1a cells (Fig. 1). The numbers of CD5$^+$, IgM$^+$ cells ranged between 4% and 7% of spleen cells regardless of treatment, suggesting that the depletion of B-1 cells is a local phenomenon restricted to the peritoneal cavity. There was an increase in the fraction of IgM$^+$, IgD$^+$ conventional splenic B cells in pristane vs. PBS treated mice and the level of surface IgD expression increased substantially (Fig. 1). This may suggest that local exposure to pristane in the peritoneal cavity promotes the expansion of follicular-type B cells (23) in the spleen. However, further studies will be needed to evaluate the significance of this population of cells in the autoimmune process.

Peritoneal B-1 Cell Depletion Is not Mediated by IFNγ

Pristane induces production of IFNγ in the peritoneal cavity, presumably by CD4$^+$ Th1 cells (20). Th1 cytokines in turn have been reported to deplete B-1 cells (2). Therefore, it was of interest to examine whether pristane-induced cytokine production mediates the observed depletion of peritoneal B-1 cells. To test this hypothesis, 3 month old female BALB/cJ IFNγ+/+ and IFNγ-/- mice housed under specific pathogen free conditions were treated with pristane. Unexpectedly, no difference was found between the cytokine-targeted mice and wild type controls. Peritoneal B-1 cells were depleted by pristane treatment as efficiently in IFNγ deficient animals as in controls (Fig. 3).

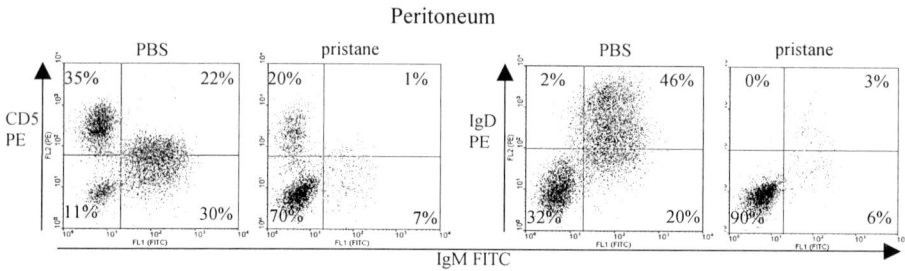

Fig. 3. Peritoneal cells from BALB/cJ IFNγ-/- mice 2 weeks after treatment with either pristane or PBS were stained with FITC-conjugated anti-IgM plus PE-conjugated anti-CD5 or IgD and analyzed by two color FACS. Diagrams from representative mice show percentages of cells within a lymphocyte gate. Percentages for each quadrant are based on 20,000 events per sample.

To assess whether absolute numbers of peritoneal B-1 cells were affected by pristane, total numbers of peritoneal cells expressing the CD5, IgM and IgD markers were determined (Table 1). The total number of CD5$^+$, IgM$^+$ cells was reduced dramatically from 0.78 X 10^6 in the PBS-treated mice to 0.14 X 10^6 in the pristane-treated group. However, the marked T cell influx (indicated by an increased number of CD5$^+$, IgM$^-$ cells) seen in conventionally housed 8-10 month old BALB/c ByJ mice was not evident in 3 month old BALB/cJ mice housed under pathogen free conditions. Similarly, the BALB/cJ IFNγ -/- mice housed under specific pathogen free conditions failed to exhibit an influx of CD5$^+$, IgM$^-$ T cells despite having markedly reduced numbers of CD5$^+$, IgM$^+$ B-1 cells. The total numbers of unstained (CD5-, IgM-) cells in the lymphocyte gate increased in both IFNγ +/+ and -/- mice, possibly representing the influx of an additional cell subset, such as NK cells. The lack of recruitment of CD5+ T cells to the peritoneal cavity by pristane is likely to reflect differences in age and/or microbial environment between the BALB/c ByJ and the BALB/cJ mice. However a strain related difference causing aberrant T cell recruitment cannot be excluded.

Table 1. Total Peritoneal Cell Counts in Pristane and PBS Treated Mice

IFNγ	Treatment (n)	Total number of cells positive X 10^{-6}			
		CD5 +, µ -	CD5 +, µ +	CD5 -, µ -	CD5 -, µ +
+/+	PBS (2)	0.96	0.78	0.30	1.26
+/+	Pristane (6)	0.63	0.14	1.62	0.41
-/-	PBS (2)	0.60	0.78	0.24	0.72
-/-	Pristane (6)	0.45	0.05	1.89	0.23

IFNγ	Treatment (n)	Total number of cells positive X 10^{-6}			
		δ +, µ -	δ +, µ +	δ -, µ -	δ -, µ +
+/+	PBS (2)	0.04	1.36	1.13	0.68
+/+	Pristane (6)	0.03	0.20	2.25	0.32
-/-	PBS (2)	0.02	0.70	0.92	0.71
-/-	Pristane (6)	0.07	0.10	2.37	0.16

Peritoneal cells from mice 2 weeks after treatment were analyzed by two-color flow cytometry for either IgM (µ) plus CD5 or IgM plus IgD (δ) staining. The mean cell numbers (% positive X total cell count within the lymphocyte gate) for each phenotype are shown.

Pristane had no effect on the absolute numbers of splenic B-1 cells in either BALB/cJ or BALB/cJ IFNγ -/- mice (not shown). These results may suggest that pristane induces a cytokine other than IFNγ that mediates depletion of B-1 cells from the peritoneal cavity. It is interesting, in this regard, that IL-12 promotes the loss of peritoneal, but not splenic, B-1 cells (2). Alternatively, pristane may diminish expression of molecules necessary for the maintenance of B-1 cells such as CD19 or CR2 (3,4,5). Finally, a direct toxic effect of pristane on B cells cannot be excluded.

Discussion

The role of B-1 cells in the pathogenesis of lupus remains unclear despite intensive study. This subset is prominent in NZB/W mice and may play an important role in the pathogenesis of spontaneous lupus-like disease in this strain (7,6,8). In addition, mutant mice with an expanded B-1 cell subset develop some features reminiscent of lupus including autoantibodies (24,25). In contrast, IgG anti-DNA/chromatin antibodies can be produced in the absence of peritoneal B-1 cells in B6/*lpr* mice, suggesting that conventional B cells are responsible for most of the autoantibodies in this strain (10). The explanation for this apparent discrepancy probably lies in the fact that autoantibodies can be produced by both B-1 and conventional B cells (7).

Autoantibody production induced by pristane occurs in two phases. In the initial phase (2-4 weeks after pristane injection) there is production of IgM anti-ssDNA and increased total IgM and IgG3 levels (13). In the late phase (4-6 months after pristane) there is production of IgG anti-nRNP/Sm and IgG anti-DNA/chromatin along with increased total IgG1, 2a and 2b. Since B-1 cells are responsible for most of the serum IgM and produce IgM anti-ssDNA antibodies (6,26,27), this subset could be instrumental in the initial phase of pristane-induced autoimmunity. However, a dramatic reduction in peritoneal B-1 cells was apparent 2 weeks after pristane-treatment coinciding with the peak of IgM anti-ssDNA antibody production (13,18), making it somewhat less likely that B-1 cells are involved in this early phase of autoantibody production. Nevertheless, in view of the fact that total IgM levels are increased at 2 weeks in pristane-treated mice, the possibility that B-1 cells are sequestered in an extra-peritoneal site as a consequence of pristane treatment cannot be excluded.

The late phase of autoantibody production is characterized by the development of non-polyreactive, IgG class anti-nRNP/Sm, Su, ribosomal P, and NF45/NF90 autoantibodies (13). The IgM^+/IgD^{high} B cells that appear at increased frequency in the spleens of pristane-treated mice (Fig. 1) have a phenotype consistent with that of follicular-type B cells (23), and could represent autoreactive conventional B cells undergoing clonal expansion and somatic mutation. It will be of interest to characterize these cells further.

References

1. Kantor AB Herzenberg LA (1993) Origin of murine B cell lineages. Annu Rev Immunol 11:501-538
2. Vogel LA, Lester TL, Van Cleave VH, and Metzger DW (1996) Inhibition of murine B-1 lymphocytes by interleukin-12. Eur J Immunol 26:219-223
3. Engel P, Zhou L, Ord DC, Sato S, Koller B, and Tedder TF (1995) Abnormal B lymphocyte development, activation, and differentiation in mice that lack or overexpress the CD19 signal transduction molecule. Immunity 3:39-50
4. Sato S, Ono N, Steeber DA, Pisetsky DS, and Tedder TF (1996) CD19 regulates B lymphocyte signaling thresholds critical for the development of B-1 lineage cells and autoimmunity. J Immunol 157:4371-4378
5. Carroll MC (1998) The role of complement and complement receptors in induction and regulation of immunity. Annu Rev Immunol 16:545-568
6. Hayakawa K, Hardy RR, Honda M, Herzenberg LA, and Steinberg AD (1984) Ly-1 B cells: functionally distinct lymphocytes that secrete IgM autoantibodies. Proc Natl Acad Sci USA 81:2494-2498
7. Casali P Notkins AL (1989) CD5+ B lymphocytes, polyreactive antibodies and the human B-cell repertoire. Immunol Today 10:364-368

8. Jiang Y, Hirose S, Hamano Y, Kodera S, Tsurui H, Abe M, Terashima K, Ishikawa S, and Shirai T (1997) Mapping of a gene for the increased susceptibility of B-1 cells to Mott cell formation in murine autoimmune disease. J Immunol 158:992-997
9. Mohan C, Morel L, Yang P, and Wakeland EK (1997) Genetic dissection of systemic lupus erythematosus pathogenesis: Sle2 on murine chromosome 4 leads to B cell hyperactivity. J Immunol 159:454-465
10. Reap EA, Sobel ES, Cohen PL, and Eisenberg RA (1993) Conventional B cells, not B-1 cells, are responsible for producing autoantibodies in lpr mice. J Exp Med 177:69-78
11. Potter M Wiener F (1992) Plasmacytomagenesis in mice: model of neoplastic development dependent upon chromosomal translocations. Carcinogenesis 13:1681-1697
12. Potter M, Wax JS, Hansen CT, and Kenny JJ (1999) BALB/c.CBA/N mice carrying the defective Btk(xid) gene are resistant to pristane-induced plasmacytomagenesis. Int Immunol 11:1059-1064
13. Satoh M, Kumar A, Kanwar YS, and Reeves WH (1995) Antinuclear antibody production and immune complex glomerulonephritis in BALB/c mice treated with pristane. Proc Natl Acad Sci USA 92:10934-10938
14. Satoh M, Richards HB, and Reeves WH (1999) Pathogenesis of autoantibody production and glomerulonephritis in pristane-treated mice: an inducible model of SLE. In Lupus: Molecular and Cellular Pathogenesis. G.M.Kammer and G.C.Tsokos, editors. Humana Press, Totowa, NJ. 399-416
15. Hilbert DM, Kopf M, Mock BA, Kohler G, and Rudikoff S (1995) Interleukin 6 is essential for in vivo development of B lineage neoplasms. J Exp Med 182:243-248
16. Richards HB, Satoh M, Shaw M, Libert C, Poli V, and Reeves WH (1998) IL-6 dependence of anti-DNA antibody production: evidence for two pathways of autoantibody formation in pristane-induced lupus. J Exp Med 188:985-990
17. Byrd LG, McDonald AH, Gold LG, and Potter M (1991) Specific pathogen-free BALB/cAn mice are refractory to plasmacytoma induction by pristane. J Immunol 147:3632-3637
18. Hamilton KJ, Satoh M, Swartz J, Richards HB, and Reeves WH (1998) Influence of microbial stimulation on hypergammaglobulinemia and autoantibody production in pristane-induced lupus. Clin Immunol Immunopathol 86:271-279
19. Richards HB, Satoh M, Shaheen VM, Yoshida H, and Reeves WH (1999) Induction of B cell autoimmunity by pristane. Curr Top Microbiol Immunol 246:387-392
20. McDonald AH Degrassi A (1993) Pristane induces an indomethacin inhibitable inflammatory influx of CD4+ T cells and IFN-gamma production in plasmacytoma-susceptible BALB/cAnPt mice. Cell Immunol 146:157-170
21. Westhoff CM, Whittier A, Kathol S, McHugh J, Zajicek C, Shultz LD, and Wylie DE (1997) DNA-binding antibodies from viable motheaten mutant mice. Implications for B cell tolerance. J Immunol 159:3024-3033
22. Hayakawa K, Hardy RR, and Herzenberg LA (1986) Peritoneal Ly-1 B cells: genetic control, autoantibody production, increased lambda light chain expression. Eur J Immunol 16:450-465
23. Amano M, Baumgarth N, Dick MD, Brossay L, Kronenberg M, Herzenberg LA, and Strober S (1998) CD1 expression defines subsets of follicular and marginal zone B cells in the spleen: beta 2-microglobulin-dependent and independent forms. J Immunol 161:1710-1717
24. Lanier LL (1998) NK cell receptors. Annu Rev Immunol 16:359-393
25. Pani G, Siminovitch KA, and Paige CJ (1997) The motheaten mutation rescues B cell signaling and development in CD45-deficient mice. J Exp Med 186:581-588
26. Hayakawa K, Hardy RR, Parks DR, and Herzenberg LA (1983) The "Ly-1 B" cell subpopulation in normal immunodefective, and autoimmune mice. J Exp Med 157:202-218
27. Forster I Rajewsky K (1987) Expansion and functional activity of Ly-1+ B cells upon transfer of peritoneal cells into allotype-congenic, newborn mice. Eur J Immunol 17:521-528

VI

B-1 Cells and Mucosal Immunity

B-1 Cells and the Intestinal Microflora

N.A. Bos, J. J. Cebra[1] and F.G.M. Kroese
Department of Histology and Cell Biology, University of Groningen, Oostersingel 69/1, 9713 EZ Groningen, The Netherlands and [1]Department of Biology, University of Pennsylvania, PA 19104-6018, USA

Introduction

The production of IgA in the murine intestine is induced by the presence of intestinal bacteria, as evidenced by the almost complete absence of IgA plasma cells in germfree mice [1]. Monoassociation of germfree mice with a single species of commensal bacteria can induce a strong mucosal IgA response [2,3]. Intestinal bacteria can induce a germinal center reaction in the Peyer's patches, leading to the production of IgA that is specific for the inducing bacteria. But next to this specific IgA, a long lasting production of IgA, not reacting with the original bacteria, is often seen, even after waning of the germinal center reaction. This 'non-specific' IgA might reflect the 'natural' IgA antibodies found in human secretions [4]. Recently, natural serum IgM antibodies have been shown to be important in control of early viral and bacterial distribution and disease [5]. Whether 'natural' IgA plays a similar role in controlling the intestinal bacteria is unknown.

In conventionally reared mice with a stable microflora, the vast majority of bacteria in the small intestine is coated with IgA, while this percentage is only 20-40% in the caecum, large intestine and feces [6]. The effect of this IgA coating on the composition of intestinal bacterial populations is largely unknown. Animals totally lacking Ig expression (μMT mice) mice show very little changes in their normal gut flora compared to wildtype littermates when they are associated with a limited set of bacteria [7]. Such mice, however, are very susceptible to opportunistic infections [8]. These findings may suggest that IgA is not involved in the establishment of the gut flora. However, the possibility that compensatory mechanisms might take over the role of IgA, can not be excluded. For instance, in IgA knockout mice mucosal IgM and IgG responses are elevated [9]. In animals lacking the poly Ig-receptor, which is responsible for the selective transport of dimeric IgA across the epithelium into the lumen of the gut, there is so much IgA accumulating in the serum that leakage of all serum proteins (including IgA) into the gut occurs [10]. Directly testing the role of IgA on the composition of the gut flora can be done by reconstitution of B cells into the above mentioned knockout mice after (mono-) association with different known microorganisms.

There is accumulating evidence that, in addition to Peyer's patch derived conventional B cells (B-2 cells), also peritoneally derived B-1 cells contribute significantly to the generation of IgA secreting cells in the murine intestine [11,12]. The exact contribution of B-1 cells to the pool of intestinal IgA plasma cells in normal, untreated animals is still not known. Also little is known about the differentiation process of peritoneal sIgM$^+$ B-1 cells to IgA secreting cells. B-1 cells may switch inside the peritoneal cavity, since Cα germline transcripts are found in the peritoneal B-1 cell subset [13,14]. Furthermore, relatively high proportions of peritoneal T cells are in an 'activated' state. These T cells may provide crucial help in isotype switching of B-1 cell at this anatomical location [14].

B-1 Cells Are Selectively Retained in the Peritoneal Cavity and Both B-1 and B-2 Cells Can Contribute to the Pool of Murine IgA Plasmablasts over Long Periods.

We assessed maintenance and dissemination of B-1 vs. B-2 cells in and from the peritoneal cavity (PeC) by transferring PeC cells (as source of B-1 cells) and Peyer's patch (PP) B cells (as source of B-2 cells) from sets of congenic mice that differ in Ig allotype into conventionally reared C.B17- scid mice. Our findings were that PeC-derived B-1 cells were selectively retained in the peritoneal cavity compared to PP-derived B-2 cells (Figure 1).

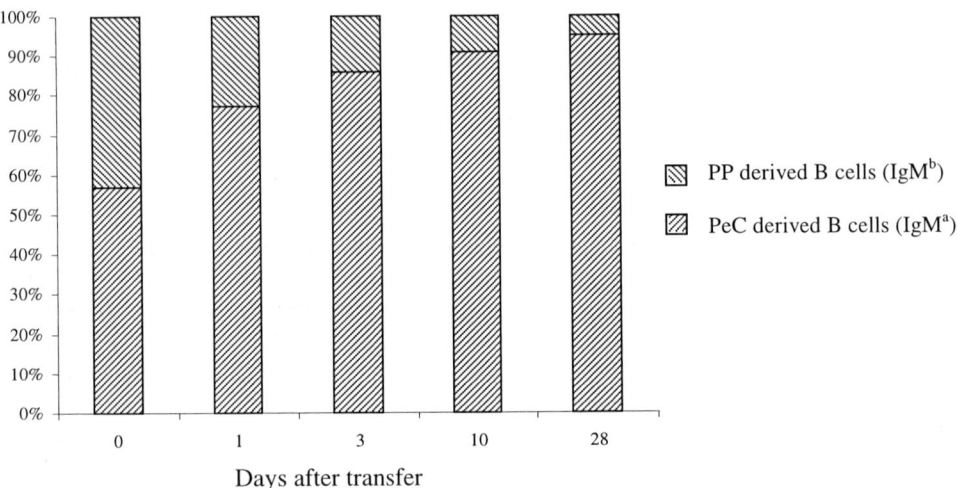

Figure 1: Contribution of peritoneal cavity cells and Peyer's patches cells to IgM$^+$ B cells in the peritoneal cavity after transfer of equal amounts of 2.5 x 10^6 cells from each donor into C.B17 scid mice. Data represent percentages of IgM$^+$ B cells within the peritoneal cavity derived from each donor as determined by flowcytometric analysis with IgM-allotype specific monoclonal antibodies. Each time point represents analysis of 1-2 mice and the experiment has been repeated four times starting with different ratios of donor cells with similar results.

When the contribution of the two cell type sources to IgA plasma cells in the mesenteric lymph nodes and the lamina propria of the small intestine of these repopulated C.B17 scid mice were analyzed a more equal contribution over long periods of both B-1 and B-2 cells was observed (Figure 2).

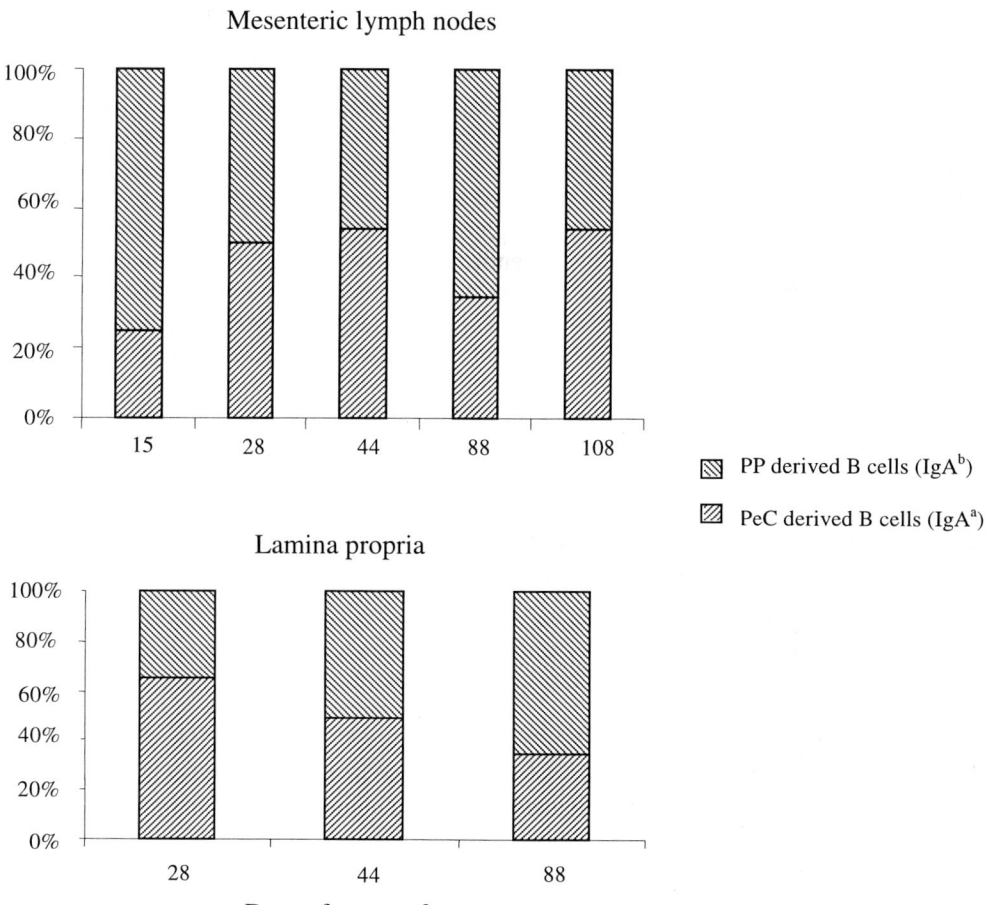

Figure 2: contribution of peritoneal cavity cells and Peyer's patch cells to IgA plasma cells in mesenteric lymph nodes and lamina propria of the small intestine after transfer into CB 17 scid mice. Data represent percentage of IgA$^+$ B cells on cytospots of mesenteric lymph node and lamina propia of the small intestine cells derived from each donor. The ratio of PP to PeC cells was 3:1 in the original inoculum with a total of 5×10^6 cells per mouse. Each time point represent analysis of 1-2 mice and the experiment has been repeated four times starting with different ratios of donor cells with similar results

In the mesenteric lymph nodes we could show by BrdU incorporation many recently divided IgA-positive B cells from both sources (data not shown). We have used this observation to make IgA-producing hybridomas from the mesenteric lymph nodes of animals after repopulation with only PeC cells [15].

We do not know how and why B-1 cells are retained within the peritoneal cavity in contrast with B-2 cells. Differential expression of "homing" receptors, other adhesion molecules or chemokine receptors on B-1 and B-2 cells might be involved. The unique microenvironment of the peritoneal cavity might contribute to the production of IgA by B-1 cells. For example, B-1 cells constitutively express IL-5 receptor and B-1 cells seem to be dependent on IL-5 for switching to IgA production [16]. The finding that peritoneal T cells in contrast to splenic T cells produce IL-5, support the hypothesis that peritoneal T cells might be involved in local differentiation to IgA producing B-1 cells [14].

In the aforementioned reconstitution experiments, few B-2 cells could be recovered from most lymphoid tissues at any time point. This raises the question of the fate and whereabouts of those B-2 cells that provided the long-term contributions to the pool of IgA plasma cells in the mesenteric lymph nodes and lamina propria. Possibly, the distribution of B-2 cells throughout the entire lymphoid system and their lack of self-renewal capacity may explain why they are difficult to recover in appreciable numbers compared with the self-renewing B-1 cells that remain sequestered in the peritoneal cavity. Alternatively, the long-term contribution by B-2 cells to the pool of IgA plasma cells may mainly have been provided by long lived, IgA memory cells within the PP inocula. The half-life of IgA plasma cells at the mucosal sites is still a matter of debate. Originally, it was calculated that most IgA plasma cells have a half-life of about 5 days, although few IgA plasma cells seemed to be long-lived [17]. Recent data suggest that the existence of long living plasma cells, although in these experiments this has not been shown at mucosal sites [18,19]. Whether B-1 and B-2 cell derived IgA plasma cells have a different life span is an important issue to resolve in order to understand their physiological contribution.

The Majority of IgA Coated Fecal Bacteria Are Coated by B-1 cell Derived IgA and Association of Germfree Mice with Different Bacteria Results in Different Contributions of B-1 and B-2 Derived IgA

The specificity repertoire of natural IgM antibodies secreted by B-1 cells appears to be biased towards microorganism coat antigens and are thought to play a major role in the first line of defense of the body [5]. We have evidence at the monoclonal and polyclonal level that B-1 cell derived IgA can bind to bacteria that belong to the indigenous microflora. We have shown by flowcytometry that IgA from either B-1 cell derived hybridomas or silicone gel induced peritoneal plasmacytomas (probably B-1 cell derived) react with fecal bacteria ([15] and Kroese, Bos and Potter, unpublished observations). Transfer of peritoneal cells, as source of B-1 cells, into C.B17 scid mice results in coating of fecal bacteria with IgA and long term IgA-coating coincided with appearance of PeC derived IgA plasma cells in the lamina propria of the gut [6]. To estimate the contribution of B-1 and B-2 cells to the IgA coating of the normal gut flora, we analyzed fecal samples from conventionally reared mice that were neonatally treated with allotype specific anti-IgM antibodies and simultaneously given peritoneal cavity (B-1) cells from congenic donors differing in Ig allotype (in collaboration with Dr. N. Baumgarth, Stanford). In these neonatal chimeric mice we observed that among all IgA coated fecal bacteria the majority were coated with PeC (B-1) derived IgA (Figure 3).

In another set of experiments, germfree SCID mice were given mixtures of FACS sorted B-1 cells and bone marrow (B-2) cells, from congenic pairs of mice differing in Ig allotype. These mice were monoassociated with the bacterium *Oochrobactrum anthropi* (OA) at the time of reconstitution. Monoassociation with OA results in a low level of mucosal IgA production as measured in fragment cultures from the small intestine. Analysis of fecal bacteria (OA only) revealed that the majority of bacteria-bound IgA antibodies were of B-1 cell origin (Figure 3). When these monoassociated mice were superinfected with *Morganella morganii* (MM) for another 28 days, the mucosal IgA production raised, and the relative contribution of B-2 cells to the production of IgA that coated fecal bacteria rose considerably. This was also reflected in the serum where after OA monoassociation only B-1 cell derived IgA could be detected while after secondary association with MM the ratio of B-1 to B-2 cell derived IgA was 39 to 61%.

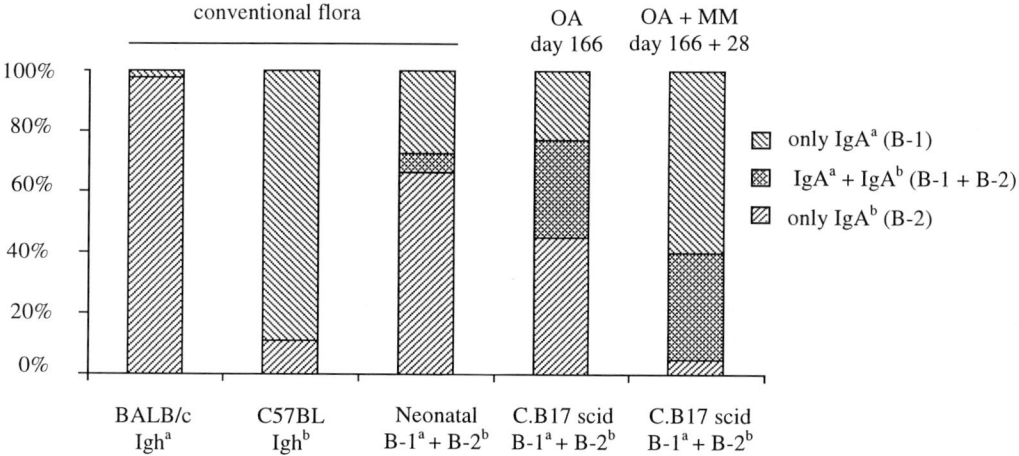

Figure 3. Relative contribution of B-1 and B-2 derived IgA to IgA coating of fecal bacteria. Percentages represent flowcytometric analysis of relative contribution of B-1 (Igh^a) and B-2 (Igh^b) derived IgA to in vivo coating of fecal bacteria detected by double staining with anti-IgA^a (Hy16) and anti-IgA^b (HISM2) monoclonal antibodies. Fecal samples of conventionally reared neonatal chimeric mice and of formally germfree C.B17 scid mice given mixtures of sorted B-1 cells (Igh^a) and bone marrow cells (Igh^b) which were monoassociated *Oochrobactrum anthropi* (OA) for 166 days and of mice that were subsequently associated with *Morganella morganii* (MM) for 28 days. Fecal samples of conventionally reared BALB/c (Igh^a) and C57BL (Igh^b) mice are shown as controls for allotype stainings.

Remarkably, most bacteria were coated exclusively by B1 or B2 cell derived IgA and only a minority was coated by IgA antibodies of both origins. Several reasons can be envisaged for this differential staining patterns. Firstly, it might reflect different epitopes recognized by the different sets of IgA antibodies. In the conventionally reared animals this can be explained by the diversity of the bacterial flora. But even in monoassociated mice the expression of epitopes by bacteria can be diverse, as we have shown by staining of monocultured bacterial species with IgA monoclonal antibodies [14]. Secondly, competition for available epitopes could explain exclusion of some IgA antibodies. The size of the repertoire and the affinities of the different sets of IgA antibodies might also contribute to the competition for available epitopes.

Manipulation of IgA Leads to Changes in the Gut Microflora

In order to obtain direct evidence for the role of IgA in influencing the composition of the gut flora, we injected B-1 cell derived IgA hybridoma cells as backpack tumor into conventionally reared C.B17 scid mice and analyzed IgA coating of the fecal bacteria. The percentage of intestinal and fecal bacteria that were *in vivo* coated with the injected monoclonal antibody became 35-40% within two weeks and remains stable afterwards. Since in normal mice this monoclonal IgA antibody can only stain about 5% of the fecal flora, this suggested changes in the gut flora. We also monitored the composition of the gut flora by FACS analysis of bacterial population after 16S rRNA *in situ* hybridization. With the Bact338 probe 70-80% of the fecal bacteria from control C.B17 scid mice were positive. This percentage decreased to 35-40% at 4-5 weeks after injection of the hybridoma cells. Similar decreases were seen with another probe (Erec482), while the percentage of Bfra602/Bdis656 positive fecal bacteria showed an increase. This results show that *in vivo* treatment with B-1 cell derived monoclonal IgA antibodies leads to great alterations in the composition of the normal gut flora ([14] and Bos et al. submitted). B-2 cell derived IgA antibodies have not been tested yet for their effect on the gut flora. Furthermore, the impact of regulation of the composition of the gut flora of polyclonal (B-1 and B-2 cell derived) IgA, present under physiologically conditions, still needs to be determined.

IgA may also play a role in the distribution of certain bacteria along the intestinal tract. Segmented filamentous bacteria (SFB) are only present in high numbers in the small intestine shortly after weaning, both in SFB monoassociated mice as well in mice with a conventional flora containing SFB [20]. SFB are strongly anchored to the epithelial cells of the small intestine. They induce a very strong mucosal IgA response [3]. However, in athymic nude mice the number of SFB in the small intestine remains the same until 12 weeks, while in their heterozygotic littermates there is rapid decrease of SFB in the small intestine in that same period. Since nude mice are impaired in their IgA production, this might suggest some role of mucosal IgA in controlling the presence of SFB in the small intestine [20].

Epitope Specificity and Affinity Might Determine the Effect of Different IgA Antibodies on the Intestinal Bacteria

We have hypothesized previously that B-1 cell derived IgA and B-2 cell derived IgA may exert differential roles in the humoral immune response in the intestine [6]. Briefly, low affinity IgA antibodies derived from B-1 cells could play a role in maintaining a stable microflora, whereas high affinity IgA antibodies (originating from B-2 cells) could result in 'immune exclusion'. We want to extent this hypothesis by speculating that bacterial properties determine the magnitude of the IgA response and the involvement of B-1 and B-2 cells. Bacteria such as *O. anthropi,* an aerobic, gram-negative bacterial strain, which grows poorly in the intestinal tract and almost does not translocate. After monoassociation with OA, there is almost no germinal center reaction within the Peyer's patches. The little IgA production is mostly produced by B-1 cells. How B-1 cells are induced to this IgA production is still unknown.

Some bacteria such as SFB, a strictly anaerobic bacterium can only live within the gastro-intestinal tract. They will die immediately after translocating the epithelial lining of the gut. In the Peyer's patches a long lasting presence of a low number of germinal centers is found, when animals are monoassociated with SFB [3]. By the firm attachment to the gut epithelial cells, SFB might trigger the mucosal immune system, leading to a great "background/natural" IgA response (B-1 derived?), but only minimal specific, high affinity IgA. The presence of IgA ("background/natural" and/or "specific") might exclude SFB bacteria from the small intestine, by preventing adherence to the intestinal lining.

Normal commensal bacteria, such as Morganella morganii (MM), a facultative, anaerobic gram-negative bacterial strain, might behave as opportunistic pathogenic bacteria, because of their ability to penetrate the host. They induce on top of the "background" IgA a strong germinal center reaction within the Peyer's patches, which results in a high level of specific IgA. The occurrence of this specific IgA strongly correlates with the moment that translocation of the bacteria does not occur anymore. The resulting long lasting IgA production consist of both "background" as well as specific IgA, which is in a dynamic equilibrium with the continuous presence of these bacteria in the gut [2]. The "background/natural" IgA that is being induced by SFB cannot prevent the mucosal IgA response by MM, since mice that were first monoassociated with SFB and then with MM, showed a similar mucosal IgA response as after monoassociation with MM alone [3]. Whether "background/natural" IgA induced by one bacterial strain can influence other bacteria that colonize the gut will dependent on the level of 'cross-reactivity' between these bacteria. Apparently, there is little cross-reactivity between SFB and MM, which was also reflected in the low level of MM-specific IgA that was found after SFB monoassociation [3].

We hypothesize that B-1 cell derived IgA antibodies in general will be of lower affinity and recognize epitopes which are less effective in immunoexclusion. The normal gut flora will be in a stable equilibrium between growth of the bacteria and production of natural IgA. High affinity IgA is being produced by B-2 cells in specific germinal center reactions towards bacteria that can penetrate the host despite the production of the natural

IgA. With this layered IgA immune response the host is able to maintain a stable gut flora, while more pathogenic bacteria are rapidly excluded from the gut (Figure 4).
The efficiency of "immune exclusion" by IgA is dependent on the epitope-specificity and affinity of the produced IgA, as a regulated response depending on the properties of the involved bacteria,
Evidence for this hypothesis awaits characterization of the involved epitopes and testing of poly- or monoclonal IgA antibodies, which differ in specificity and affinity in animals with a defined gut flora, for their effectiveness in controlling the gut flora.

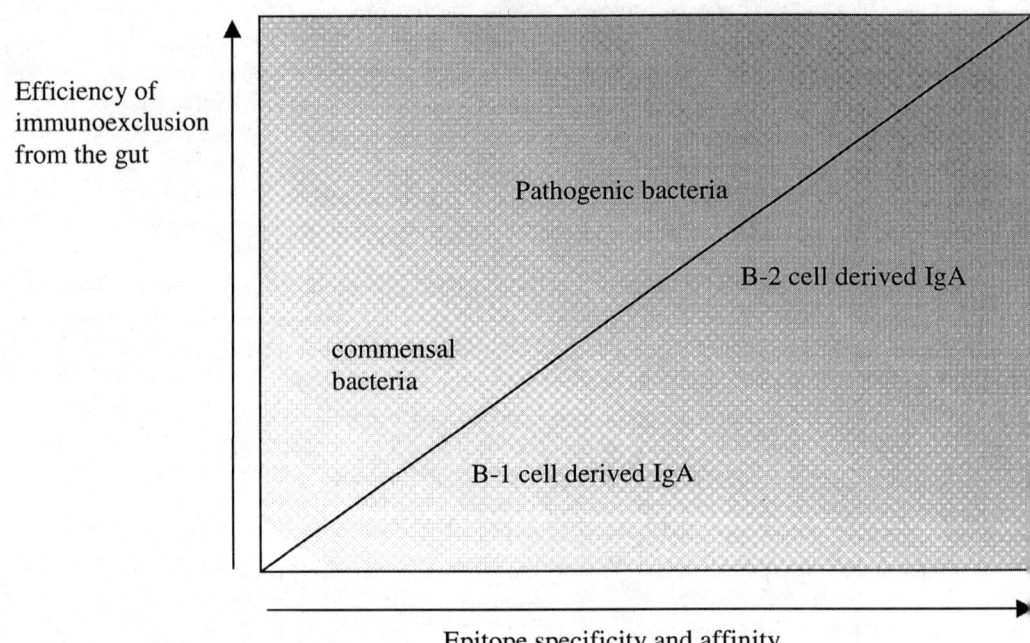

Figure 4. Affinity hypothesis: We hypothesize that commensal bacteria induce IgA with epitope specificities and affinities that have a low efficiency of excluding intestinal bacteria from the gut, while more pathogenic bacteria induce IgA with a higher affinity which is more efficient in excluding bacteria from the gut.

Adknowledgments

We wish to thank E. R. Cebra, J.C.A.M Bun and D.J. de Groot for excellent technical assistance and Dr. M. Potter (NIH, Bethesda) and Dr. N. Baumgarth (Stanford University, Stanford) for sharing unpublished information. This work was in part supported by NATO grant CRG 961213 (NAB and JJC) and NIH grant AI37108 (JJC).

References

1. Bos NA, Meeuwsen CG, Wostmann BS, Pleasants JR, Benner R (1988) The influence of exogenous antigenic stimulation on the specificity repertoire of background immunoglobulin-secreting cells of different isotypes. Cell.Immunol. 112:371-380.
2. Shroff KE, Meslin K, Cebra JJ (1995) Commensal enteric bacteria engender a self-limiting humoral mucosal immune response while permanently colonizing the gut. Infect.Immun. 63:3904-3913.
3. Talham GL, Jiang HQ, Bos NA, Cebra JJ (1999) Segmented filamentous bacteria are potent stimuli of a physiologically normal state of the murine gut mucosal immune system. Infect.Immun. 67:1992-2000.
4. Quan CP, Berneman A, Pires R, Avrameas S, Bouvet J-P (1997) Natural polyreactive secretory immunoglobulin A autoantibodies as a possible barriere to infection in humans. Infect. Immun. 65:3997-4004.
5. Ochsenbein AF, Fehr T, Lutz C, Suter M, Brombacher F, Herngartner H, Zinkernagel RM (1999) Control of early viral and bacterial distribution and disease by natural antibodies. Science 286:2156-2159.
6. Kroese FGM, de-Waard R, Bos NA (1996) B-1 cells and their reactivity with the murine intestinal microflora. Semin.Immunol. 8:11-18.
7. Marcotte H, Lavoie MC (1996) No apparent influence of immunoglobulins on indigenous oral and intestinal microbiota of mice. Infect.Immun. 64:4694-4699.
8. Marcotte H, Levesque D, Delanay K, Bourgeault A, de-la-Durantaye R, Brochu S, Lavoie MC (1996) Pneumocystis carinii infection in transgenic B cell-deficient mice. J.Infect.Dis. 173:1034-1037.
9. Harriman GR, Bogue M, Rogers P, Finegold M, Pacheco S, Bradley A, Zhang Y, Mbawuike IN (1999) Targeted deletion of the IgA constant region in mice leads to IgA deficiency with alterations in expression of other Ig isotypes. J.Immunol. 162:2521-2529.
10. Johansen FE, Pekna M, Natvig Noderhaug I, Haneberg B, Hietala MA, Kraji P, Betsholz C, Brandtzaeg P (1999) Absence of epithelial immunoglobulin A transport, with increased mucosal leakiness, in polymeric immunoglobulin receptor/ secretory componenet deficient mice. J.Exp.Med. 190:915-921.
11. Kroese FGM, Butcher EC, Stall AM, Lalor PA, Adams S, Herzenberg LA (1989) Many of the IgA producing plasma cells in murine gut are derived from self-replenishing precursors in the peritoneal cavity. Int.Immunol. 1:75-84.
12. Kroese FGM, Ammerlaan WA, Kantor AB (1993) Evidence that intestinal IgA plasma cells in mu, kappa transgenic mice are derived from B-1 (Ly-1 B) cells. Int.Immunol. 5:1317-1327.
13. De Waard R, Dammers PM, Tung JW, Kantor AB, Wilshire JA, Bos NA, Herzenberg LA, Kroese FGM (1998) Presence of germline and full-length IgA RNA transcripts among peritoneal B-1 cells. Dev.Immunol. 6:81-87.
14. Kroese FGM, Bos NA (1999) Peritoneal B-1 cells switch in vivo to IgA and these IgA antibodies can bind to bacteria of the normal intestinal microflora. Curr.Top.Microbiol.Immunol. 246:343-349.

15. Bos NA, Bun JCAM, Popma SH, Cebra ER, Deenen GJ, Cammen MJFvd, Kroese FGM, Cebra JJ (1996) Monoclonal Immunoglobulin A derived from peritoneal B cells is encoded by both germ line and somatically mutated VH genes and reactive with commensal bacteria. Infect. Immun. 64:616-623.
16. Bao S, Beagley KW, Murray AM, Caristo V, Matthaei KI, Young IG, Husband AJ (1998) Intestinal IgA plasma cells of the B1 lineage are IL-5 dependent. Immunology 94:181-188.
17. Mattioli CA, Tomasi-TB J (1973) The life span of IgA plasma cells from the mouse intestine. J.Exp.Med. 138:452-460.
18. Manz RA, Lohning M, Cassese G, Thiel A, Radbruch A (1998) Survival of long-lived plasma cells is independent of antigen. Int.Immunol. 10:1703-1711.
19. Slifka MK, Antia R, Whitmire JK, Ahmed R (1998) Humoral immunity due to long-lived plasma cells. Immunity. 8:363-372.
20. Snel J, Hermsen CC, Smits HJ, Bos NA, Eling WMC, Cebra JJ, Heidt PJ (1998) Interactions between gut-associated lymphoid tissues and indigenous, filamentous bacteria in the small intestine of mice. Can.J.Microbiol. 44:1177-1182.

Mechanism of B1 cell differentiation and migration in GALT

Sidonia Fagarasan, Reiko Shinkura, Tadashi Kamata, Fumiaki Nogaki, Koichi Ikuta and Tasuku Honjo
Department of Medical Chemistry, Kyoto University Faculty of Medicine, Japan;

Two subsets of B cells-designated B1 and B2 cells have been identified, based upon their anatomical location, developmental origin, cell surface markers, antibody repertoire and self-renewal capacity (1-4). B1 cells are dominant in the peritoneal cavity (PEC) and the size of the B1 cell population is kept constant due to their self-renewal capacity. This activity is necessary for antibody production (5,6). More than 20 years ago, Husband and Gowans suggested a link between PEC cells and antigen-specific B cells in the lamina propria (LP) of the rat small intestine (7). It is generally accepted that many of the IgA plasma cells in the intestinal LP are derived from B2 cell precursors in Peyer's patches (PP)(7-11). More recent studies demonstrated that aproximately half of the IgA plasma cells in the LP are derived from PEC cells and belong to the B1 cell lineage, suggesting that frequent migration of these cells may take place between PEC and LP of the gut (12,13).

aly/aly mice are characterized by the systemic absence of lymph nodes (LN) and PP, disorganized splenic architecture and immunodeficiency, because they have a point mutation in the gene encoding NF-κB-inducing kinase (NIK)(14,15). The lack of lymphoid organogenesis in *aly/aly* mice is caused by the defect in non-bone marrow (BM) derived cells, because the *aly*-type mutation was found to affect LTβR signaling pathway (15) and LTβR is exclusively expressed by non-lymphoid cells (16,17). However, there are several features of *aly/aly* mice which suggest that they have also a defect in BM-derived cells: disturbed thymic architecture, impaired T cell functions, reduced numbers of B cells in BM, spleen, LN and peripheral blood (14,18). Moreover, spleen, mesenteric lymph nodes (MLN) and PP remain atrophic when *aly/aly* BM cells are transferred to irradiated wild-type mice, suggesting the presence of a homing defect in *aly/aly* lymphocytes (14). Another constant feature of *aly/aly* mice suggesting a migration defect is that their PEC contains more B1 cells than PEC of normal or *aly/+* mice.

In this study, we investigated the homing capacity of PEC cells from *aly/aly* and *aly/+* mice. We found that PEC cells from *aly/aly* mice have a defect in homing to other lymphoid tissues, and this defect was more severe regarding their migration to the gut-associated lymphatic tissue (GALT) system. *In vivo* migration defect correlated with *in vitro* decrease chemotactic activity of SLC (secondary lymphoid-tissue chemokine) and BLC (B lymphocyte chemoattractant) on *aly/aly* PEC cells. The defective chemotactic response of *aly/aly* PEC lymphocytes was not due to the lack of chemokine or their receptors but to a defect in signaling pathway through the chemokine receptors. We found that the *aly* mutation of the NIK gene blocks signaling from the receptors for SLC, providing the first evidence that NIK is involved in signal transduction through seven-transmembrane protein receptors.

Abundant B1 in PEC and Complete Absence of B Cells in LP of *aly/aly* Mice.

A constant feature of *aly/aly* mice is that their PEC contains more B1 cells than PEC of *aly/+* mice (Fig. 1A).Since many IgA plasma cells are derived from PEC B1 cells (12,13) we examined LP of *aly/aly* and *aly/+* mice. We found that LP of *aly/aly* mice did not contain either $B220^+IgM^+$ or $B220^-IgA^+$ cells which we found in LP of *aly/+* mice (Fig. 1B,C).

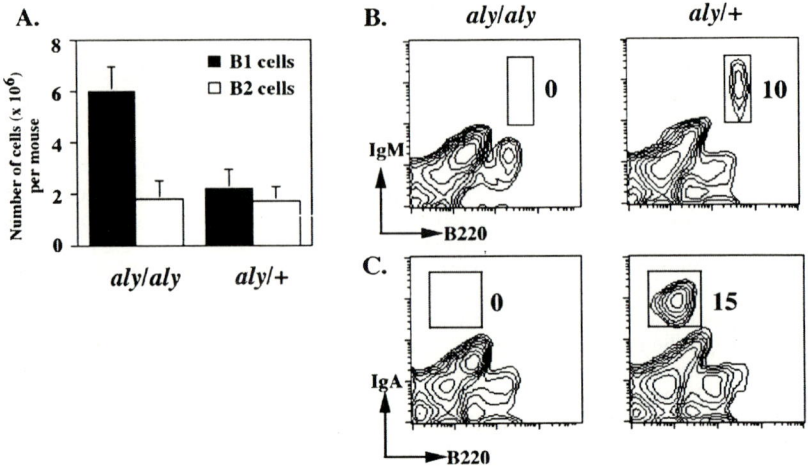

Fig. 1. More B1 cells in PEC and absence of B cell populations in LP of *aly/aly* mice. (A) Total numbers of PEC cells in *aly/aly* and *aly/+* mice, as calculated by (% of IgM$^+$Mac-1$^+$ B1 cells or IgM$^+$Mac-1$^-$ B2 cells)x (number of viable cells). (B,C) Flow cytometric analysis of LP cells, stained with FITC-anti B220 and PE-anti IgM or IgA. The numbers are the % of B cells (B220$^+$IgM$^+$) and plasma cells (B220$^-$IgA$^+$).

FACS analysis revealed that B220$^+$IgM$^+$ cells are small lymphocytes bearing all surface markers of mature B cells (data not shown). The light scatter profile showed that B220$^-$IgA$^+$ cells are larger than B220$^+$IgM$^+$ cells.

To characterize the B220$^-$IgA$^+$ we sorted these cells and examined by May-Grunwald Giemsa and cytoplasmic IgA staining. As shown in Fig 2 A. B220$^-$IgA$^+$ cells display a typical plasma cell morphology, almost all of them contained IgA in their cytoplasm (Fig.2B) and also secrete IgA as detected by the ELISPOT assay (Fig. 2C). The surface

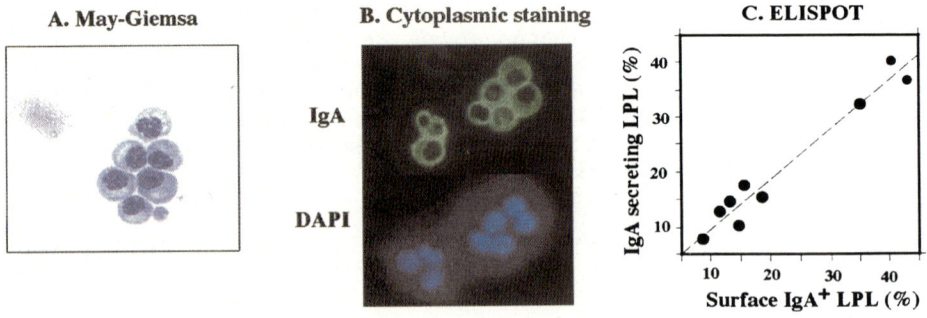

Fig. 2. Characterization of B220$^-$IgA$^+$ cells in LP. Sorted B220$^-$IgA$^+$ cells, stained (A) for MGG and (B) cytoplasmic IgA and DAPI to visualize nuclei. (C). A significant correlation between the frequency of surface IgA$^+$ cells detected by FACS and IgA secreting cells detected by ELISPOT.

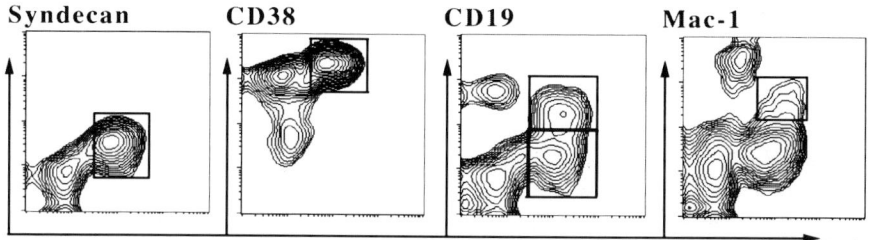

Fig. 3. Phenotypic analyses of IgA plasma cells in LP. Cells were stained for IgA in combination with Syndecan-1, CD38, CD19 and Mac-1.

phenotypes of the B220$^-$IgA$^+$ plasma cells were found to be syndecan^{low}, CD38^{hi}, but IgD$^-$, CD5$^-$, CD23$^-$ (Fig. 3 and data not shown). Interestingly, a subpopulation of B220$^-$IgA$^+$ plasma cells also express CD19 and Mac-1 (Fig. 2D). We speculate that this population, B220$^-$IgA$^+$CD19$^+$Mac1$^+$ are immature plasma cells, as they incorporate more BrdU than B220$^-$IgA$^+$CD19$^-$Mac1$^-$ cells (data not shown) which further differentiate *in situ* into mature plasma cells, with only IgA on their surface. In conclusion, *aly/aly* mice are missing not only mature B cells but also IgA plasma cells in the LP, in spite that they have many B1 cells in PEC.

aly/aly PEC Cells Fail to Generate IgA Plasma Cells in the GALT System of RAG-2$^{-/-}$ Mice.

The unique features of *aly/aly* mice mentioned above, led us to suspect that *aly/aly* PEC cells may have a migration defect to LP although the lack of IgA plasma cells in LP can be partially explained by the lack of PP, which normally contains the B2 cell precursors of plasma cells (7-11). To verify our hypothesis, we transferred PEC cells from *aly/aly* and *aly/+* mice into PEC of RAG-2$^{-/-}$ mice and analyzed the lymphoid tissues 3 and 6 wk after the transfer. No B220$^-$IgA$^+$ cells could be detected in LP or MLN of RAG-2$^{-/-}$ mice transferred with *aly/aly* PEC cells. By contrast, *aly/+* peritoneal B cells injected into RAG-2$^{-/-}$ mice migrated to MLN and intestinal LP, where they gave rise to IgA plasma cells (Fig. 4A, B). In agreement with these results, PEC of RAG-2$^{-/-}$ mice injected with *aly/aly* lymphocytes contained more B1 cells than PEC of RAG-2$^{-/-}$ mice injected with *aly/+* lymphocytes (Fig. 4 C), in spite of the fact that we injected approximately the same number of cells.

The absence of IgA plasma cells reflects a homing defect to the GALT system and not a defect in class switching because *aly/aly* lymphocytes can switch to IgA-producing cells by *in vitro* culture in the presence of cytokines (unpublished data). FACS results were further confirmed and extended by staining of fixed cytocentrifuge preparations of LP and MLN lymphocytes for IgM, IgA and IgG. As shown in Table 1, reconstitution of IgA or IgG plasma cells in LP of the small intestine was not observed when PEC cells from *aly/aly* were transferred to RAG-2$^{-/-}$ mice, whereas LP of RAG-2$^{-/-}$ mice injected with *aly/+* PEC cells contained many IgA plasma cells and a few IgG plasma cells. No IgM, IgA and IgG plasma cells could be identified in MLN of RAG-2$^{-/-}$ mice injected with *aly/aly* PEC cells. On the contrary, MLN of RAG-2$^{-/-}$ mice injected with *aly/+* PEC contained IgA, IgM and IgG plasma cells, and the percentage and the numbers of IgA plasma cells were at least two times higher than that in LP (Fig. 4 and data not shown). This observation suggests that PEC B cell proliferation and differentiation to IgA plasma cells may take place in the MLN, although we cannot exclude the possibility that this phenomenon is restricted to the RAG-2$^{-/-}$ environment which is probably devoid of normal regulatory mechanisms.

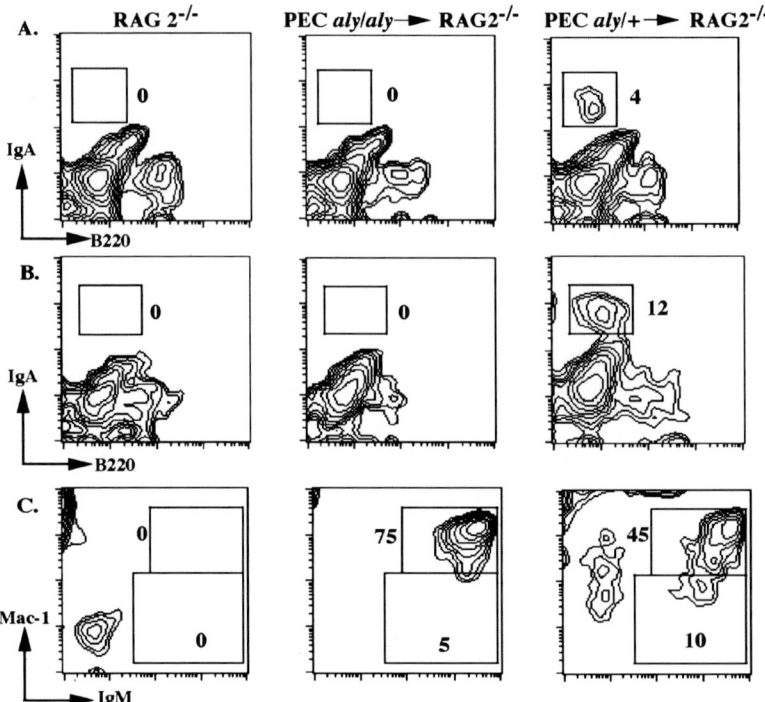

Fig. 4. Transfer of PEC cells from *aly/aly* mice fails to generate IgA plasma cells in GALT of RAG-2$^{-/-}$ mice. FACS analyses of (A) LP, (B) MLN and (C) PEC from RAG-2$^{-/-}$ mice and RAG-2$^{-/-}$ mice injected with ~ 10^7 PEC cells from *aly/aly* and *aly/+* mice, 6 weeks previously. The numbers are the percentages of IgA plasma cells (B220$^-$IgA$^+$), B1 (IgM$^+$Mac1$^+$) and B2 (IgM$^+$Mac1$^-$) cells.

Table 1. Impaired generation of plasma cells in RAG-2$^{-/-}$ mice transferred with *aly/aly* PEC cells.

	% IgM		% IgA		% IgG	
	aly/aly	*aly/+*	*aly/aly*	*aly/+*	*aly/aly*	*aly/+*
LPL 3 wk	0	0	0	2.33 ± 0.81	0	< 1
LPL 6 wk	0	0	0	5.66 ± 1.36	0	< 1
MLN 3 wk	0	< 1	0	7.86 ± 1.20	0	< 1
MLN 6 wk	0	3.66 ± 3.12	0	11.66 ± 3.93	0	< 1
Spleen 3 wk	< 1	2.27 ± 0.55	0	< 1	0	1.66 ± 0.57
Spleen 6 wk	< 1	7.10 ± 3.35	0	1.37 ± 0.47	< 1	3.33 ± 1.15

Fixed cytocentrifuge preparations of LP, MLN and spleen from RAG-2$^{-/-}$ mice injected with PEC cells from *aly/aly* and *aly/+* mice were stained for cytoplasmic IgM, IgA and IgG and counterstained with DAPI. The percentages are plasma cells in total cells, as mean and standard error.

Migration Defect of *aly/aly* PEC Cells is not Restricted to GALT

At 3 wk after transfer, the spleen or BM of RAG-2$^{-/-}$ mice transferred with *aly/aly* peritoneal B cells contained very few IgM or IgG and no IgA plasma cells (Table 1 and data not shown). On the contrary, spleen of RAG-2$^{-/-}$ mice injected with *aly/+* peritoneal B cells contained many IgM or IgG plasma cells and less but significant number of IgA plasma cells, while their BM contained less than 0.5% each of plasma cells expressing various isotypes (Table 1 and data not shown). The presence of high numbers of IgM-secreting B cells in RAG-2$^{-/-}$ mice transferred with *aly/+* PEC cells and yet reduced numbers of B220$^+$IgM$^+$ cells (1/3 of that in normal mice) in spleen (data not shown) suggests that peritoneal B cells may contain precursors to plasma cells or differentiate rapidly to plasma cells after transfer.

The difference in the numbers of plasma cells between RAG-2$^{-/-}$ mice injected with *aly/aly* and *aly/+* PEC cells was also reflected in the serum levels of immunoglobulins (Ig) (data not shown). In RAG-2$^{-/-}$ mice, 6 wk after transfer of *aly/aly* PEC cells, the serum titers of IgM and IgG were very low as compared with those injected with *aly/+* PEC cells, and the IgA levels were not detected, even after 6 wk. These results demonstrated that B1 cells were unable to secrete Ig *in situ* , suggesting that the migration capacity of B1 cells is closely related with their ability to produce antibodies.

Both Chemokines and their Receptors are Expressed in *aly/aly* Cells

Chemokines are chemotactic cytokines involved in the constitutive homing of lymphocytes to their proper location within lymphoid organs. The role for chemokines in B cell migration was shown in mice with targeted disruption of the BLR1 (Burkitt's lymphoma receptor, known also as CXCR5) which lack functional germinal centers in the spleen, caused by impaired lymphocyte recirculation (19). The ligand for BLR1, BLC has been shown to promote migration of B lymphocytes (20,21). Another chemokine, SLC stimulates the chemotaxis of naive T, memory T and B cells (22-25). CCR7-/- mice revealed that this receptor (one of the receptors for SLC) is important for T, B and dendritic cells migration to the proper environments, to initiate an antigen-specific immune response and to establish a functional architecture of the secondary lymphoid tissues (26).

To explore the cause of migration defect of *aly/aly* PEC cells, we measured mRNA expression levels of chemokines and their receptors. We first studied the mRNA expression of SLC, ELC (EBV-induced molecule 1 ligand chemokine) (27,28) and BLC, in the spleen and small intestine, to which PEC cells were shown to migrate in RAG-2$^{-/-}$ mice. SLC mRNA was expressed in the spleen and small intestine of RAG-2$^{-/-}$ mice and *aly/aly* mice although at lower levels than those in *aly/+* mice (data not shown). ELC mRNA expression levels were lower in spleen and absent or lower in small intestine of both RAG-2$^{-/-}$ and *aly/aly* mice, as compared with those in *aly/+* mice. BLC mRNA expression was drastically reduced in spleen and absent in small intestine of RAG-2$^{-/-}$ and *aly/aly* mice. Therefore, among the chemokines constitutively expressed in secondary lymphoid organs, SLC mRNA was found most abundantly expressed in RAG-2$^{-/-}$ mice, suggesting that this chemokine could be important for homing of injected lymphocytes to the lymphoid tissues of RAG-2$^{-/-}$ mice.

We also found that both PEC and spleen cells of *aly/aly* and *aly/+* mice expressed similar levels of CCR7 and BLR1 (data not shown). LPS stimulation did not affected the CCR7 and slightly up-regulated the BLR1 expression levels in *aly/aly* PEC cells and splenic cells from *aly/aly* as well as *aly/+* mice. On the other hand, LPS stimulation increased the mRNA expression levels of both CCR7 and BLR1 in *aly/+* PEC cells (data not shown).

Reduced Chemotactic Activities of SLC and BLC on Resting and LPS Activated *aly/aly* Peritoneal Lymphocytes.

To test if *aly/aly* cells respond to chemokines, chemotaxis assays were performed with PEC cells and splenocytes, using two chemokines: SLC and BLC. As shown in Figure 5A, peritoneal B1 cells from *aly/aly* mice, did not respond to SLC at all, whereas those from *aly/+* responded to SLC. *aly/aly* peritoneal B2 and T cells also showed very week chemotactic responses to SLC, as compared with those from *aly/+* mice (Figure 5 B, C). Surprisingly, splenic B and T cells of *aly/aly* mice showed a normal chemotactic response to SLC (Fig. 5 D, E).

BLC induced a very weak chemotactic response in peritoneal B1 and B2 cells of *aly/aly* mice. On the other hand, BLC induced a strong chemotactic response in peritoneal B1 and B2 cells and splenic B cells of *aly/+* mice as well as splenic B of *aly/aly* mice (data not shown).LPS activation was found to enhance the chemotactic response to SLC and BLC of peritoneal but not of splenic B cells (Fig. 5 and data not shown). Although *aly/aly* peritoneal B lymphocytes showed an increase in the chemotactic response to SLC after LPS stimulation, their responses were still weaker than those of non-stimulated *aly/+* peritoneal B cells. By contrast, LPS stimulation augmented the BLC responsiveness of *aly/aly* PEC B cells to similar levels as non-stimulated *aly/+* PEC B cells, in agreement with upregulation of BLR1 mRNA by LPS stimulation (data not shown).The chemotactic response of splenic cells to SLC and BLC was very similar between *aly/aly* and *aly/+* mice, indicating that chemotactic responses of splenic and PEC cells might be regulated differently.

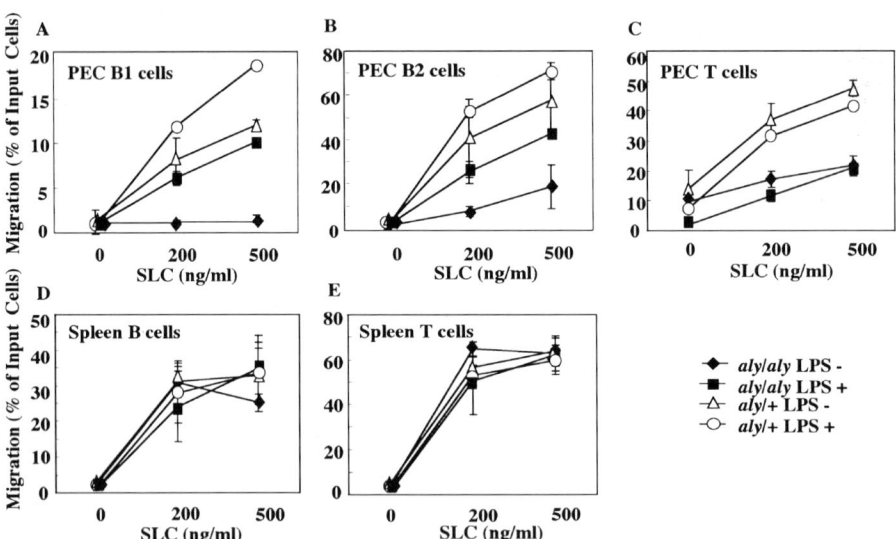

Fig. 5. Decreased chemotactic activity of SLC on resting and LPS stimulated cells from *aly/aly* and *aly/+* mice. The number of input and migrated cells of each subtype were determined by immunostaining and flow cytometry. Results are expressed as the percentage of input cells of each subtype migrating to the lower chamber of a transwell filter. PEC cells and splenic cells were incubated with media alone (LPS-) or LPS (20µg/ml) for 3 hours. Data points represent the mean and standard error for experiments performed in triplicate.

NIK is Downstream of SLC Receptor Signaling

aly/aly PEC cells failed to migrate to the GALT system as well as to spleen. The *in vivo* migration defect of *aly/aly* PEC cells correlates well with *in vitro* defective chemotactic response of *aly/aly* PEC cells to SLC and BLC. This migration defect cannot be explained by the lack of chemokines or their receptors, suggesting that the signal downstream of chemokine receptors might be affected in *aly/aly* mice. Therefore, we examined whether chemokine-induced NF-κB activation is affected by the *aly* mutation of NIK. We stimulated PEC cells from *aly/aly* and *aly/+* mice with SLC and the activation of NF-κB was determined by the gel shift assay of nuclear extracts, 15 or 30 min after stimulation. As shown in Fig. 6, SLC stimulation did not induce activation of NF-κB in *aly/aly* PEC, but increased the levels of NF-κB in the nuclei of *aly/+* PEC. The same stimulation did not affect the amount of Oct-1 transcription factor in the nuclear extract of either *aly/aly* or *aly/+* PEC cells (data not shown). Although the SLC-induced activation of NF-κB was not observed in nuclei of *aly/aly* PEC cells, the NF-κB complexes were present even before SLC stimulation. What was different from *aly/+* PEC cells was not the constitutive level, but rather the species of the NF-κB complexes (data not shown).

These results clearly demonstrate that NIK acts downstream the signaling pathway of the receptors for SLC, leading to activation of NF-κB and that the *aly*-type mutation in the NIK gene affects this signaling pathway, resulting in migration defect of *aly/aly* PEC cells. Although we do not know which chemokines are involved in PEC cell migration *in vivo*, the fact that SLC was expressed in RAG-2$^{-/-}$ mice and that this expression was sufficient for homing of normal PEC cells, suggests that SLC may be at least one of important chemokines for PEC cell migration. Regardless which chemokines or chemokine receptors are involved in the PEC cell migration, it is likely that NIK is involved in their signaling pathways.

The present study offers an explanation for the complex and severe *aly* phenotype, showing that the migration defect of lymphocytes to the proper places is also responsible for the immunologic abnormalities in *aly/aly* mice.

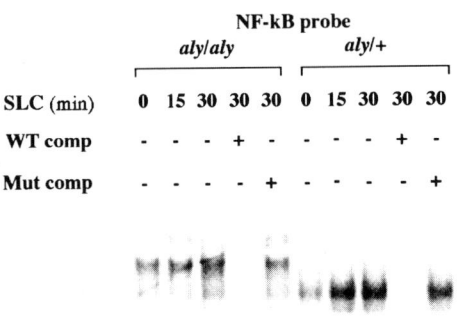

Fig. 6. Impaired activation of NF-κB by SLC in *aly/aly* PEC cells. Nuclear extracts from unstimulated and SLC (500ng/ml) stimulated PEC cells were preincubated with unlabeled (WT comp) or mutant NF-κB probe (Mut. comp) and then incubated with ^{32}P-labeled NF-κB probe, followed by separation on polyacrylamide gel and analyses by autoradiography.

Acknowledgment

We thank Dr. Jason G. Cyster (UCSF) for providing us chemokines and critical discussion and Dr. Kazuo Kinoshita for computerized artwork.

References

1. Hayakawa, K., R.R. Hardy, D.R. Parks, and L.A. Herzenberg. (1983). The "Ly-1 B" cell subpopulation in normal immunodefective, and autoimmune mice. *J Exp Med* 157, no. 1:202-18.
2. Hayakawa, K., R.R. Hardy, and L.A. Herzenberg. (1985). Progenitors for Ly-1 B cells are distinct from progenitors for other B cells. *J Exp Med* 161, no. 6:1554-68.
3. Hardy, R.R., and K. Hayakawa. (1994). CD5 B cells, a fetal B cell lineage. *Adv Immunol* 55:297-339.
4. Kantor, A.B., A.M. Stall, S. Adams, K. Watanabe, and L.A. Herzenberg. (1995). De novo development and self-replenishment of B cells. *Int Immunol* 7, no. 1:55-68.
5. Sidman, C.L., L.D. Shultz, R.R. Hardy, K. Hayakawa, and L.A. Herzenberg. (1986). Production of immunoglobulin isotypes by Ly-1+ B cells in viable motheaten and normal mice. *Science* 232, no. 4756:1423-5.
6. Herzenberg, L.A., A.M. Stall, P.A. Lalor, C. Sidman, W.A. Moore, and D.R. Parks. (1986). The Ly-1 B cell lineage. *Immunol Rev* 93:81-102.
7. Husband, A.J., and J.L. Gowans. (1978). The origin and antigen-dependent distribution of IgA-containing cells in the intestine. *J Exp Med* 148, no. 5:1146-60.
8. Craig, S.W., and J.J. Cebra. (1971). Peyer's patches: an enriched source of precursors for IgA-producing immunocytes in the rabbit. *J Exp Med* 134, no. 1:188-200.
9. Craig, S.W., and J.J. Cebra. (1975). Rabbit Peyer's patches, appendix, and popliteal lymph node B lymphocytes: a comparative analysis of their membrane immunoglobulin components and plasma cell precursor potential. *J Immunol* 114, no. 1 Pt 2:492-502.
10. Tseng, J. (1981). Transfer of lymphocytes of Peyer's patches between immunoglobulin allotype congenic mice: repopulation of the IgA plasma cells in the gut lamina propria. *J Immunol* 127, no. 5:2039-43.
11. Tseng, J. (1984). A population of resting IgM-IgD double-bearing lymphocytes in Peyer's patches: the major precursor cells for IgA plasma cells in the gut lamina propria. *J Immunol* 132, no. 6:2730-5.
12. Kroese, F.G., E.C. Butcher, A.M. Stall, P.A. Lalor, S. Adams, and L.A. Herzenberg. (1989). Many of the IgA producing plasma cells in murine gut are derived from self-replenishing precursors in the peritoneal cavity. *Int Immunol* 1, no. 1:75-84.
13. Bos, N.A., J.C. Bun, S.H. Popma, E.R. Cebra, G.J. Deenen, M.J. van der Cammen, F.G. Kroese, and J.J. Cebra. (1996). Monoclonal immunoglobulin A derived from peritoneal B cells is encoded by both germ line and somatically mutated VH genes and is reactive with commensal bacteria. *Infect Immun* 64, no. 2:616-23.
14. Miyawaki, S., Y. Nakamura, H. Suzuka, M. Koba, R. Yasumizu, S. Ikehara, and Y. Shibata. (1994). A new mutation, aly, that induces a generalized lack of lymph nodes accompanied by immunodeficiency in mice. *Eur J Immunol* 24, no. 2:429-34.
15. Shinkura, R., K. Kitada, F. Matsuda, K. Tashiro, K. Ikuta, M. Suzuki, K. Kogishi, T. Serikawa, and T. Honjo. (1999). Alymphoplasia is caused by a point mutation in the mouse gene encoding Nf-kappa b-inducing kinase. *Nat Genet* 22, no. 1:74-7.

16. Crowe, P.D., T.L. VanArsdale, B.N. Walter, C.F. Ware, C. Hession, B. Ehrenfels, J.L. Browning, W.S. Din, R.G. Goodwin, and C.A. Smith. (1994). A lymphotoxin-beta-specific receptor. *Science* 264, no. 5159:707-10.
17. Ware, C.F., T.L. VanArsdale, P.D. Crowe, and J.L. Browning. (1995). The ligands and receptors of the lymphotoxin system. *Curr Top Microbiol Immunol* 198:175-218.
18. Shinkura, R., F. Matsuda, T. Sakiyama, T. Tsubata, H. Hiai, M. Paumen, S. Miyawaki, and T. Honjo. (1996). Defects of somatic hypermutation and class switching in alymphoplasia (aly) mutant mice. *Int Immunol* 8, no. 7:1067-75.
19. Forster, R., A.E. Mattis, E. Kremmer, E. Wolf, G. Brem, and M. Lipp. (1996). A putative chemokine receptor, BLR1, directs B cell migration to defined lymphoid organs and specific anatomic compartments of the spleen. *Cell* 87, no. 6:1037-47.
20. Gunn, M.D., V.N. Ngo, K.M. Ansel, E.H. Ekland, J.G. Cyster, and L.T. Williams. (1998). A B-cell-homing chemokine made in lymphoid follicles activates Burkitt's lymphoma receptor-1. *Nature* 391, no. 6669:799-803.
21. Legler, D.F., M. Loetscher, R.S. Roos, I. Clark-Lewis, M. Baggiolini, and B. Moser. (1998). B cell-attracting chemokine 1, a human CXC chemokine expressed in lymphoid tissues, selectively attracts B lymphocytes via BLR1/CXCR5. *J Exp Med* 187, no. 4:655-60.
22. Nagira, M., T. Imai, K. Hieshima, J. Kusuda, M. Ridanpaa, S. Takagi, M. Nishimura, M. Kakizaki, H. Nomiyama, and O. Yoshie. (1997). Molecular cloning of a novel human CC chemokine secondary lymphoid-tissue chemokine that is a potent chemoattractant for lymphocytes and mapped to chromosome 9p13. *J Biol Chem* 272, no. 31:19518-24.
23. Hedrick, J.A., and A. Zlotnik. (1997). Identification and characterization of a novel beta chemokine containing six conserved cysteines. *J Immunol* 159, no. 4:1589-93.
24. Hromas, R., C.H. Kim, M. Klemsz, M. Krathwohl, K. Fife, S. Cooper, C. Schnizlein-Bick, and H.E. Broxmeyer. (1997). Isolation and characterization of Exodus-2, a novel C-C chemokine with a unique 37-amino acid carboxyl-terminal extension. *J Immunol* 159, no. 6:2554-8.
25. Gunn, M.D., K. Tangemann, C. Tam, J.G. Cyster, S.D. Rosen, and L.T. Williams. (1998). A chemokine expressed in lymphoid high endothelial venules promotes the adhesion and chemotaxis of naive T lymphocytes. *Proc Natl Acad Sci U S A* 95, no. 1:258-63.
26. Forster, R., A. Schubel, D. Breitfeld, E. Kremmer, I. Renner-Muller, E. Wolf, and M. Lipp. (1999). CCR7 coordinates the primary immune response by establishing functional microenvironments in secondary lymphoid organs. *Cell* 99, no. 1:23-33.
27. Yoshie, O., T. Imai, and H. Nomiyama. (1997). Novel lymphocyte-specific CC chemokines and their receptors. *J Leukoc Biol* 62, no. 5:634-44.
28. Yoshida, R., M. Nagira, T. Imai, M. Baba, S. Takagi, Y. Tabira, J. Akagi, H. Nomiyama, and O. Yoshie. (1998). EBI1-ligand chemokine (ELC) attracts a broad spectrum of lymphocytes: activated T cells strongly up-regulate CCR7 and efficiently migrate toward ELC. *Int Immunol* 10, no. 7:901-10.

VII

B-1 Cells and Antibodies

Early and Natural Antibodies in Non-mammalian Vertebrates

M. F. Flajnik and L. L. Rumfelt
Department of Microbiology and Immunology, University of Maryland at Baltimore School of Medicine, 655 West Baltimore St. (13-009), Baltimore, MD 21201-1559.

Introduction

In mammals, there is good evidence that the types of antibodies produced early in development are different from those found at later stages [reviewed in 1]. Furthermore, the lymphocytes that express such antibodies comprise a unique subpopulation dubbed B1 cells, believed by most investigators to arise from early dedicated stem cells. B1 cells have unusual properties such as lifelong self-renewal in defined microenvironments, and most importantly, an "intentional" self-reactivity and polyreactivity of their immunoglobulin (Ig) receptors. Recent data would even argue that self antigens must interact with cell-surface Ig to perpetuate the B1 cell population(s) [2]. The function of B1 cells is not known, but they might be important for homeostasis of the immune system, innate protection against pathogens, and/or the mopping up of debris inside the body [see this issue!]. The discovery of B1 cells, γ/δ T cells, and two populations of B2 or conventional B cells, prompted the Herzenbergs in 1989 to propose the "Layering Hypothesis [3,4]" in which lymphocytes with "innate" receptors (B1 cell Ig, γ/δ TCR, and perhaps NK cells) arise early in ontogeny and those with "adaptive" receptors (B2 cell Ig, α/β TCR) develop later. It was further suggested by Janeway that "innate" antibodies and TCRs might precede adaptive antigen receptors in phylogeny, perhaps even before the emergence of the adaptive immune system [5]. Studies of non-mammalian vertebrates could help to determine whether these propositions have validity, first to examine whether such different populations of lymphocytes exist in most vertebrates, and secondly to determine whether some non-mammalian vertebrates or invertebrates might have only one type of receptor of the innate class.

Here we will concentrate on studies of natural and early antibodies carried out in two ectothermic (cold-blooded) creatures, the amphibian *Xenopus* and the cartilaginous fish (sharks) [reviewed in 6]. We survey these few studies and provide evidence that, indeed, antibodies expressed in early life are qualitatively and quantitatively different from those found in the later, mature Ig repertoire.

Xenopus antibodies

Of the major classes of vertebrates including cartilaginous fish, bony fish, amphibians, reptiles, birds and mammals, the amphibians mark an important transitional taxon regarding the immune system. They are the first animals to have a "translocon" (i.e. mouse and human-like) organization of their heavy (H) and light (L) chain genes, and are also the first class of animals that undergoes the heavy chain switch [reviewed in 6]. Of greatest appeal to most immunologists, however, is the transformation of the immune system when tadpoles metamorphose into adult frogs [reviewed in 7].

Tadpole antibodies
Du Pasquier demonstrated long ago that tadpoles, having fewer than one million lymphocytes, are capable of making a heterogeneous response to hapten antigens [8]. Du Pasquier and his colleagues later showed in isogeneic animals that the specific tadpole antibody repertoire is more restricted and different from the one expressed in adult life [9]. In large part, this change in repertoire appears to be due to the lack of N-region addition to the junctions of H chain rearranging gene segments during tadpole life [10]. The repertoire does not change until the mature tadpole B cells turn over at metamorphosis and presumably a new wave of differentiating cells appears that seeds the adult lymphoid tissues [7]; tadpoles that are inhibited from undergoing metamorphosis for up to one year (with agents that block thyroid function) maintain this early repertoire [11]. The lack of N-regions is correlated with low TdT expression in the immunocompetent larvae [12]. Hsu has suggested that an extensive tadpole antibody repertoire may not be advantageous due to potential cross-reactivities of antibodies induced in tadpoles on adult-specific self antigens expressed at the metamorphic climax [12].

In frogs, and in fact all vertebrates so far examined apart from the cartilaginous fish, there is a large number of V_H gene families [6,13,14]. The earliest antibodies produced in *Xenopus* are of the V_{H1} family, the equivalent of the V_{H3} or V_{H7183} (group III) genes of man and mouse respectively [15]. The tadpole then proceeds to express the full complement of V_H genes found in the adult (but without N-additions in the V-D-J junctions). In addition, a particular D segment is used in a preferential reading frame early in development. This DJ junction shows homology-based joining in 3 nucleotides of these two segments, but it is generated in mature cells by virtue of selection since many genomic sequences with other frames of this particular join can be detected in the early B cells. Thus, it indeed appears to be true that this gene segment is selected for some reason early in ontogeny, and that at least some portion of the tadpole CDR3 and V_H repertoire is under a different type of selection as compared to adults.

A CD5 candidate has been uncovered in *Xenopus* [16]. Surprisingly it is not expressed by B cells (spleen, intestine and peripheral blood were tested), but is found on thymocytes and T cells. If B cells are stimulated in the presence of T cells they can be induced to express the putative CD5; the molecule's function and cDNA have yet to be uncovered. However, the preliminary studies would argue that CD5 is not a marker for B cell subpopulations in *Xenopus*.

All bony fish species investigated also have a large number of V_H families (at least eleven, ref 17). No study of their differential usage during ontogeny has been attempted, but is now possible with the discovery of primary and secondary lymphoid tissues in several species [18,19]. The highest expression of RAG and TdT outside of the thymus is in the pronephros (head kidney), long suspected to be a primary lymphoid tissue in this species. Analysis of V_H families in developing B cells obviously will be of great interest.

Shark antibodies

Sharks are members of the vertebrate class cartilaginous fish (Chondricthyes: sharks, skates, and rays) which is the oldest to have an adaptive immune system rooted on Ig, TCR and major histocompatibility complex (MHC). There are at least three Ig isotypes in representative species, IgM [20], NAR [21], and IgW (also referred to as IgNARC and IgX, refs 22-24). Members in each of these gene families are found in the so-called "cluster" configuration discovered by Litman and colleagues [25] with one V, one-three D, one J and one set of C exons in each cluster. It is estimated that there are between 100-200 IgM clusters [26], and many fewer NAR [21] and IgNARC [23] clusters in the species that have been examined. There is no switching between the isotypes, and it is hypothesized (but far from proven) that each isotype delineates a separate subpopulation of B cells [27]. There are also at least three light (L) chain isotypes, all in the cluster-type configuration as well with between 5-100 genes depending on the isotype and the particular species [reviewed in 28]. In most immunization studies antigen-specific IgM responses are not mammalian-like, with increases neither in affinity nor in titer of the secondary responses. The roles of NAR or IgW in antigen-specific responses have not been tested yet.

Natural antibodies
IgM was the first isotype to be discovered about 35 years ago and until recently it was believed to be the only isotype in sharks [20]. It is found in multimeric (19S) and monomeric forms (7S) in approximately equal amounts [29]. IgM levels are very high in sharks, making up at least 50% of the serum protein; thus,

in addition to its role in defense IgM is believed to be important for osmoregulation as well. Marchalonis and his colleagues have shown that "sticky" shark IgM antibodies exist that bind to a variety of antigens [30,31]. They have suggested that there may be a correlation between the expression of such antibodies in sharks and autoimmune mammals. The significance of such auto- and polyreactivity is not known, but the antibodies only make up a small percentage of the high levels of IgM found in the blood.

Leslie and Clem performed the most detailed experiment in the study of these natural antibodies [32]. They carefully separated the multimeric (19S) and monomeric (7S) IgM molecules from unimmunized animals and measured reactivity to DNP. The 19S and 7S DNP-specific antibodies isolated by precipitation were incapable of neutralizing DNP-phage, but at least the 19S antibodies were able to bind to DNP immunoabsorbents. Surprisingly, the natural 19S DNP-specific antibodies agglutinated both sheep and chicken erythrocytes, and this (poly)reactivity was not inhibited by free DNP hapten. The authors suggested that these low affinity, polyreactive antibodies might have a role as "carriers" for small molecules in the body. Marchalonis also suggested that shark natural antibodies to thyroglobulin were polyreactive [30,31]. The general conclusion is that a percentage of shark IgM antibodies can show the same type of auto- and polyreactivity as described for some mouse and human B1 cell gene products.

"Germline-joined" shark antibody genes
All shark IgM genes analyzed to date are in the aforementioned cluster organization with a V, 2 Ds and a J segment in each cluster [25,33]. Litman and his colleagues showed long ago that up to one-half of these 100-200 gene clusters are either completely or partially "germline-joined." However, in the shark species that he studied, the horned shark, he found no expression of such genes in adults, i.e. the conventional, rearranging genes always "win." He and his colleagues did not investigate expression during early development, so the question became two-pronged: Is the usage of V_H genes different in early ontogeny as compared to adults, and might the germline-joined genes be expressed preferentially early in development?

We recently addressed these questions by accident in the nurse shark when we began to study expression of all of the secreted antigen receptors (IgM 19S and 7S, NAR, IgW) in neonatal sharks. We found one particular Ig cluster to be expressed in high amounts in immunoprecipitation experiments with our panel of monoclonal antibodies. To our surprise, this shark antibody was unlike *bona fide* IgM in that it contains three constant (C) rather than four C domains found in IgM. The CH2 domain is deleted from this molecule, similar what occurred to mammalian IgG over evolutionary time [34]. Most interestingly, this V gene is completely "germline-joined," with unambiguous V, D, and J segments. This gene is expressed in the spleen and epigonal organ of neonatal animals, and the protein does not appear to be transferred from the mother via the yolk. Expression wanes in the spleen, but is still detectable in the epigonal organ, a tissue in which we also observe relatively high levels of RAG and TdT expression strongly suggesting that it is a primary lymphoid tissue in cartilaginous fish. We envision two possibilities for cells producing this neonatal Ig that are not mutually exclusive: 1) there is continual production of B cells positive for this "neonatal" antibody throughout life, possibly because this Ig gene cluster has a transcriptional advantage over the other conventional clusters, or 2) the epigonal organ serves as a reservoir for these (self renewing?) cells as well as providing a primary lymphoid tissue, i.e. it may serve a function similar to the mouse peritoneal cavity.

Primordial or derived?

Which of these characteristics of the frog and shark immune systems were found in the common ancestor of all the vertebrates and which are specific to the particular species? The lack of N-regions in the junctions of frog Ig genes and the expression of V_{H3}-like genes early in ontogeny certainly seem to be ancestral traits [35,36]. However, the persistence of the N-less antibodies throughout the entire life of the immunocompetent tadpole suggests an added complexity, perhaps related to the avoidance of autoimmunity at metamorphosis.

The shark natural antibodies surely seem to be similar to their mammalian counterparts and therefore this feature is also an old one. However, because IgM makes up about half of the serum protein in shark plasma, perhaps a higher relative level of these "sticky" natural antibodies affects the generation of adaptive immune responses and may partially explain the lack of a true memory response.

We hypothesize that the shark neonatal antibody described here for the first time has been selected for a function similar to that of the mammalian B1 cell-produced antibodies. That such an antibody is expressed early in development fits well with the "Layering Hypothesis" and suggests that the common ancestor of all vertebrates expressed different types of antibodies during the fetal and adult lives. However, the shark cluster-type organization may provide an advantage in selecting for particular germline-joined clusters whose products are useful in defense and/or homeostasis; in mammals, it seems we are "stuck" with preferential rearrangement and selection for the "innate" antibodies expressed from the single IgH locus. Chalk one up for the sharks!

Acknowledgments

Our work is supported by NIH grant RR06603. Discussions with Marilyn Diaz on the advantages of the shark Ig gene clusters over the mouse/human "translocon" organization were stimulating..

References

1. Youinou P, Jamin C, Lydyard PM (1999) CD5 expression in human B cell populations. Immunol Today 20:312-316
2. Hayakawa K, Asano M, Shinton SA, Gui M, Allman D, Stewart CL, Silver J, Hardy RR (1999) Positive selection of natural autoreactive B cells. Science 285:113-116
3. Herzenberg LA, Herzenberg LA (1989) Toward a layered immune system. Cell 59:953-954
4. Herzenberg LA, Kantor AB, Herzenberg LA (1992) Layered evolution in the immune system: a model for the ontogeny and development of multiple lymphocyte lineages. Ann N Y Acad Sci 651:1-9
5. Janeway CA (1992) The immune system evolved to discriminate infectious self from noninfectious self. Immunol Today 13:11-16
6. Du Pasquier L, Flajnik MF (1999) Origin and evolution of the vertebrate immune system In: Paul WE (ed) Fundamental Immunology, Fourth Edition. Lippincott-Raven, Philadelphia, pp 605-650
7. Rollins-Smith LA (1998) Metamorphosis and the amphibian immune system. Immunol Rev 166:221-230
8. Du Pasquier L (1970) Ontogeny of the immune response in animals having less than one million lymphocytes: the larvae of the toad *Alytes obstetricans*. Immunology 19:353-362
9. Du Pasquier L, Blomberg B, Bernard CC (1979) Ontogeny of immunity in amphibians: changes in antibody repertoires and appearance of adult major histocompatibility complex antigens in *Xenopus*. Eur J Immunol 9:900-906
10. Schwager J, Buerckert N, Courtet M, Du Pasquier L (1991) The ontogeny and diversification of the immunoglobulin heavy chain locus in *Xenopus*. EMBO J 10:2461-2470
11. Hsu E, Du Pasquier L (1992) Changes in the amphibian antibody repertoire are correlated with metamorphosis and not with age or size. Dev Immunol 2:1-6
12. Lee A, Hsu E (1994) Isolation and characterization of the *Xenopus* terminal deoxynucleotidl transferase. J Immunol 152:4500-4507
13. Haire RN, Ohta Y, Litman RT, Amemiya CT, Litman GW (1990) Eleven distinct V_H gene families and additional patterns of sequence variation suggest

a high degree of immunoglobulin gene complexity in a lower vertebrate, *Xenopus laevis*. J Exp Med 171:1721-1737
14. Schwager J, Buerckert N, Courtet M, Du Pasquier L (1989) Genetic basis of the antibody repertoire in *Xenopus*: analysis of the Vh diversity. EMBO J 8:2989-3001
15. Mussmann R, Courtet M, Du Pasquier L (1998) Development of the early B cell population in *Xenopus*. Eur J Immunol 28:2947-2959
16. Jurgens JB, Gartland LA, Du Pasquier L, Horton JD, Gobel TW, Cooper MD (1995) Identification of a candidate CD5 homologue in the amphibian *Xenopus laevis*. J Immunol 155:4218-4223
17. Andersson E, Matsunaga T (1998) Evolutionary stability of the immunoglobulin heavy chain variable region gene families in teleosts. Immunogenetics 47:272-277
18. Hansen JD (1997) Characterization of rainbow trout terminal deoxynucleotidyl transferase structure and expression. TdT and RAG1 co-expression define the trout primary lymphoid tissues. Immunogenetics 46:367-375
19. Hansen JD, Zapata AG (1998) Lymphocyte development in fish and amphibians. Immunol Rev 166:199-220
20. Marchalonis JJ, Edelman GM (1965) Phylogenetic origins of antibody structure. I. Multichain structure of immunoglobulins in the smooth dogfish (*Mustelus canis*). J Exp Med 122:601-618
21. Greenberg AS, Avila D, Hughes M, Hughes A, McKinney EC, Flajnik MF (1995) A new antigen receptor gene family that undergoes rearrangement and extensive somatic diversification in sharks. Nature 374:168-173
22. Bernstein RM, Schluter S, Shen S, Marchalonis JJ (1996) A new high molecular weight immunoglobulin class from the carcharhine shark: implications for the properties of the primordial immunoglobulin. Proc Natl Acad Sci 90:2385-2388
23. Greenberg AS, Hughes AL, Guo J, Avila D, McKinney EC, Flajnik MF (1996) A novel "chimeric" antibody class in cartilaginous fish: IgM may not be the primordial immunoglobulin. Eur J Immunol 26:1123-1129
24. Anderson MK, Strong SJ, Litman RT, Luer CA, Amemiya CT, Rast JP, Litman GW (1999) A long form of the skate IgX exhibits a striking resemblance to the new shark IgW and IgNARC genes. Immunogenetics 49:56-67
25. Hinds KR, Litman GW (1986) major reorganization of immunoglobulin V_H segmental elements during vertebrate evolution. Nature 320:546-549
26. Kokubu F, Hinds K, Litman RT, Shamblott MJ, Litman GW (1987) Extensive families of constant region genes in a phylogenetically primitive vertebrate indicate an additional level of immunoglobulin complexity. Proc Natl Acad Sci 84:5868-5872
27. Flajnik MF, Rumfelt LL (2000) The immune system of cartilaginous fish. Curr Top Microbiol Immunol, in press

28. Litman GW, Anderson MK, Rast JP (1999) Evolution of antigen-binding receptors. Annu Rev Immunol 17:109-147
29. Clem LW, Small PA (1967) Phylogeny of immunoglobulin structure and function. I. Immunoglobulins of the lemon shark. J Exp Med 125:893-920
30. Marchalonis JJ, Hohman VS, Thomas C, Schluter SF (1993) Antibody production in sharks and humans: a role for natural antibodies. Dev Comp Immunol 17:41-53
31. Marchalonis JJ, Schluter SF, Bernstein RM, Hohman VS (1998) Antibodies of sharks: revolution and evolution. Immunol Rev 166:103-122
32. Leslie GA, Clem LW (1970) Reactivity of normal shark immunoglobulins with nitrophenyl ligands. J Immunol 105:1546-1552
33. Kokubu F, Litman RT, Shamblott MJ, Hinds K, Litman GW (1988) Diverse organization of immunoglobulin V_H loci in a primitive vertebrate. EMBO J 7:3413-3422
34. Frazer JK, Capra JD (1999) Immunoglobulins: structure and function. In: Paul WE (ed) Fundamental Immunology, Fourth Edition, Lippincott-Raven, Philadelphia, pp37-74
35. Perlmutter RM, Kearney JF, Chang SP, Hood LE (1985) Developmentally controlled expression of immunoglobulin VH genes. Science 227:1597-1601
36. Feeney AJ (1990) Lack of N regions in fetal and neonatal mouse immunoglobulin V-D-J junctional sequences. J Exp Med 172:1377-1390

Polyreactive Antibodies and Polyreactive Antigen-Binding B (PAB) Cells

A.L. Notkins, M.D.
Experimental Medicine Section, Oral Infection and Immunity Branch, National Institute of Dental and Craniofacial Research, National Institutes of Health, 30 Convent Drive, Bethesda, Maryland 20892

Polyreactive Antibodies.

The development of hybridoma technology in the late 1970s made it possible to study the fine specificity of individual antibody (Ab) molecules. The majority of these Abs showed high specificity and affinity for immunizing antigens (Ags). In the early 1980s, working independently, studies from our laboratory at the National Institutes of Health [1-3] and S. Avrameas' Laboratory at the Institut Pasteur [4,5] showed that some of these monoclonal Abs had the capacity to bind to a variety of different Ags. By immortalizing human peripheral B lymphocytes with Epstein-Barr Virus (EBV) we found that 10-15% of the cells made broadly reactive Abs which we now refer to as polyreactive Abs. At first we thought that these Abs were autoantibodies because they bound to a number of different organs and cell types. It soon became apparent, however, that they were not true autoantibodies because they bound equally well to a variety of foreign Ags. **Fig. 1** shows by ELISA, several typical monoclonal polyreactive Abs that react to different degrees with a variety of self and foreign Ags [6]. The majority of polyreactive Abs are IgM, but some are IgG and IgA [7,8]. In contrast to polyreactive Abs, monoreactive Abs bind only to the immunizing Ag and not to the other Ags in the panel **(Fig. 1)**.

The affinity of polyreactive Abs for most Ags is very low [6,9,10]. As seen in **Table 1**, the dissociation constants (K_d) of polyreactive Abs vary with the Ag and range from 10^{-3} to 10^{-7} (mol/l). In contrast, monoreactive Abs bind to the immunizing Ag with K_d ranging from 10^{-7} to 10^{-11} and not to the other Ags in the panel.

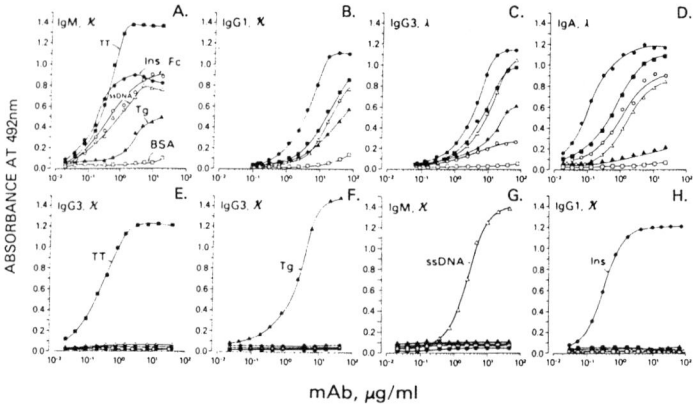

Fig. 1. Dose-dependent binding of four monoclonal polyreactive Abs and four monoclonal monoreactive Abs to different Ags as measured by ELISA [6].

A variety of studies have been carried out in our laboratory and other laboratories to determine the structural or molecular basis by which polyreactive Abs are able to bind to so many different antigens, whereas monoreactive Abs are unable to bind to these antigens. We now know that many of the V_H and V_L genes of polyreactive Abs are in or near germline configuration, but it should be emphasized that the V_H and V_L genes of a number of polyreactive Abs also show moderate to extensive somatic mutations [11-13]. The Fab fragments of these Abs maintain polyreactivity [14], and based on gene shuffling experiments and site-directed mutagenesis, the CDR3 region is particularly important [15,16]. The precise mechanism underlying polyreactivity, however, still remains unclear. One possibility is that the confirmational flexibility [17,18] at the Ag-binding site of polyreactive Abs is greater than that of monoreactive Abs, thereby allowing a greater number of Ags

DISSOCIATION CONSTANT (K_d, mole/l) OF HUMAN MAbs
FOR DIFFERENT ANTIGENS

MAbs	Selecting Antigen	Ligand			
		Ins	Tg	ssDNA	TT
IgG1 (k)	Ins	3.3×10^{-6}	5.5×10^{-7}	1.0×10^{-6}	4.0×10^{-7}
IgA (λ)	ssDNA	1.2×10^{-5}	$>5.0 \times 10^{-3}$	2.0×10^{-7}	3.3×10^{-6}
IgG1 (k)	Ins	5.0×10^{-7}	—	—	—
IgG1 (k)	Tg	—	9.6×10^{-10}	—	—
IgG1 (λ)	ssDNA	—	—	2.0×10^{-10}	—
IgM (k)	ssDNA	—	—	8.0×10^{-9}	—
IgG3 (k)	TT	—	—	—	3.0×10^{-11}

to bind but at lower affinity. Still another possibility is a diverse array of binding site structures at the combining site [19]. Crystallographic studies involving the binding of different Ags to the same polyreactive molecule might help elucidate the mechanism of polyreactivity.

The biological role of polyreactive Ab also is still unclear. If monoclonal polyreactive or monoclonal monoreactive Abs are injected into mice, the polyreactive Abs are cleared many times faster than the monoreactive Abs presumably because they bind to self-antigens [20]. As seen in **Fig. 2**, the half life of monoreactive IgM, IgA and IgG is 35, 26 and 280 hours respectively, while that

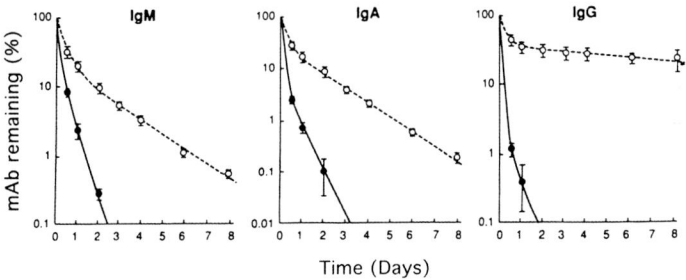

Fig. 2. Clearance of intravenously administered human monoreactive and polyreactive Abs in SCID mice [20].

of polyreactive IgM, IgA and IgM is 8, 8 and 10 hours respectively. **Fig. 3** shows that polyreactive Abs are present in the circulation, but not in free form. Incubation of sera with a panel of Ags results in little if any Ag-binding [21]. However, if the

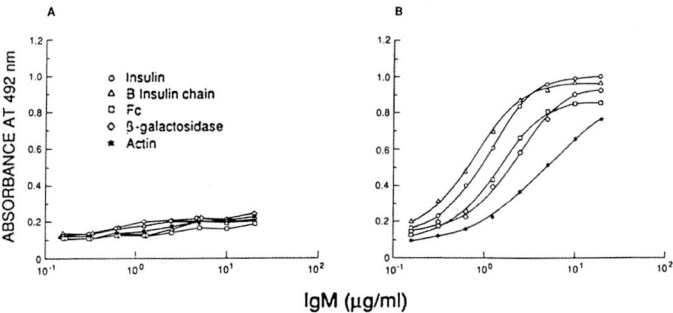

Fig. 3. Polyreactive Abs in the blood. Equal concentrations of (A) IgM from human sera and (B) affinity-purified IgM from the same sera [21]. Ag-binding determined by ELISA.

IgM in the sera is first affinity-purified, which results in dissociation of low-affinity Ags, the affinity-purified IgM shows substantial polyreactivity (**Fig. 3**). Adding back as little as a 1 in 10 dilution of normal sera to the affinity-purified IgM results in complete loss of Ag-binding and masks the polyreactive Ab [21].

Early in our work, we thought that polyreactive Abs might play a role in the first line of defense against foreign Ags [10]. This idea is now less appealing based on the fact that polyreactive Abs are of low affinity, poor immune percipitators, rapidly cleared from the circulation and exist in a masked form bound to self-Ags. Over the last few years we have focused on the issue of biological function, and are beginning to believe that perhaps what is important is not secreted polyreactive Abs, but the cells that make these Abs.

Polyreactive Antigen-Binding B Cells

If the cells that make polyreactive Abs are anything like the cells that make monoreactive Abs, one would expect polyreactive Ig molecules to be expressed on

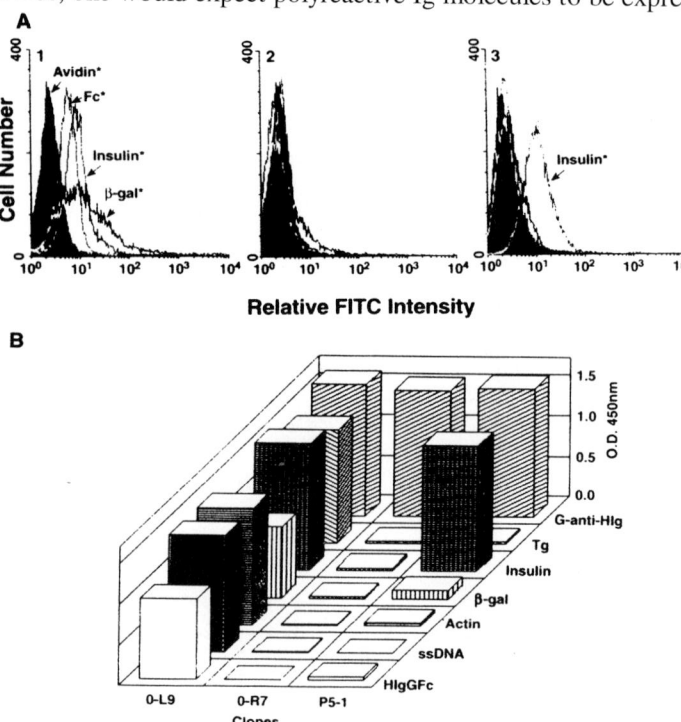

Fig. 4. (A) Binding of three different antigens to B cell clones producing (1) polyreactive Ig, (OL-9), (2) non-reactive Ig (O-R7) and (3) monoreactive Ig (P5-1). (B) Supernatants from the three clones shown in panel A were tested for reactivity with different Ags by ELISA [22].

the surface of these cells. To see if, in fact, this was the case, we studied by FACS analysis the binding of a panel of FITC-labelled Ags to the polyreactive Ab-producing hybridoma, O-L9. As seen in **Fig. 4**, Fc, insulin, and β-gal all bound to this hybridoma and the degree of binding varied depending on the Ag used [22]. In constrast, the insulin-specific monoclonal hybridoma, P5-1, bound only insulin and none of the other Ags. We refer to these polyreactive Ag-binding B cells as PAB cells.

To see if PAB cells actually exist in the peripheral circulation, we chose three FITC-labelled Ags: insulin, Fc and β-gal. All three Ags bound to varying degrees to peripheral B cells. We then sorted these cells into Ag-binding and non-Ag-binding cells and then transformed them with EBV. The cells then were cloned in microtiter wells by limiting dilution and the number of wells that made polyreactive Ab was determined. As seen in **Fig. 5**, five-to-ten times as many wells containing Ag-binding cells made polyreactive Abs as compared to the wells containing non-Ag-binding cells and most were IgM producers [22].

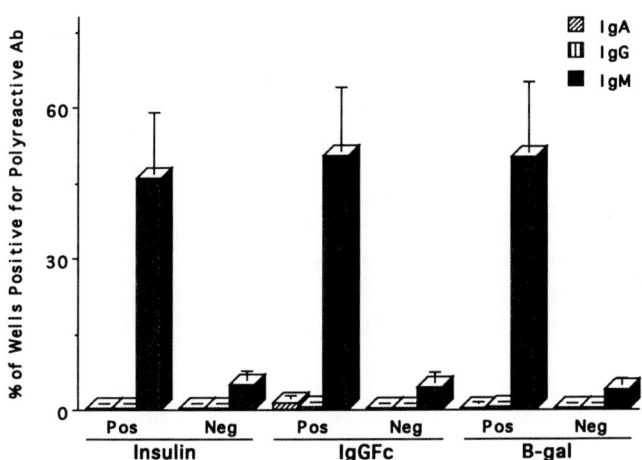

Fig. 5. Percentage of Ag-binding and non-Ag-binding human B lymphocytes making polyreactive Ab and their corresponding Ig isotype. Cells were first positively or negatively sorted based on Ag-binding. Sorted cells then were immortalized with EBV and supernatants tested by ELISA [22].

To examine the role of CD5 in polyreactivity we obtained $CD5^+$ peripheral B cells and then sorted them into fractions that bound Ags (i.e., insulin, Fc, β-gal) and fractions that did not bind Ags. We then transformed the cells with EBV, cloned by limiting dilutions and determined the percentage of wells that made polyreactive Ab [22]. As seen in **Fig. 6**, five-to-six times more wells from the $CD5^+$ Ag-binding groups made polyreactive Ab as compared to the wells from the $CD5^+$ non-

Ag-binding groups. Thus, Ag-binding appears to be a better marker for identifying B cells that make polyreactive antibody than CD5[+] or other cell surface markers currently used to characterize the B-1 subset [23]. Moreover, since some CD5[-] cells also can make polyreactive Abs [22,24,25] it would appear that PAB cells may overlap with, but do not necessarily belong to the B-1 subset.

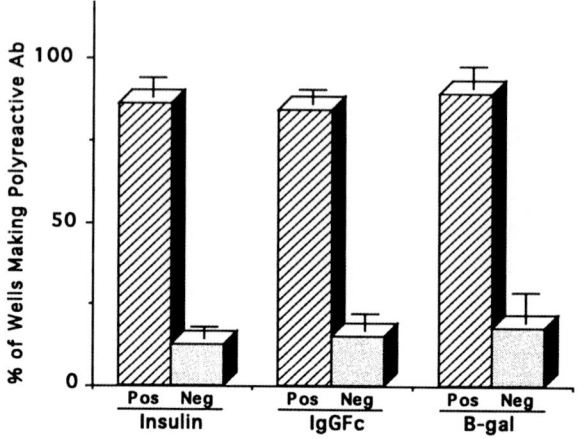

Fig. 6. Percentage of Ag-binding CD5[+] and non-Ag-binding CD5[+] cells making polyreactive Ab. Cells were positively and negatively selected based on Ag-binding, transfected with EBV and supernatants tested for Ag-binding by ELISA [22].

To see if Ags that bound to the polyreactive receptor could be processed, that is internalized and degraded into peptides, ^{125}I-lablelled Ags were used [26]. Fc, insulin and thyroglobulin all bound to the surface of these cells and were internalized within 1 hour. Degraded ^{125}I-labelled peptides, in the acid-soluble fraction, also were apparent within 1 hour and increased over the next 2 hours. The efficiency of binding and processing varied with the Ag and, as expected, was not as great as with control Ags (i.e., anti-IgM) which bound with high affinity [26].

The fact that PAB cells could bind and process Ags raised the possibility that these cells also could present Ags to T cells. Experiments in progress (Z. Wang, et al.) indicate that this is the case. T cells obtained from Ag-immuninzed animals proliferated in the presence of that Ag and a PAB cell hybridoma. Similarly, an Ag-specific T cell hybridoma proliferated in the presence of that Ag and PAB cells from normal animals. However, in the absence of B or T cell hybridomas, PAB cells from normal animals failed to stimulate T cells from Ag-immunized animals. The reason for this becomes apparent when one looks at the profile of surface markers, especially B7-1 and B7-2, on PAB cells sorted by the binding of β-gal, insulin or Fc (**Fig. 7**). In all three cases, there was no expression or at most very minimal expression of the co-stimulatory molecules B7-1 and B7-2. To see if low affinity Ag-binding would up-regulate B7, PAB cells from normal individuals were incubated with Fc. The binding of Fc failed to up-regulate B7, whereas high affinity Ag binding (i.e., anti-IgM), lead to the rapid up-regulation of B7 [26].

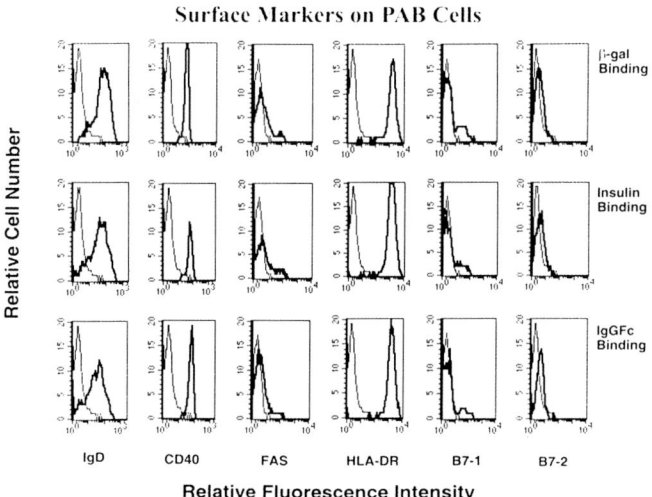

Fig. 7. Cells were sorted based on Ag-binding and then analyzed for B cell surface markers [26].

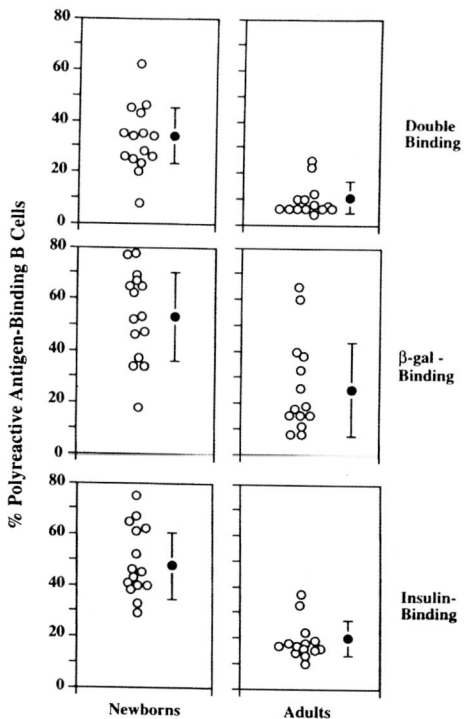

Fig. 8. Frequency of polyreactive Ag-binding B cells in the blood of newborn infants and adults [27].

To see if the percentage of PAB cells in normal subjects varied with age, we used the binding of β-gal, insulin or both as a marker for PAB cells [27]. **Fig. 8** shows that approximately 49% of cord blood B cells bound insulin, 54% bound β-gal and 38% bound both insulin and β-gal. In contrast, only 21% of adult blood B cells bound insulin, 28% bound β-gal and 10% bound both insulin and β-gal. Thus, PAB cells are present in high numbers in the peripheral circulation of both newborns and adults, but appear to be the predominant B cell type in the newborn.

The demonstration that PAB cells can bind, process and present many different Ags to T cells, but do so in the absence of the important co-stimulatory molecules B7-1 and B7-2 argues that under ordinary circumstances PAB cells do not stimulate T cells. Thus, PAB cells have all the properties required for inducing and maintaining immunological tolerance [28-32] and are particularly suited to pick-up the myriad of endogenous Ags to which the host might make an autoimmune response. Thus, it is tempting to speculate, that whereas high affinity monoreactive Ag-binding B cells provide defense against foreign invaders, low affinity polyreactive Ag-binding PAB cells may provide defense against autoimmune self-destruction by inducing and maintaining immunological tolerance.

References

1. Haspel MV, Onodera T, Prabhakar BS, McClintock PR, Essani K, Ray UR, Yagihashi S and Notkins AL (1983) Multiple organ-reactive monoclonal autoantibodies. Nature. 304:73-76
2. Satoh J, Prabhakar BS, Haspel MV, Ginsberg-Fellner F, Notkins AL (1983) Human monoclonal autoantibodies t at react with multiple endocrine organs. N Engl J Med 309: 217-20
3. Prabhakar BS, Saegusa J, Onodera T and Notkins AL (1984) Lymphocytes capable of making monoclonal antibodies that react with multiple organs are a common feature of the normal B cell repertoire. J. Immunol 133:2815-2817
4. Dighiero GP, Lymberi J-C, Mazie S, Rouyre GS, Butler-Browne RG, Whalen and Avrameas S (1983) Murine hybridomas secreting natural monoclonal antibodies reacting with self antigens. J Immunol 131:2267-2272
5. Ternynck T, and Avrameas S (1986) Murine natural monoclonal autoantibodies A study of their polyspecifities and their affinities. Immunol Rev 94: 99-112.
6. Nakamura M, Burastero SE, Ueki Y, Larrick JW, Notkins AL, and Casali P (1988) Probing the normal and autoimmune B cell repertoire with Epstein-Barr Virus. Frequency of B cells producing monoreactive high affinity autoantibodies in patients with Hashimoto's disease and systemic lupus erythematosus. J Immunol 141: 4165-4172
7. Nakamura M, Burastero SE, Notkins AL and Casali P (1988) Human monoclonal rheumatoid factor-like antibodies from CD5 (Leu-1)[+] B cells are polyreactive. J. Immunol 140: 4180-4186
8. Casali P, Burastero SE, Nakamura M, Inghirami G, Notkins AL (1987) Human lymphocytes making rheumatoid factor and antibody to ssDNA belong to Leu-1[+] B cell subset. Science 236: 77-81
9. Burastero SE, Casali P, Wilder RL and Notkins AL (1988) Monoreactive high affinity and polyreactive low affinity rheumatoid factors and produced by CD5[+] B cells from patients with rheumatoid arthritis. J. Exp Med 168: 1979-1992
10. Casali P and Notkins AL (1989) Probing the human B-cell repertoire with EBV: polyreactive antibodies and CD5[+] B lymphocytes. Annu Rev Immunol 7: 513-35

11. Sanz I, Casali P, Thomas JW, Capra J.D., Notkins AL (1989) Nucleotide sequences of eight human natural autoantibody VH regions reveals apparent restricted use of VH families. J Immunol 142: 4054-61
12. Harindranath H, Goldfarb S, Ikematsu H, Burastero SE, Wilder RL, Notkins AL and Casali P (1991) Complete sequence of the genes encoding the VH and VL regions of low- and high affinity monoclonal IgM and IgA1 rheumatoid factors produced by $CD5^+$ B cells from a rheumatoid arthritis patient. Int Immunol. 3:865-875
13. Harindranath N, Ikematsu H, Notkins AL and Casali P (1993) Structure of the VH and VL segments of polyreactive and monoreactive human natural antibodies to HIV-1 and Escherichia coli beta-galactosidase. Int Immunol 5: 1523-33
14. Cheung SC, Takeda S, and Notkins AL (1995) Both V_H and V_L chains of polyreactive IgM antibody are required for polyreactivity: Expression of Fab in E. coli. Clin and Exp Immunol 101: 383-386
15. Ichiyoshi Y, and Casali P (1994) Analysis of the structural correlates for antibody polyreactivity by multiple reassortments of chimeric human immunoglobulin heavy and light chain V segments. J Exp Med 180:885-895
16. Deng Y and Notkins AL (2000) Molecular determinants of polyreactive antibody binding: HCDR3 and cyclic peptides. Clin Exp Immunol 119: 69-76
17. Kramer A, Keitel T, Winkler K, Stocklein W, Hohne W, Schneider-Mergener J (1997) Molecular basis for the binding promiscuity of an anti-p24 (HIV-1) monoclonal antibody. Cell 91:799-809
18. Wedemayer GJ, Patten PA, Wang LH, Schultz PG, Stevens RC (1997) Structural insights into the evolution of an antibody combining site. Science 276:1665-9
19. Ramsland PA, Guddat LW, Edmundson AB, Raison RL (1997) Diverse binding site structures revealed in homology models of polyreactive immunoglobulins. J Commut Aided Mol Des 11:453-61
20. Sigounas G, Harindranath N, Donadel G, and Notkins AL (1994) Half-life of polyreactive antibodies. J Clin Immunol 14: 134-40
21. Sigounas G, Kolaitis N, Monell-Torrens E, and Notkins AL (1994) Polyreactive IgM antibodies in the circulation are masked by antigen binding. J Clin Immunol 14:375-381
22. Chen ZJ, Wheeler J and Notkins AL (1995) Antigen-binding B cells and polyreactive antibodies. Eur J Immunol 25:579-586
23. Kantor A (1991) A new no enclature for B cells. Immunol Today 12: 388
24. Kasaian MT, H Ikematsue and Casali P (1992) Identification and analysis of a novel human surface CD5⁻ B lymphocyte subset producing natural antibodies. J Immunol 148: 2690-702
25. Vernino LA, Pisetsky DS, Lipsky PE (1992) Analysis of the expression of CD5 by human B cells and correlation with functional activity. Cell Immunol 139: 185-97
26. Chen ZJ, Shimizu F, Wheeler J and Notkins AL (1996) Polyreactive antigen-binding B cells in the peripheral circulation are IgD^+ and B7⁻. Eur J Immunol 26:2916-2923
27. Chen ZJ, Wheeler CJ, Shi W, Wu AJ, Yarboro CH, Gallagher M and Notkins AL (1998) Polyreactive antigen-binding B cells are the predominant cell type in the newborn B cell repertoire. Eur J Immunol 28:989-994
28. Gilbert KM and Weigle WO (1994) Tolerogenicity of resting and activated B cells. J Exp Med 179: 249-258
29. Eynon EE, and Parker DC (1992) Small B cells as antigen-presenting cells in the induction of tolerance to soluble protein antigens. J. Exp. Med 175:131-138
30. Fuchs, EJ and Matzinger P (1992) B cells turn off virgin but not memory T cells. Nature 258:1156-1159
31. Eynon EE and Parker DC (1993) Parameters of tolerance induction by antigen targeted to B lymphocytes. J Immunol 151: 2958-64
32. Cassell DJ and Schwartz RH (1994) A quantitative analysis of antigen-presenting cell function: activated B cells st mulate naive GD4 T cells but are inferior to dendritic cells in providing costimulation. J Exp Med 180: 1829-40

A B-Cell Superantigen that Targets B-1 Lymphocytes

G. J. Silverman[1], S. Cary[1], M. Graille[2], V.E. Curtiss[1], R. Wagenknecht[1], L. Luo[1], D. Dwyer[1], C. Goodyear[1], A.L. Corper[3], E.A. Stura[2] and J.-B. Charbonnier[2]

[1]Department of Medicine, University of California, San Diego, La Jolla, CA / USA 92093-0663
[2]Département d'Ingénierie et d'Etudes des Protéines, CEA C.E. Saclay, 91191 Gif-sur-Yvette Cedex, France [3]Department of Immunology, The Scripps Research Institute, 10555 North Torrey Pines Road, La Jolla CA / USA 92037

Introduction

Superantigens are microbial products that target large proportions of the host lymphocyte pool by virtue of highly adapted interactions with conserved and highly represented variable region determinants of the lymphocyte antigen receptor. The structural basis for these interactions and their biologic consequences have been well described for several bacterial and viral products that specifically interact with V_β sites on the antigen receptors of T lymphocytes (TcR). These studies have demonstrated interactions with framework determinants that results in binding and stimulation of more than 5% of the host's lymphocytes, resulting in altered host immunity.

While the activities of natural proteins with comparable interactions through the antigen receptors of the B lymphocytes (BcR) have not been as extensively investigated, there is now substantial evidence that protein A of *Staphylococcus aureus* (SpA) has the properties of a superantigen for B lymphocytes. As outlined in the following sections, recent studies have defined the unconventional molecular basis for the BcR binding interactions of this bacterial toxin, and challenge studies in murine models have demonstrated that SpA can induce deletion of large V_H family-defined clonal sets resulting in altered host immunity. Susceptible clonal sets are represented within B-1 cell populations, and the suppressive effects on these B-1 clones appears to result in longterm adverse effects on host immune responses, even after only a limited interval of superantigen exposure.

SpA: A model B-cell superantigen

SpA is a 42kD membrane protein produced by most clinical isolates of this common pathogen. The extra-membrane portion of SpA is composed of five tandem repeats of 56-61 amino acid domains, that share greater than 85% protein sequence homology. Each of these domains is now known to possess two completely separate immunoglobulin (Ig) binding specificities: a specificity for the Fc gamma portion of most IgG antibodies (Fcγ); and a separate specificity for a conserved site contributed by the V_H region within the Ig that is responsible for antigen binding (Fab). Based on studies that correlated primary amino acid sequence with binding activity, this SpA specificity has been shown to be restricted to the products of the human V_H3 family, and its analogs in other mammalian species. In the mouse, the analogous VH families, S107, J606, 7183, and DNA4/ V_H10, commonly encode for SpA binding [1]. In vitro stimulation involving SpA has been shown to select for human V_H3-expressing B lymphocytes. In the past year, several important features regarding the molecular and cellular aspects of the B-cell superantigen activities of SpA have been elucidated, and an overview is provided in the following sections.

Overview on the structural basis of antigen recognition

To appreciate the unique structural features of SpA-antibody interaction, one must understand the fundamental basis for conventional antigen binding by an antibody. Antigen recognition is mediated by a combining site in the Fab fragment involving contributions from both the variable heavy (V_H) and variable light (V_L) domains. Within each variable region, there are three non-contiguous linear intervals of greatest variability, that have been termed hypervariable regions or complementarity determining regions (CDR). Separating these CDR are other sequence intervals, termed framework regions (FR), that are more highly conserved. In general, at both a nucleotide and a protein sequence level, the sequences for these framework regions are highly conserved amongst the products of a V gene family. In structural studies, the H and L chains of Ig have been shown to fold into a heterodimeric compressed β barrel structure. Herein, the CDR represent β loops that are juxtaposed at one end of the antibody to form a composite surface forming the classic antigen binding site (reviewed in [2,3]). The FR subdomains fold into relatively rigid β strands that maintain this overall Ig structure [4-6]. Amongst the CDR, the somatically generated heavy chain CDR3 (HCDR3) shows the greatest variation in length and sequence. Comparison of the crystal structures of a number of antibodies suggests that the CDRs, with the exception of HCDR3, adopt a limited number of conformations or "canonical structures" [7,8]. Analysis of the structures of the many antigen-antibody co-complexes solved to date indicate that in each antibody the residues from only 3-5 of the six CDR loops actually contribute contact sites to the interaction. However, in every case examined to date the HCDR3 makes important contributions to the binding of a conventional antigen. Thereby, in the human and murine antibody system a near unlimited potential range of antigen binding sites are created by accessing many different inherited antibody minigenes into combinatorial rearrangements that are modified by several additional somatic mechanisms to further structurally diversify the accessible repertoire of receptors. Hence, at any one time the immune repertoires contain a great diversity (i.e. estimated 10^5-10^6) of somatically generated conventional antigen binding sites, which is each generally represented relatively infrequently.

The extraordinarily high frequency of binding interactions between SpA and human monoclonal antibodies provided the first evidence of the special molecular features of the Fab-binding interactions of SpA [9;10]. In these surveys, most rearrangements of human genes from the large human V_H3 family were found to possess this binding activity, and despite extensive surveys not a single antibody from any of the other six human V_H families has exhibited this activity. Notably, in the several cases in which a V_H3 antibody is devoid of SpA binding activity, recent data suggest that this results from non-permissive replacement mutations affecting V_H region contact sites. Correlative binding studies with antibody primary sequences and heavy-light chain recombination experiments indicate that the SpA contact site on an antibody maps to the portion of the V_H region contributed solely by the V_H gene segment -- relative SpA activity is little effected by V_L usage or by the somatically generated heavy chain CDR3 that plays such a critical role in the binding of conventional ligands. These findings implicated the solvent exposed V_H FR1 and FR3 subdomains, which are the most highly conserved intervals among members of the V_H3 family, while there was no conservation in the CDR1 and CDR2 that vary greatly in length and composition among different V_H3 members. The SpA binding site was anticipated to involve a conformational surface, potentially requiring contacts from residues in both FR1 and FR3, as this binding activity is absent in denatured V_H3 antibodies or isolated H chains, and great loss of activity was documented in chimeric antibodies with shuffled V_H region subdomains [11]. The high frequency of this binding activity in the human repertoire is also rationalized by the unconventional structural correlates for Fab-mediated binding of SpA, which requires only the contributions from highly conserved framework intervals that are present in all members of the large V_H3 family.

Crystallographic analysis of the SpA-BcR binding interaction

The molecular basis for the Fab-mediated binding interaction of SpA has recently been determined in the structure of a co-crystal of SpA domain D and a human V_H3 IgM Fab solved to 2.7 Å [12]. In these studies, domain D was shown to fold into the same highly stable, three α helical bundle previously reported for domains B and E, and for the synthetic domain Z. The antibody in the complex is a V3-30-encoded IgM rheumatoid factor from a B-cell line isolated from synovium of a rheumatoid arthritis patient. Interactions with the Fab are formed through sidechain atoms in four V_H region β strands; B, C", D, and E, that are involved in bonds with residues in the helix II and helix III of domain D. Notably, the contact surface in the Fab is completely remote from the light chain, and it is also distant from the V_H region complementarity determining region (CDR) loops that create the conventional antigen binding site. In addition, neither the HV4 β loop of the V_H region FR3 nor the heavy and light chain constant regions are involved. These findings explain how an antibody can simultaneously interact with both protein A and a conventional antigen without competition [13].

In these crystallographic studies, each of the critical V_H region residues responsible for SpA binding derive from codons in the germline V_H gene segment and none are in the somatically generated HCDR3. In the human repertoire, most of these contact residues occur only in Ig from the human V_H 3 family, explaining the V_H family restriction for the binding of SpA. These findings explain why this "preimmune" binding specificity, linked to V_H3 genes that represent about half of inherited genes, is expressed so frequently by human peripheral B cells [14].

Amongst the V_H residues responsible for SpA binding, only position 57 commonly displays variation among germline V_H3 genes. At this site, there are three natural variations; isoleucine threonine or lysine, but each is permissive of SpA binding. The identification of the direct involvement of position 57 of the C" strand of FR3, as well as position 82a of the E strand of FR3, as a contact site in the co-crystal provides the structural explanation for mutagenesis studies demonstrating that non-conservative amino acid replacements at these sites abolish SpA binding activity [15;16].

Each of the five extra-membrane domains of SpA has been reported to bind V_H3 Ig [17;18], and our finding that the residues in domain D responsible for V_H3 interactions are also conserved in the other natural Ig-binding domains suggests that all of these domains bind Fab in a similar fashion. These findings therefore provide the structural explanation for reports that the interaction of native protein A with native pentameric IgM is 2-3 orders of magnitude greater than for monovalent Fab, as the increase of binding activity with IgM is likely due to the benefit of enhanced avidity [17;19;20].

Structural comparisons with known T-cell superantigens

Many aspects of the domain D-Fab interaction emulate the general features reported for co-complexes of TcR V_β regions and staphylococcal enterotoxins that have the biologic properties of T-cell superantigens [21-25]. In crystallographic studies, Mariuzza and coworkers have demonstrated that these toxins bind to the V_β region of the TcR, outside of the conventional peptide binding groove. About 25% of the contacts derive from the V_β FR1 and FR3 subdomains, a spatial relationship with some similarity to that of the SpA- Fab interaction. However, enterotoxin contacts generally derive from the V_β gene segment codons, and hydrogen bonds are made with α carbon backbone atoms from the CDR1, CDR2 and HV4 V_β loops of the TcR. This finding is distinct from the involvement of BcR side-chain atoms that are responsible for SpA binding. Hence, in the TcR the occurrence of variations in the conformations and spatial relationships of the β strands and the loops of a V_β region of the TcR, and not which specific amino acids are at these contact positions, determines whether an interaction with an enterotoxin can occur.

The structures of these staphylococcal products also differ. While a SpA domain is a triple α helical bundle structure, staphylococcal enterotoxins are composed of both α helical portions and β strands that each contribute contacts for interactions with the TcR. Hence, despite the great similarity of the TcR V_β regions and BcR V_H regions, these toxins hav evolved very different structural strategies to interact with their respective lymphocyte targets. This concept is even more intriguing in light of the fact that the same staphylococcal isolates often express both types of virulence factors.

B-cell superantigen-induced lymphocyte supra-clonal deletion

In recent reports, to provide the structural foundation for interpreting in vivo on the consequences of in vivo exposure of the murine immune system to SpA, we sought to identify the inherited murine V_H genes associated with binding activity. In the murine system, inherited V_H gene families display a much greater diversity, especially within the framework regions that predominantly define these families. Families that display greater structural homology are grouped into one of three subgroups or clans [26]. In surveys of the SpA binding activity of panels of murine monoclonal antibodies, we confirmed that murine clan III analogues of human V_H3 genes, that include S107, J606, 7183 and DNA4, all commonly convey SpA binding activity [1]. In these murine V_H families, there is conservation of almost every critical contact residue identified in the V_H 3-SpA complex, and in DNA4 and certain 7183 genes that have non-conservative variations at these contact sites, antibodies encoded by these genes commonly have weaker SpA binding activities. Other structurally related murine clan III families, like V_H11 and X24, are associated with even greater diversity at these critical positions (unpublished observations), and antibodies from these V_H families did not have detectable activity in our assays. In studies of a large number of clan I and II murine and human antibodies, not a single antibody has exhibited Fab-mediated SpA binding activity [1;19]. These correlative studies demonstrate the association of clan III genes with SpA binding activity, and also indicate that genes from certain clan III V_H families, and in particular the small S107 V_H family, are more commonly associated with stronger SpA binding interactions.

In our initial flow cytometric studies, we found that following a short duration of exposure to neonatal BALB/c mice to SpA (i.e. 100 µg in saline by intraperitoneal injection every other day for the first two weeks of life), there is a marked reduction in the representation in the spleen and bone marrow of B cells capable of Fab-mediated SpA binding interactions [27]. Based on binding studies with a novel clan III-specific chicken monoclonal antibody, termed LJ-26 [28], the induced defect was limited to the repertoire from clan III V_H families [28;29]. B cells expressing Ig recognized by markers for the S107 family demonstrated the greatest induced reductions [28;29], but the overall magnitude of this induced defect suggests that B cells expressing additional clan III V_H families are also likely to be affected by SpA treatment.

We also sought to better understand the functional properties of the SpA molecule responsible for its superantigen properties. For this purpose, we compared the immunobiologic effects of different forms of SpA domains that display varying levels of Fc and Fab binding activities. We therefore made comparisons between native pentameric SpA, and a chemically modified form without Fc binding activity, and a monomeric form of SpA (i.e. domain D), and genetically modified oligovalent forms with attenuated Fab binding activities and without Fc binding activity [29]. These studies documented that the native pentameric form of SpA, which has the strongest Fab binding activity, induced the most severe in vivo immunodefects. Importantly, Fc binding activity did not correlate with these defects.

The special susceptibility of B-1 cells

In our initial report, we demonstrated that susceptible neonatal B cells can be deleted by exposure to a form of SpA [27], and subsequently found that mature splenic B cells in adult mice are also sensitive to SpA-induced deletion [29]. To ensure that the measured decrement in SpA binding B cells was not due to SpA-induced down-regulation of membrane associated BcR, splenocytes were evaluated immediately ex vivo, and also following 24 and 48 hr of in vitro culture. These studies confirmed the specific supra-clonal deletion of susceptible clan III-expressing conventional B cells. Hence, the murine B cells deleted by in vivo exposure are determined predominantly by the BcR V_H family expression.

To determine how long the influence of SpA on clonal representation persisted after the last SpA dose, we examined splenocytes and bone marrow cells at different times after last neonatal treatment. We found that in BALB/c mice the frequency of B cells capable of Fab-mediated SpA binding activity began to revert within 3 days of the last dose, and appeared to normalize by 7-10 days post-treatment [27]. However, these findings seemed to present a paradox because, despite the evidence of normalization in the composition of the conventional splenic B-cell pool (i.e. mature B-2 cells), a specific SpA-induced defect still persisted in the representation of natural clan III-encoded IgM-secreting cells in the spleen, and to a lesser extent in the bone marrow, of these SpA-treated mice. Moreover, following neonatal treatment these immune defects were long-lasting, and were uniformly documented even in mice more than 13 months after last immune exposure [27]. These findings led us consider whether the persistent defect was due to the deletion of B-1 cells, which represent only a minor component of splenic B cells (discussed below).

To better define the cellular targets responsible for the persistent induced immunodeficit, we also performed RT-PCR Southern blots [29]. When we surveyed commonly expressed murine V_H families, we found severe decreases in the level of splenic VH S107-μ rearrangement transcripts, while levels of other clan III families, J606, 7183 and X24, and the levels of clan I families; J558 and VGAM3, and the clan II family Q52, did not demonstrate significant changes. Using a real time kinetic cDNA measurement method, we confirmed that neonatal exposure resulted in 20-40 fold decrease in S107-μ levels, while exposure of adult mice to proportionately higher doses induced a mean 90% decrease in S107-μ levels compare to control treated mice. Using V region-specific serologic markers, a concurrent selective loss of circulating S107-encoded IgM was also documented in these mice [29]. These findings documented that B cells expressing S107-encoded BcR were the major targets for SpA-induced immunosuppression, but the absence of persistent changes in the frequency of conventional B cells suggested that the persistent immunodefect reflects the influence of SpA upon a special B-cell population responsible for natural immunity.

To better understand the consequences of immune exposure, we sought a transgenic Ig murine system that would allow further characterization of the cellular mechanisms of SpA induced immunosuppression. Based on the identification of a specific induced defect in S107 encoded Ig expression, we evaluated the response of transgenic Ig mice expressing the T15i "knock-in" gene [28]. The T15 B-cell clone produces a natural antibody that recognizes phosphorylcholine, and which is essential for protective immune defense from disseminated pneumococcal infection [30-32]. In these antibodies, the V_H region is encoded by a canonical S107.1-DFL16.1-J_H1 rearrangement, which also conveys high affinity SpA binding activity [1;33]. Furthermore, B cells expressing the exact T15 V_H – V_L encoded antibodies have been rigorously demonstrated in adoptive transfer studies to reside predominantly (if not solely) within the B-1 pool [34-37]. In T15i mice, generated in the laboratory of Dr. Klaus Rajewsky, the T15 V_H gene is linked to the IgM^a allotype, and co-expressed with diverse endogenous light chains [38;39]. Notably, in the peritoneal cavity the great majority of T15 V_H-expressing B cells have the phenotype of B-1b cells (i.e. B220$^+$, Mac-1$^+$, CD5$^-$), while expression of the T15 V_H region in the spleen is predominantly in cells with the phenotype more akin to conventional follicular B-cell (B220$^+$, IgD^{hi}, CD5$^-$).

We confirmed that splenic T15 V_H^+ B cells were deleted immediately after treatment, but when examined months after the last SpA exposure, the levels of splenic conventional B cells (B220$^+$ IgD^{hi}) expressing the T15 V_H transgene were similar to those of control treated mice [29]. Significantly, in the peritoneal B-1 pool, there were persistent dramatic decrements in the levels of these superantigen-binding B cells. The only plausible explanation for the post-treatment normalization in the spleen of potentially reactive conventional mature B cells is that they are replenished by the B-lymphoneogenesis that continues in the bone marrow throughout life. In contrast, the B-1 pool has been reported to be established during the first 4-6 weeks of life, and thereafter these B cells are self-sustaining [40-44]. Hence, after depletion of T15 V_H-expressing peritoneal B-1 cells this set cannot later be replenished. We speculate that treatment with SpA deletes these susceptible B-1 cells, resulting in a persistent "hole" in the B-1 repertoire because this anatomic site is filled with other competing non-reactive B-1 cell clones. Using either V_H specific serologic markers, similar outcome was also demonstrated after SpA treatment of mice with polyclonal B cell compartments. These studies indicate that any B cell expressing an appropriate V_H gene may be susceptible to negative selection by this B-cell superantigen, but that B-1 cells are specially prone to the induction of long lasting immune defects.

The sensitivity of certain B-1 clones to SpA-induced effects has consequences for host immunity, as B-1 cells are responsible for the production of natural antibodies and protection from viral and bacterial pathogens [45;46]. We have found that neonatal treatment can permanently decrease circulating levels of natural antibodies to phosphorylcholine [29], which have been shown to contribute to immune defense to virulent pneumococcal infections and other microbial pathogens [47]. Moreover, neonatal treatment results in specific tolerance to both thymus-independent type 2 (TI-2) and T dependent forms of phosphorylcholine, while responses to $\alpha 1,3$ dextran, an immunogen that elicits a response dominated by a clan I/J558 family-encoded B-cell clone, remain intact. These studies confirm the special susceptibility of immune defenses linked to V_H family-defined B-1 cell clones to this bacterial toxin.

Evolutionary implications for SpA activity

Staphylococcus aureus is a common pathogen for many species, but this oligomeric protein, SpA, is not essential for the cell cycle of the bacterium and plays no known role in its metabolism. Consequentally, the highly refined Ig-binding activities of this virulence factor can only be viewed as relevant to its capacity to impair host defenses. While in vitro and in vivo membrane BcR crosslinking agents can induce the proliferation of sensitive B cells, other BcR cross-linking stimuli induce apoptosis [48]. The parameters that determine the outcome are not well defined, but it appears that the repetitive oligomeric structure of the Ig-binding domains of the secreted form of SpA have been optimized for the negative selection of reactive host B lymphocytes.

From a structural perspective, the V_H surface defined by the SpA binding site has been highly conserved in the immune system of diverse species, and analogues of clan III V_H genes are represented in species as primitive as the shark [6;26]. Moreover, antibodies with near perfect SpA binding motifs in their V_H framework regions are identifiable in nearly every characterized mammalian species, and the Ig from all mammals evaluated to date display Fab-mediated SpA binding activity (G. Silverman, manuscript in preparation). Moreover, this same V_H clan III-specific motif is also identifiable in the antibody genes such as the $V_H 1$ family of the amphibian, *Xenopus laevi* [49], which we have found also has Ig with the SpA binding activity. Current evidence suggests that the efficient T-dependent B cell responses associated with germinal center reactions only developed with the appearance of mammals, and this became a new layer overlayed on pre-existing facets of the adaptive immune system [50]. Therefore, it is conceivable that the ancient microbial pathogen, *Staphylococcus aureus*, first developed SpA as a virulence factor adapted to hosts that possess only very primitive immune systems. While the functional capacity of the immune system of non-mammalian species has not been extensively investigated, it is plausible that B-cell compartment of species like *Xenopus* include the equivalent of B-1 cells, which express evolutionarily selected binding specificities that aid host immune defenses. Therefore it is plausible that this microbial pathogen has developed a virulence factor that targets primordial and evolutionarily conserved immune defenses within the B-cell compartment of the adaptive immune system. In our current studies, we hope to test these hypotheses by examining immune responses of other species to SpA. Our studies also seek to better understand the molecular properties and limitations of immune interactions with variant SpA proteins. Our longterm goals are to better understanding the functional capacities of the different components of the B-cell compartment, and exploring the possibility of harnessing B-cell superantigens as therapeutic agents.

References

1. Cary, S., M.R. Krishnan, T. Marion, and G.J. Silverman. 1999. The murine clan V_H III related 7183, J606 and S107 and DNA4 families commonly encode for binding to a bacterial B cell superantigen. *Mol. Immunol.* 36:769-776.

2. Wilson, I. A., R.L. Stanfield. 1993. Antibody-antigen interactions. *Current Biology* 4: 857-867.

3. MacCallum, R. M., A.C. Martin, and J.M. Thornton. 1996. Antibody-antigen interactions: contact analysis and binding site topography. *J Mol Biol* 262: 732-45.

4. Tutter, A. and R. Riblet. 1988. Selective and neutral evolution in the murine Igh-V locus. *Curr.Top.Microbiol.Immunol.* 137:107-115.

5. Kirkham, P. M. and H.W. Schroeder, Jr. 1994. Antibody structure and the evolution of immunoglobulin V gene segments. *Semin.Immunol.* 6: 347-360.

6. Kirkham, P.M., F. Mortari, J.A. Newton, and H.W.J. Schroeder. 1992. Immunoglobulin VH clan and family identity predicts variable domain structure and may influence antigen binding. *EMBO J.* 11:603-609.

7. Chothia, C. and A.M. Lesk. 1987. Canonical structures for the hypervariable regions of immunoglobulins. *J.Mol.Biol.* 196:901-917.

8. Chothia, C., A.M. Lesk, A. Tramontano, M. Levitt, S.J. Smith-Gill, G. Air, S. Sheriff, E.A. Padlan, D. Davies, and W.R. Tulip. 1989. Conformations of immunoglobulin hypervariable regions. *Nature* 342:877-883.

9. Sasso, E.H., G.J. Silverman, and M. Mannik. 1991. Human IgA and IgG F(ab')2 that bind to staphylococcal protein A belong to the VHIII subgroup. *J.Immunol.* 147:1877-1883.

10. Sasso, E.H., G.J. Silverman, and M. Mannik. 1989. Human IgM molecules that bind staphylococcal protein A contain VHIII H chains. *J.Immunol.* 142:2778-2783.

11. Potter, K.N., Y. Li, and J.D. Capra. 1996. Staphylococcal protein A simultaneously interacts with framework region 1, complementarity-determining region 2, and framework region 3 on human VH3-encoded Igs. *J Immunol* 157:2982-8.

12. Graille, M., E. A. Stura, A. L. Corper, B. J. SuttonM. J. Taussig, J.-B. Charbonnier, and G. J. Silverman. Crystal structure of a *Staphylococcus aureus* protein A domain complexed with the Fab fragment of a human IgM antibody: structural basis for recognition of B-cell receptors and superantigen activity. Proc. Natl. Acad. Sci. (USA) 97:(10) 5399-5404.

13. Young, W.W., Y. Tamura, D.M. Wolock, and J.W. Fox. 1984. Staphylococcal protein A binding to the Fab of mouse monoclonal antibodies. *J.Immunol.* 133:3163-3166.

14. Silverman, G.J., M. Sasano, and S.B. Wormsley. 1993. Age-associated changes in binding of human B lymphocytes to a VH3-restricted unconventional bacterial antigen. *J.Immunol.* 151:5840-5855.

15. Randen, I., K.N. Potter, Y. Li, K.M. Thompson, V. Pascual, O. Forre, J.B. Natvig, and J.D. Capra. 1993. Complementarity-determining region 2 is implicated in the binding of staphylococcal protein A to human immunoglobulin VHIII variable regions. *Eur.J.Immunol.* 23:2682-2686.

16. Zhang, M., A. Majid, P. Bardwell, C. Vee, and A. Davidson. 1998. Rheumatoid factor specificity of a VH3-encoded antibody is dependent on the heavy chain CDR3 region and is independent of protein A binding. *J Immunol* 161: 2284-9.

17. Roben, P., A. Salem, and G.J. Silverman. 1995. VH3 antibodies bind domain D of staphylococcal protein A. *J.Immunol.* 154:6437-6446.

18. Jansson, B., M. Uhlen, and P.A. Nygren. 1998. All individual domains of staphylococcal protein A show Fab binding. *FEMS.Immunol.Med.Microbiol.* 20:69-78.

19. Sasano, M., D.R. Burton, and G.J. Silverman. 1993. Molecular selection of human antibodies with an unconventional bacterial B cell superantigen. *J.Immunol.* 151:5822-5839.

20. Silverman, G.J., R. Pires, and J.P. Bouvet. 1996. An endogenous sialoprotein and a bacterial B cell superantigen compete in their VH family-specific binding interactions with human Igs. *J.Immunol.* 157:4496-4502.

21. Malchiodi, E.L., E. Eisenstein, B.A. Fields, D.H. Ohlendorf, P.M. Schlievert, K. Karjalainen, and R.A. Mariuzza. 1995. Superantigen binding to a T cell receptor beta chain of known three-dimensional structure. *J.Exp.Med.* 182:1833-1845.

22. Li, H., A. Llera, E. L. Malchiodi, and R. A. Mariuzza. 1999. The structural basis of T cell activation by superantigens. *Annu Rev Immunol* 17: 435-66.

18. Fields, B. A., E. L. Malchiodi, H. Li, X. Ysern, C. V. Stauffacher, P. M. Schlievert, K. Karjalainen, and R. A. Mariuzza. 1996. Crystal structure of a T-cell receptor beta-chain complexed with a superantige. *Nature* 384: 188-92.

19. Li, H., A. Llera, D. Tsuchiya, L. Leder, X. Ysern, P.M. Schlievert, K. Karjalainen, and R.A. Mariuzza. 1998. Three-dimensional structure of the complex between a T cell receptor beta chain and the superantigen staphylococcal enterotoxin B. *Immunity* 9:807-16.

25. Li, H., A. Llera, and R.A. Mariuzza. 1998. Structure-function studies of T-cell receptor-superantigen interactions. *Immunol Rev* 163:177-86.

26. Schroeder, H.W.J., J.L. Hillson, and R.M. Perlmutter. 1990. Structure and evolution of mammalian VH families. *Int.Immunol.* 2:41-50.

27. Silverman, G.J., J.V. Nayak, K. Warnatz, S. Cary, H. Tighe, and V.E. Curtiss. 1998. The dual phases of the response to neonatal exposure to a VH family-restricted staphylococcal B-cell superantigen. *J.Immunol.* 161:5720-5732.

28. Cary, S., J. Lee, R. Wagenknecht, and G. J. Silverman. 2000. Characterization of superantigen induced clonal deletion with a novel clan III-restricted avian monoclonal antibody: Exploiting evolutionary distance to create antibodies specific for a conserved VH region surface. *J. Immunol. 164*: 4730-4741

29. Silverman, G.J., S. Cary, R. Aguilar, D. Dwyer, L. Linda, R. Wagenknecht, and V. E. Curtiss. A B-cell superantigen induced persistent "hole" in the B-1 repertoire. submitted.

30. Briles, D.E., M. Nahm, K. Schroer, J. Davie, P. Baker, J. Kearney, and R. Barletta. 1981. Antiphosphocholine antibodies found in normal mouse serum are protective against intravenous infection with type 3 streptococcus pneumoniae. *J.Exp.Med.* 153:694-705.

31. Briles, D. E., C. Forman, S. Hudak, and J. L. Claflin. 1982. Antiphosphorylcholine antibodies of the T15 idiotype are optimally protective against Streptococcus pneumoniae. *J.Exp.Med.* 156: 1177-1185.

32. Briles, D. E., J. L. Claflin, J. L., K. Schroer, and C. Forman. 1981. Mouse IgG3 antibodies are highly protective against infection with Streptococcus pneumoniae. *Nature* 294: 88-90.

33. Seppala, I., M. Kaartinen, S. Ibrahim, and O. Makela. 1990. Mouse Ig coded by VH families S107 or J606 bind to protein A. *J.Immunol.* 145:2989-2993.

34. Masmoudi, H., T. Mota-Santos, F. Huetz, A. Coutinho, and P.A. Cazenave. 1990. All T15 Id-positive antibodies (but not the majority of VHT15+ antibodies) are produced by peritoneal CD5+ B lymphocytes. *Int.Immunol.* 2:515-520.

35. Kaplan, D.R., J. Quintans, and H. Kohler. 1978. Clonal dominance: loss and restoration in adoptive transfer. *Proc.Natl.Acad.Sci.U.S.A.* 75:1967-1970.

36. Augustin, A.A., M.H. Julius, and H. Cosenza. 1977. Change in the idiotypic pattern of an immune response following syngeneic hemopoetic reconstitution of lethally irradiated mice. *In* Regulation of the Immune System. E. Sercarz, L. Herzenberg, and C. Fox, editors. Academic Press, New York. 195-199.

37. Wemhoff, G. A. and J. Quintans. 1987. Alterations of idiotypic profiles: the cellular basis of T15 dominance in BALB/c mice. *J Mol Cell Immunol* 3: 307-20.

38. Taki, S., M. Meiering, and K. Rajewsky. 1993. Targeted insertion of a variable region gene into the immunoglobulin heavy chain locus. *Science* 262:1268-1271.

39. Taki, S., F. Schwenk, and K. Rajewsky. 1995. Rearrangement of upstream DH and VH genes to a rearranged immunoglobulin variable region gene inserted into the DQ52-JH region of the immunoglobulin heavy chain locus. *Eur.J.Immunol.* 25:1888-1896.

40. Kantor, A.B., A.M. Stall, S. Adams, K. Watanabe, and L.A. Herzenberg. 1995. De novo development and self-replenishment of B cells. *Int.Immunol.* 7:55-68.

41. Hayakawa, K., R.R. Hardy, and L.A. Herzenberg. 1985. Progenitors for Ly-1 B cells are distinct from progenitors for other B cells. *J.Exp.Med.* 161:1554-1568.

42. Hayakawa, K., R.R. Hardy, A.M. Stall, and L.A. Herzenberg. 1986. Immunoglobulin-bearing B cells reconstitute and maintain the murine Ly-1 B cell lineage. *Eur.J.Immunol.* 16:1313-1316.

43. Forster, I. and K. Rajewsky. 1987. Expansion and functional activity of Ly-1+ B cells upon transfer of peritoneal cells into allotype-congenic, newborn mice. *Eur.J.Immunol.* 17:521-528.

43. Lalor, P.A., L.A. Herzenberg, S. Adams, and A.M. Stall. 1989. Feedback regulation of murine Ly-1 B cell development. *Eur.J.Immunol.* 19:507-513.

45. Herzenberg, L.A. and A.B. Kantor. 1993. B-cell lineages exist in the mouse. *Immunol. Today.* 14:79-83.

46. Ochsenbein A.F., T. Fehr, C. Lutz, M. Suter, F. Brombacher, H. Hengartner, and R.M. Zinkernagel. 1999. Control of early viral and bacterial distribution and disease by natural antibodies. *Science* 286:2156-2159.

47. Harnett, W. and M.M. Harnett. 1999. Phosphorylcholine: friend or foe of the immune system. *Immunology Today* 20: 125-129.

48. Parry, S.L., M.J. Holman, J. Hasbold, and G.G.B. Klaus. 1994. Plastic-immobilized anti-mu or anti-delta antibodies induce apoptosis in mature murine B lymphocytes. *Eur.J.Immunol.* 24:974-979.

49. Mussmann, R., M. Courtet, and L. Du Pasquier. 1998. Development of the early B cell population in Xenopus. *Eur J Immunol* 28:2947-59.

50. Herzenberg, L.A. 1989. Toward a layered immune system. *Cell* 59:953-954.

Myeloma Proteins that Bind Hsp65 (GroEL) are Polyreactive and are Found in High Incidence in Pristane Induced Plasmacytomas

M. Potter, G. Jones, W. DuBois, K. Williams, and E. Mushinski
Laboratory of Genetics, National Cancer Institute, National Institutes of Health, Bethesda, MD 20892 USA

Abstract

The myeloma proteins produced by 44 plasmacytomas (PCTs) recently induced by pristane in BALB/cAnPt and closely related PCT susceptible congenic strains of mice were isolated chromatographically and screened against a panel of 10 protein, nucleic acid and lipid antigens. This sample was highly unusual because 82% of the proteins had IgG isotypes. Nine of the proteins bound to Hsp65 (GroEL), and all of these were polyreactaive. Twenty-one of the myeloma proteins were polyreactive and bound two or more antigens in the panel, and five were monoreactive. Thus, an antigen binding activity was determined for 59% of these myeloma proteins.

Introduction

The term 'myeloma protein' indicates a 4-chain immunoglobulin (Ig) monomer composed of 2 light (L) and 2 heavy (H) chains that is secreted by a plasma cell tumor such as a multiple myeloma in humans, an immunocytoma in the rat or a plasmacytoma (PCT) in the mouse. Myeloma proteins are monoclonal immunoglobulins and, because they possess the structural properties of homogeneous antibody molecules [1], it is appropriate to ask what kinds of antigen binding activities they have. An extensive literature exists on the antigen binding activities of both human and mouse antigen binding myeloma proteins (for reviews see [2] and [3], respectively). The great majority of reports of antigen binding activity of myeloma proteins, however, are anecdotal insofar as none can account for more than a few percentage of cases in a series tested or screened. Experimental plasma cell tumor induction systems offer the potential for identifying relevant antigens, as these tumors are induced in genetically homogeneous (inbred) animals that are raised under similar environmental conditions. Relevant antigens are naturally occurring potential immunogens that can be demonstrated in the environment of the host organism. They are usually products of infectious microorganisms, antigens that are ingested. They can be generated autogenously from tissue and cellular destruction.

Previous attempts to identify antigens for mouse myeloma proteins have relied heavily on chance matches of myeloma protein and antigens available in the laboratory such as polysaccharides from various microbes and plants, synthetic antigens, e.g., chemically conjugated proteins and antigens used in clinical tests. The total number of active proteins in a series of myeloma proteins ranged between 5 and 10% [3]. The testing of autoantigens as targets for mouse myeloma proteins has been less intensively pursued. Schubert et al. in 1970 [4] described three myeloma proteins that bound to DNA, and more recently Diaw et al. [5] have described additional examples. Most of the work in the mouse system was done using precipitation in agar and before sensitive solid phase binding assays such as the ELISA method came into popular use.

In the present study we have attempted to screen a set of 44 myeloma proteins isolated from 55 primary PCTs induced by i.p. three (0.2-0.5 ml) injections of the paraffin oil, pristane (2,6,10,14-tetramethylpentadecane). The action of pristane is based on its property of being a poorly metabolized oil. Pristane in the peritoneal space causes a dramatic and sustained influx of macrophages and neutrophils that attempt to phagocytose the small oil droplets and surround the larger ones. Many of the complexes of cells and oil become deposited on peritoneal surfaces leading to the formation of the oil granuloma [6]. This new tissue is a chronic inflammatory tissue and plays an important role in plasmacytomagenesis by providing a micro-environment for the developing PCT. Various stages of PCT development are found morphologically in the oil granuloma [7].

In our conventional colony the incidence of PCTs at 300 days ranges from 50-70% in different experiments. These tumors were induced in BALB/cAnPt or BALB/cAnPt.DBA/2. PCT susceptible congenic strains and were collected during a two to three month period from eight different ongoing experiments. Pristane also induces the development of a rheumatoid arthritis in 20% of BALB/cAnPt more recently this has been 30+% and 70% of BALB/cJax mice in our laboratory [8] and a wide range of incidences in other inbred strains [9]. Heat shock proteins of bacterial origin have been implicated in the pathogenesis of autoimmune arthritis in rats (see review [10] by van Eden) and pristane induced arthritis (PIA) in mice [11]. The first association of a mycobacterial Hsp65 protein was made in the rat model [10]. Thompson and Elson and their collaborators in Bristol have shown that PIA in mice also is a T-cell driven disease process and that the critical antigen is a member of the Hsp60 class, Hsp65 of mycobacterial origin [12]. Elson et al. [11] proposed that microbial Hsp60 antigens lead to a process of epitope spreading in pristane injected mice whereby specific T cells that react with the murine homologue of Hsp60, i.e., Hsp58 can become activated. Further, it is postulated that when mouse anti-Hsp65 T cells are activated in joints they contribute indirectly to joint destruction [13]. These workers have also found that arthritic mice have elevated titers of Hsp65 binding antibodies in their serum [14,13]. In addition, the pristane associated epitope spreading and development of arthritis can be prevented in these mice by preimmunizing the mice with the mycobacterial Hsp65 or the immunodominant peptide derived from positions 261-271 in this protein. These findings led to testing myeloma proteins derived from PCTs induced in pristane injected mice for antigen binding activity to various Hsps. Heat shock proteins fall into four multi-membered classes of proteins, the Hsp60s, 70s, 90s and the low molecular weight forms [15] and within each the classes are structurally highly conserved across species from microbes to humans. The Hsp60 class is the most prevalent in microbial organisms. The first antigen that we used in the screen was the heat shock protein Hsp65 (GroEL) of *E. Coli* origin [16].

Heavy Chain Classes Expressed in the Myeloma Proteins

The ascites was examined electrophoretically for the presence of M-components using Beckman Paragon SPE 1% agarose gel system and the heavy chain isotype with specific heterologous antisera. A single isotype was found in 44 of the cases and the remaining 11 cases either had no predominating heavy chain isotype or presented with two or more isotypes. The screening with various antigens was carried out with the 44 monoclonal PCTs. Of these 8 were IgA, 6 were IgG1, 14 were IgG2a and 16 were IgG2b (Table 1).

Table 1. Polyreactivity among the 44 myeloma proteins

Isotype	No. myeloma protein (% of total)	No. Binding Hsp65 (GroEL)	Antigen binding activity for ten antigens		
			Poly-reactive (% of total)	Mono-reactive	None demon-strated
IgA	8 (18.1)	0	1 (2.2%)	0	7
IgG1*	6 (13.6)	1	1 (2.2%)	2**	1
IgG2a*	14 (31.8)	3	8 (18.1%)	0	4
IgG2b*	16 (36.4)	5	12 (25.0%)	3***	1
Totals	44*	9	22 (47.7%)		

*IgG1+ IgG2a + IgG2b = 81.8%
**1 with lysozyme; 1 with PC-BSA
***2 with lysozyme, 1 with actin

Surprisingly, the IgG isotypes comprised 82% of the cases. In previous studies the IgA isotype has been the predominant heavy chain class expressed by PCTs [17]. At present we cannot explain this dramatic shift. It should be noted, however, that our mouse colony is housed in a new facility at NIH, and this move is possibly associated with changes in environmental antigen exposure. We are currently examining another series of myeloma proteins to determine if this trend continues.

Binding to Hsp65 (GroEL)

Initial screening with whole ascites revealed a few reactive myeloma sera. The 44 ascites were then purified by protein A sepharose or by DEAE ion exchange chromatography. The IgG and IgA proteins eluted from these columns were much more effective in binding to the heterologous antigens on the plates. Previously, Sigounas *et al.* [18] have shown that polyreactive antibodies (see below) are masked in serum due to their ability to bind to self proteins. Accordingly, we routinely have used the isolated proteins in all of the serological reactions.

The recombinant *E. Coli* GroEL protein obtained from Sigma (Chaperonin 60) was adsorbed onto Immulon 2 HB flat bottom 96 well polystyrene plates (Dynex Corp., Chantilly, VA) and then reacted with myeloma proteins. The adsorbed myeloma proteins were detected by goat anti-mouse Ig serum 1010-05 (Southern Biotechnology Associates, Inc., Birmingham, AL) to which horse radish peroxidase (HRP) had been conjugated. The myeloma proteins were then titered beginning with a 5 µg/ml solution. The reactivity to GroEL varied with the isotype of the myeloma protein. There were no IgA's, 5IgG2b's, 3 IgG2a's and 1 IgG1. None of the proteins were monoreactive to Hsp65 (GroEL). The total Hsp65 reactors was 9 (20.5%).

The GroEL Binding Myeloma Proteins are Polyreactive

The nine myeloma proteins that reacted with GroEL were reacted with one or more nine other antigens: mycobacterial Hsp70, bovine muscle actin, chicken egg lysozyme (HEL), chicken ovalbumin, calf thymus ds DNA, polyinosinic acid, insulin, phosphorylcholine-BSA. A total of 21 of the 44 proteins (47.7%) were found to be polyreactive. This represents the highest number of myeloma proteins in any sample that have a biological activity. There was marked skewed distribution of the polyreactive samples according to isotype, as 75% were IgG2b, 25% IgG2a, 18.1% IgA and 2.2% IgG1 (Table 1). The high incidence of polyreactive myeloma proteins may be related to the unusual prevalence of IgG isotypes. We are currently testing another sample of IgG myeloma proteins from PCTs that were induced in other facilities.

Biological Implications

The ELISA assays system has not been systematically used in the mouse plasmacytomas/myeloma protein system. We chose the *E. Coli* GroEL (Hsp65, Chaperonin) protein as a starting point because of its involvement in pristane arthritis (P.I.A.).

The original objective in these experiments was to find relevant antigens to which myeloma proteins bind. While this antigen was a common target, it became clear that the strong reactors to Hsp65 (GroEL) were polyreactive to a variety of other protein, nucleic acid and lipid antigens. Most polyreactive antibodies that have been extensively studied have been derived from cell fusions of B cells with PCT cells. These B cells have come from a number of different sources including neonatal mice and humans, spleen cells from autoimmune mice, LPS stimulated B cells, $CD5^+$ B cells and B cells from Peyer's patches normal mice, and appear to be natural antibody producing B cells in both humans [19,20] and mice [21-23].

Polyreactive antibodies can bind self antigens, and a dramatic proof of this capability is the finding that IgM polyreactive antibodies in the plasma are masked, and only when the antibodies are isolated is polyreactivity demonstrable [18]. Also, the half life of IgM polyreactive antibodies is extremely short. These findings have suggested that immunoglobulin secretion was not the primary function of the polyreactive antibody producing B cells. Notkins and his group have suggested that these B cells have another action that is related to peripheral T cell tolerance [24,24]. They have found that B lymphocytes producing polyreactive antibodies fail to activate B7-1 and B7-2 antigens that play an active role in transmitting a second (proliferative) signal to T cells [24]. Nonetheless, these B cells actively process and present autoantigens to T cells but deliver a tolerance signal.

Mycobacterial Hsp65 is the specific member of the Hsp60 family that has been implicated in pristane arthritis, and the *E. Coli* Hsp65 (GroEL) has not as yet been implicated. In fact, preimmunization with Hsp65 (GroEL) does not protect mice from the subsequent development of arthritis [13]. The specificity of the anti-Hsp65 antibodies that appear in arthritic mice has not yet been determined. Further, the natural source of mycobacterial Hsp65 antigen in P.I.A. has not yet been explained. The levels of anti-Hsp65 antibodies in P.I.A. appear to be only modestly elevated [13]. The evidence from the PCT sample studied here is that the B cells producing polyreactive antibodies are responding to antigenic stimuli that is promoting isotype switching into IgG classes. The inducing agents for this drive may be endogenous autoantigens produced by cell death in the oil granuloma tissue. This study demonstrates that a high proportion of IgG myeloma proteins in BALB/cAnPt and related congenic strains produce polyreactive antibodies, some of which can bind to Hsp65 (GroEL).

References

1. Gearhart PJ, Johnson ND, Douglas R, Hood L (1981) IgG antibodies to phosphorylcholine exhibit more diversity than their IgM counterparts. Nature 291:29-34
2. Grunebaum E, Buskila D, Shoenfeld Y (1993) Serum human immunoglobulins (monoclonal gammopathies) as natural autoantibodies. In: Shoenfeld Y (ed) Natural autoantibodies: their physiological role and regulatory significance. CRC Press, Boca Raton, pp 81-108
3. Potter M (1977) Antigen-binding myeloma proteins of mice. In: Kunkel HG, Dixon FJ (eds) Advances in Immunology. Academic Press, New York, pp 141-211
4. Whittaker JA (1991) Solitary plasmacytoma. In: Delamore IW (ed) Multiple Myeloma and Other Paraproteinemias. Churchill Livingstone, Edinburgh, pp 193-203
5. Diaw L, Siwarski D, Coleman A, Kim J, Jones GM, Dighiero G, Huppi K (1999) Restricted immunoglobulin variable region (Ig V) gene expression accompanies secondary rearrangements of light chain Ig V genes in mouse plasmacytomas. J Exp Med 190:1405-1416
6. Cancro M, Potter M (1976) The requirement of an adherent cell substratum for the growth of developing plasmacytoma cells in vivo. J Exp Med 144:1554-1567
7. Potter M, Mushinski EB, Wax JS, Hartley J, Mock BA (1994) Identification of two genes on chromosome 4 that determine resistance to plasmacytoma induction in mice. Cancer Res 54:969-975
8. Potter M, Wax JS (1981) Genetics of susceptibility to pristane-induced plasmacytomas in BALB/cAn: reduced susceptibility in BALB/cJ with a brief description of pristane-induced arthritis. J Immunol 127:1591-1595
9. Wooley PH, Seibold JR, Whalen JD, Chapdelaine JM (1989) Pristane-induced arthritis. The immunologic and genetic features of an experimental murine model of autoimmune disease. Arthritis Rheum 32:1022-1030
10. van Eden W (1991) Heat-shock proteins as immunogenic bacterial antigens with the potential to induce and regulate autoimmune arthritis. Immunol Rev 121:5-28
11. Elson CJ, Barker RN, Thompson SJ, Williams NA (1995) Immunologically ignorant autoreactive T cells, epitope spreading and repertoire limitation. Immunol Today 16:71-76
12. Thompson SJ, Rook GA, Brealey RJ, Van der Zee R, Elson CJ (1990) Autoimmune reactions to heat-shock proteins in pristane-induced arthritis. Eur J Immunol 20:2479-2484
13. Thompson SJ, Francis JN, Siew LK, Webb GR, Jenner PJ, Colston MJ, Elson CJ (1998) An immunodominant epitope from mycobacterial 65-kDa heat shock protein protects against pristane-induced arthritis. J Immunol 160:4628-4634
14. Thompson SJ, Hitsumoto Y, Ghoraishian M, Van der Zee R, Elson CJ (1991) Cellular and humoral reactivity pattern to the mycobacterial heat shock protein hsp65 in pristane induced arthritis susceptible and hsp65 protected DBA/1 mice. Autoimmunity 11:89-95
15. Hartl FU (1996) Molecular chaperones in cellular protein folding. Nature 381:571-579
16. Braig K, Otwinowski Z, Hegde R, Boisvert DC, Joachimiak A, Horwich AL, Sigler PB (1994) The crystal structure of the bacterial chaperonin GroEL at 2.8 A [see comments]. Nature 371:578-586
17. Potter M (1972) Immunoglobulin-producing tumors and myeloma proteins in mice. Physiol Rev 52:631-719
18. Sigounas G, Kolaitis N, Monell-Torrens E, Notkins AL (1994) Polyreactive IgM antibodies in the circulation are masked by antigen binding. J Clin Immunol 14:375-381
19. Casali P, Schettino EW (1996) Structure and function of natural antibodies. Curr Top Microbiol Immunol 210:167-179
20. Chen ZJ, Wheeler J, Notkins AL (1995) Antigen-binding B cells and polyreactive antibodies. Eur J Immunol 25:579-586
21. Kaushik A, Lim A, Poncet P, Ge XR, Dighiero G (1988) Comparative analysis of natural antibody specificities among hybridomas originating from spleen and peritoneal cavity of adult NZB and BALB/c mice. Scand J Immunol 27:461-471

22. Dighiero G, Poncet P, Rouyre S, Mazie JC (1986) Newborn Xid mice carry the genetic information for the production of natural autoantibodies against DNA, cytoskeletal proteins, and TNP. J Immunol 136:4000-4005
23. Ternynck T, Avrameas S (1986) Murine natural monoclonal autoantibodies: a study of their polyspecificities and their affinities. Immunol Rev 94:99-112
24. Chen ZJ, Shimizu F, Wheeler J, Notkins AL (1996) Polyreactive antigen-binding B cells in the peripheral circulation are IgD+ and B7-. Eur J Immunol 26:2916-2923

VIII

B-CLL and B-Cell Tumors

Molecular Pathogenesis of B-Cell Chronic Lymphocytic Leukemia: Analysis of 13q14 Chromosomal Deletions

A. Migliazza[1], E. Cayanis[2], F. Bosch-Albareda[1], H. Komatsu[1], S. Martinotti[1], E. Toniato[1], S. Kalachikov[2], M.F. Bonaldo[3], P. Jelenc[2], X. Ye[2], A. Rzhetsky[2], X. Qu[2], M. Chien[2], G. Inghirami[4], G. Gaidano[5], U. Vitolo[6], G. Saglio[7], L. Resegotti[6], P. Zhang[2], M.B. Soares[3], J. Russo[2], S.G. Fischer[2], I.S. Edelman[2], A. Efstratiadis[2], and R. Dalla-Favera[1].
[1]Institute of Cancer Genetics, Columbia University, New York; [2]Genome Center, Columbia University, New York; [3]Interdepartmental Program in Human Molecular Genetics, The University of Iowa, Iowa City; [4]Department of Pathology, New York Medical Center, New York; [5]Department of Medical Sciences, A. Avogadro University of Eastern Piedmont, Novara, Italy; [6]Divisione di Ematologia, A.O. San Giovanni. Battista della citta' di Torino, Torino, Italy; [7]Dipartimento di Scienze Biomediche e Oncologia Umana, Ospedale San Luigi, Universita' di Torino, Orbassano, Italy.

B-cell chronic lymphocytic leukemia (B-CLL) is due to the monoclonal expansion of B lymphocytes characterized by the expression of the CD5 marker. These cells are non-proliferating and long lived, possibly because perturbed in their apoptotic program (for a review on the B-CLL biology, see Caligaris-Cappio and Hamblin, 1999). Recent studies focusing on the sequence status of the immunoglobulin variable (IgV) genes (Damle et al., 1999; Hamblin et al., 1999; Oscier et al., 1997) have shown that at least two B-CLL subsets can be identified: one would originate from a B cell clone with germline IgV sequences (possibly from the follicular mantle area around germinal centers), the second, harboring mutated IgV genes, would derive from cells that have been exposed to the somatic hypermutation machinery, possibly during the germinal center reaction. These two disease subsets also display distinct immunophenotypic and clinical features, the "unmutated" group being characterized by significantly higher CD38 expression and worse clinical outcome (Damle et al., 1999). In light of these observations, it is reasonable to anticipate that the pathogenesis of B-CLL might involve distinct pathways and cellular targets in different subsets of this disease. This chapter will try to summarize the current knowledge about the molecular pathogenesis of B-CLL. In particular, we will review current studies aimed at the identification of a tumor suppressor gene located on chromosome 13q14 that may be associated with the pathogenesis of B-CLL.

Genomic Instability and Molecular Lesions in B-CLL
Analogous to most tumors, the pathogenesis of B-CLL is thought to involve the progressive and clonal accumulation of multiple genetic lesions affecting proto-

oncogenes and tumor suppressor genes. As for other mature B-cell malignancies, the genome of B-CLL is relatively stable and not characterized by general random instability (Dalla-Favera et al., 1999). However, while in non-Hodgkin's lymphoma (NHL) the activation of proto-oncogenes is typically caused by non-random reciprocal and balanced chromosomal translocations, these lesions are rarely associated with B-CLL, suggesting a different mechanism causing genetic lesions in B-CLL. No gene has yet been found to be altered in B-CLL at significant frequency. In particular, none of the proto-oncogenes involved in the pathogenesis of other mature B cell malignancies, including cyclin D1, BCL-2, BCL-6, PAX-5 and c-MYC, are primarily altered in this malignancy (Gaidano et al., 1994; Lo Coco et al., 1994). Inactivation of the tumor suppressor gene p53 (on chromosome 17p13) and deletions/mutations of the ATM gene (on chromosome 11q22-23) have been reported in a fraction of cases, but such lesions seem to have been observed in late stages of the disease and therefore are likely to represent events associated with tumor progression (Gaidano et al., 1991; Bullrich et al., 1999; Stankovic et al., 1999; Starostik et al., 1998). Overall, chromosomal aberrations, often single, are observed in approximately half of the patients: a comprehensive review/reassessment of molecular cytogenetic data collected on a large series of B-CLL patients is given in (Dohner et al., 1999). The most common non-random chromosomal alteration associated with B-CLL is represented by chromosome 13q14 deletions (in about 50% of the patients), followed by structural aberrations of chromosome 11q (19% of the cases) and trisomy of chromosome 12 (15%). Albeit at lower frequencies, structural aberrations of chromosome 6q and 17p are also detected in B-CLL samples (9% and 8%, respectively).

Chromosome 13q14 Deletion in B-CLL and Multiple Human Cancers
Deletion of chromosome 13q14 represents an early clonal aberration in B-CLL and strongly suggests the presence, on this region, of a tumor suppressor gene whose loss or inactivation may be crucial for the leukemogenesis. In addition, this aberration is by far the most common chromosomal lesion identified in B-CLL, often appearing as the only detectable abnormality. For these reasons, we have focused our interest on this locus, with the ultimate goal of identifying the B-CLL associated tumor suppressor gene. A high-density clone-contig based physical map encompassing the deleted interval was generated. This allowed us to design a set of densely spaced markers (at approximately 30 kb intervals) that we used to study the presence and extent of genetic loss in B-CLL patients (Kalachikov et al., 1997). We identified somatic loss at 13q14 in 105/196 patients (54%), the loss being monoallelic in 81% of such cases and biallelic in the remaining 19%, suggesting in these latter cases complete inactivation of the putative suppressor gene. Together with several other reports summarized in Fig. 1 (Bouyge-Moreau et al., 1997; Bullrich et al., 1996; Corcoran et al., 1998; Kalachikov et al., 1997; Stilgenbauer et al., 1998), our study has led to the identification of a minimal region of

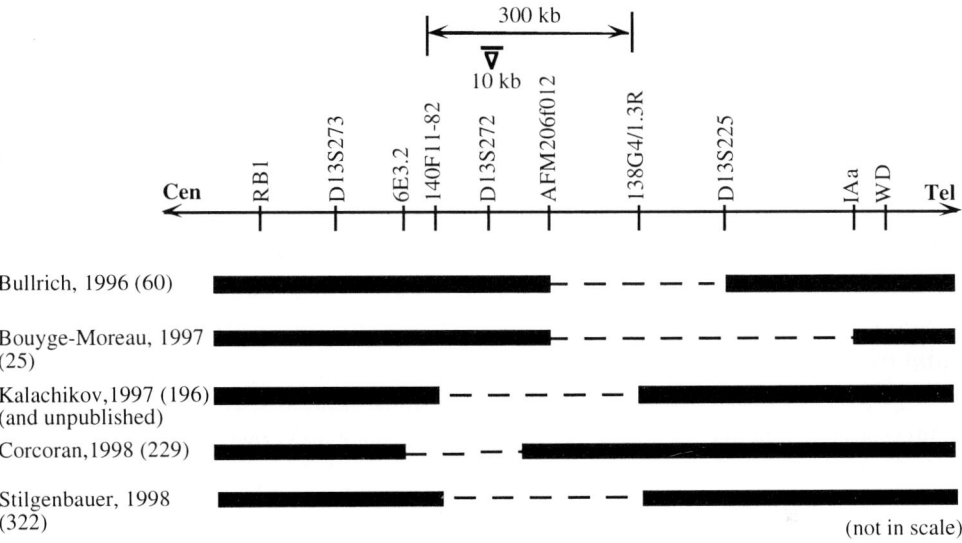

Fig. 1. Schematic representation of the genomic region involved in 13q14 deletions showing the minimal interval of deletion (MRD) identified by various groups. The analyzed region of chromosome 13q14 is represented, including the relevant molecular markers used to perform the deletion assessment. Dashed and continuous lines indicate regions of loss and retention, respectively. All together, more than 600 cases of B-CLL were characterized for genetic loss at 13q14 (the number of patients per each study is indicated in brackets). A minimal consensus region of deletion of approximately 300 kb, located between the Retinoblastoma 1 gene (RB1) and the Wilson's disease (WD) loci, could be identified, pointing to the putative candidate interval likely to include a tumor suppressor gene important for the B-CLL pathogenesis. This interval partially overlaps with the regions described in the studies from (Bouyge-Moreau et al., 1997) and (Bullrich et al., 1996) on one side, and by (Corcoran et al., 1998) on the other. A small empty arrow indicates a minimal consensus interval of only 10 kb described in Liu et al, 1997.

deletion (MRD) defined as the smallest interval of chromosomal deletion common to all the deleted samples. This chromosomal segment spans <300kb and represents the putative site of the B-CLL-associated tumor suppressor gene.

Further attention to chromosome 13q14 has been drawn more recently by several studies reporting its deletion in a variety of human tumor types in addition to B-CLL. Among lymphoid malignancies, 13q14 loss is found in NHL, of both B- and T- cell origin, with different frequencies in the various hystologic subtypes of this disease (the highest, 38-70%, reported for mantle cell lymphoma) (Liu et al., 1995; Rosenwald et al., 1999; Siu et al., 1999; Stilgenbauer et al., 1998), and in 30-76% of monoclonal gammopathies, including monoclonal gammopathy of undetermined significance and multiple myeloma (Avet-Loiseau et al., 1999a; Avet-Loiseau et al., 1999b; Cigudosa et al., 1998; Shaughnessy and Barlogie, 1999). Tumors of myeloid origin also display genetic loss at

13q14 (La Starza et al., 1998). Finally, deletion of 13q14 is found in an increasing variety of human tumors, including prostate cancer (Latil et al., 1999; Li et al., 1998; Ueda et al., 1999; Yin et al., 1999), head-and-neck-squamous cell carcinoma (Gupta et al., 1999), non-small cell lung cancer (Tamura et al., 1997), pituitary adenoma (Harada et al., 1999), hepatocellular carcinoma (Hammond et al., 1999). Although the area of chromosome 13q14 involved in the deletion awaits better refinement in most of these neoplasms, these observations may suggest the intriguing possibility that a common tumor suppressor gene may exist in this region.

Mapping of Genes on 13q14

Some progress has been made toward the identification of the B-CLL associated tumor suppressor gene that may be located on 13q14. First, since no known gene had yet been

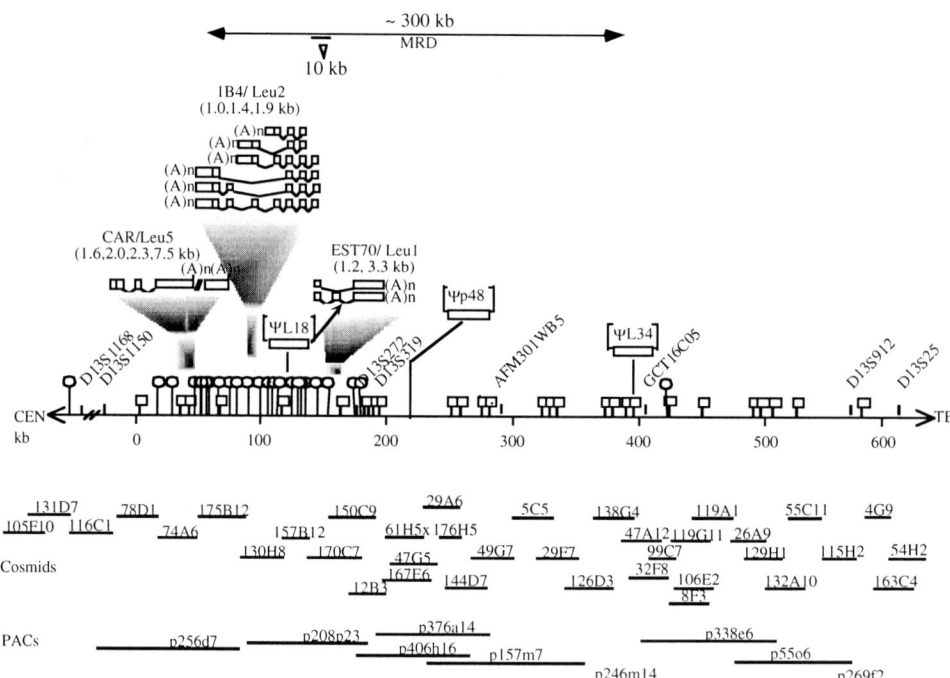

Fig. 2. Physical and transcriptional map of the B-CLL candidate interval. In the lower part of the figure, a minimum tiling-path of genomic clones (including cosmids and PACs) encompassing the region is shown. Genomic distances are shown at the bottom of the chromosome 13 line. On top of this line, exons (rectangles) and cDNA (circles) are indicated. The transcripts that were identified and characterized are blown up and the full-length transcripts exon composition is depicted. Pseudogenes are annotated with the Greek letter Ψ. On top, the arrow indicate the position of the 300 kb minimal region of deletion (MRD), as well as the mini-deletion of 10 kb from (Liu et al., 1997), respect to the genes.

mapped inside the candidate segment on 13q14, we sought to identify transcribed sequences located within or in the proximity of the identified minimal region of deletion, using a combination of techniques including exon trapping, cDNA selection and homology searches versus public databases (for this purpose, we have sequenced to 95% completion a total of 600 kb of DNA encompassing the candidate interval). The resulting transcriptional map of the B-CLL locus includes three non transcribed pseudogenes and three new transcripts (Fig. 2). Two genes, EST70 and 1B4 (Kalachikov et al., 1997) also called Leu1 and Leu2, respectively, in (Liu et al., 1997), contain only small (<70 amino acids) open reading frames and may represent non-coding RNAs (Erdmann et al., 1999). The main mRNA species of 1.2, 3.3 kb and 1.0,1.4, 1.9 kb, respectively, could be detected in various tissues including B-cell-derived cell lines; the corresponding full-length sequences were cloned, accounting for at least three and five exons, respectively. The third gene, CAR, CLL Associated RING finger, encodes for a 407 amino acid-long protein characterized by a "RING-finger-B-box-Coiled-Coil" motif (this gene was also partially characterized in (Kapanadze et al., 1998) and named Leu5). Four corresponding mRNAs were identified, of 1.6, 2.0, 2.3 and 7.5 kb, which share the same coding potential and are expressed "ubiquitously" (15 tissues tested and several cell lines), including various lymphoid-specific tissues.

Mutation and Expression Analysis of the Candidate Genes
The paradigm for complete loss of gene function for tumor suppressors in cancer implicate disruption of both copies of such genes, typically through a mechanism involving point mutations disrupting one allele and deletions affecting the second one (Knudson, 1971). For this reason, we searched for mutations affecting the identified candidate genes in patients where one copy of such genes had already been lost because of chromosomal deletion. However, we have failed to find such nucleotide alterations when we analyzed exonic as well as exon-intron junction sequences in 20 B-CLL patients (including patients with and without deletion at 13q14). Another possibility by which loss-of-function could be achieved, is by loss of expression at the transcriptional level, for example upon methylation of the remaining copy of the gene in the patients with deletion. When the expression pattern of the candidate genes was analyzed in deleted patients, we found expression of the retained (normal) allele for the Est70/Leu1 and the CAR/Leu5 genes (Fig. 3), excluding also this inactivation mechanism for these two candidates. On the contrary, no expression was detected by Northern blot analysis when the 1B4/Leu2 gene was tested in B-CLL patients (regardless of the deletion status) (Fig.3). This finding raises the possibility that lack of expression of the 1B4/Leu2 might be important for the pathogenesis of B-CLL. Alternatively, lack of expression may simply reflect the physiological absence of 1B4/Leu2 expression in the CD5+ B-cell compartment.

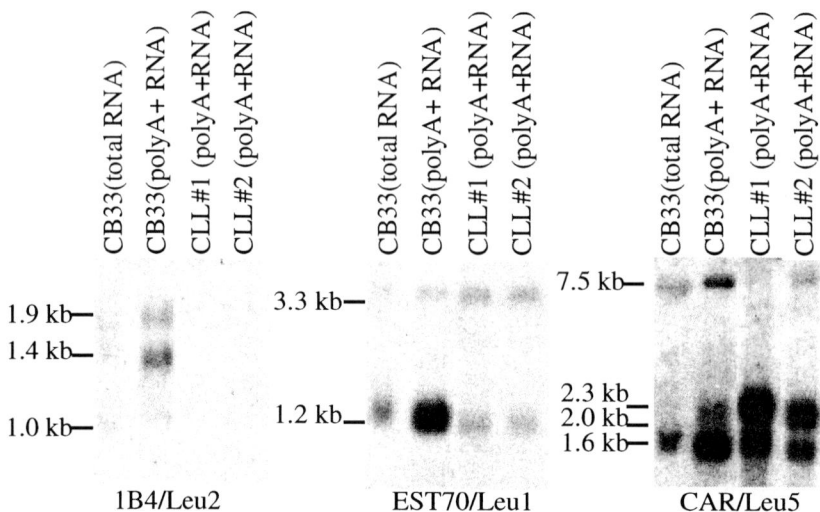

Fig. 3 Expression analysis of the B-CLL candidate genes in patients with deletion at 13q14 (CLL#1 and CLL#2). CB33 is a normal cord-blood-derived cell line immortalized by EBV infection. Twenty five micrograms of total RNA and four micrograms of polyA+ RNA were loaded on gel. The same filter was sequentially hybridized with 1B4, EST70 and CAR –specific probes (in this order). Exposure times are: six days for the 1B4 gene and two days for the EST70 and CAR genes. The normal retained allele of genes EST70 and CAR is expressed in the deleted samples. On the contrary, no expression of the 1B4 gene was detected in B-CLL samples (including cases where no 13q14 deletion was observed, not shown).

More refined expression analysis aimed at studying expression of the B-CLL candidate genes in the CD5+ B cell compartment is needed. Nevertheless, because of their location in the 13q14 deleted interval, these genes are candidates for the B-CLL-associated tumor suppressor gene.

Conclusion and Perspectives

In conclusion, somatic deletions involving a region of chromosome 13q14 and smaller than 300 kb are found in more than 50% of B-CLL patients. Furthermore, these deletions can be mono- or bi-allelic. The deleted genomic segment is candidate to contain a tumor suppressor gene whose inactivation may be relevant for the pathogenesis of B-CLL. In the attempt to discover such transcript, we have identified three genes located in the 13q14 MRD and therefore candidate to be the leukemia-associated tumor suppressor gene. However, since these genes lack mutations in B-CLL patients, they may not fit in the "two-hits" inactivation mechanism typically involving tumor suppressor genes in cancer (deletion of one allele and inactivating mutations in the second allele).

Nevertheless, the causal link between these genes (or one of them) and tumor suppression in B-CLL could be represented by their haplo-insufficiency (or reduced dosage) in the leukemia, as a result of 13q14 deletion. This scenario is well exemplified by the tumor suppressor $p27^{Kip1}$, whose levels are decreased in various human tumors by a variety of mechanisms, including hemizygous chromosomal deletions (for a review, see Clurman and Porter, 1998). Data obtained in mice confirmed $p27^{Kip1}$ to be haplo-insufficient for tumor suppression, since heterozygous animals appear to display a tumor-prone phenotype (Fero et al., 1998). We are currently attempting to test this model for the B-CLL candidate genes in mouse systems by "targeting" a region syntenic to the human 13q14 deleted segment, in the attempt to mimic the deletions encountered in the human patients.

References

Avet-Loiseau, H., T. Facon, A. Daviet, C. Godon, M.J. Rapp, J.L. Harousseau, B. Grosbois, and R. Bataille. 1999a. 14q32 translocations and monosomy 13 observed in monoclonal gammopathy of undetermined significance delineate a multistep process for the oncogenesis of multiple myeloma. Intergroupe Francophone du Myelome. *Cancer Res.* 59:4546-4550.

Avet-Loiseau, H., J.Y. Li, N. Morineau, T. Facon, C. Brigaudeau, J.L. Harousseau, B. Grosbois, and R. Bataille. 1999b. Monosomy 13 is associated with the transition of monoclonal gammopathy of undetermined significance to multiple myeloma. Intergroupe Francophone du Myelome. *Blood.* 94:2583-2589.

Bouyge-Moreau, I., G. Rondeau, H. Avet-Loiseau, M.T. Andre, S. Bezieau, M. Cherel, S. Saleun, E. Cadoret, T. Shaikh, M.M. De Angelis, S. Arcot, M. Batzer, J.P. Moisan, and M.C. Devilder. 1997. Construction of a 780-kb PAC, BAC, and cosmid contig encompassing the minimal critical deletion involved in B cell chronic lymphocytic leukemia at 13q14.3. *Genomics.* 46:183-190.

Bullrich, F., D. Rasio, S. Kitada, P. Starostik, T. Kipps, M. Keating, M. Albitar, J.C. Reed, and C.M. Croce. 1999. ATM mutations in B-cell chronic lymphocytic leukemia. *Cancer Res.* 59:24-27.

Bullrich, F., M.L. Veronese, S. Kitada, J. Jurlander, M.A. Caligiuri, J.C. Reed, and C.M. Croce. 1996. Minimal region of loss at 13q14 in B-cell chronic lymphocytic leukemia. *Blood.* 88:3109-3115.

Caligaris-Cappio, F., and T.J. Hamblin. 1999. B-cell chronic lymphocytic leukemia: a bird of a different feather. *J Clin Oncol.* 17:399-408.

Cigudosa, J.C., P.H. Rao, M.J. Calasanz, M.D. Odero, J. Michaeli, S.C. Jhanwar, and R.S. Chaganti. 1998. Characterization of nonrandom chromosomal gains and losses in multiple myeloma by comparative genomic hybridization. *Blood.* 91:3007-3010.

Clurman, B.E., and P. Porter. 1998. New insights into the tumor suppression function of P27(kip1). *Proc Natl Acad Sci U S A.* 95:15158-15160.

Corcoran, M.M., O. Rasool, Y. Liu, A. Iyengar, D. Grander, R.E. Ibbotson, M. Merup, X. Wu, V. Brodyansky, A.C. Gardiner, G. Juliusson, R.M. Chapman, G. Ivanova, M. Tiller, G. Gahrton, N. Yankovsky, E. Zabarovsky, D.G. Oscier, and S. Einhorn. 1998. Detailed molecular delineation of 13q14.3 loss in B-cell chronic lymphocytic leukemia. *Blood.* 91:1382-1390.

Dalla-Favera, R., A. Migliazza, C.C. Chang, H. Niu, L. Pasqualucci, M. Butler, Q. Shen, and G. Cattoretti. 1999. Molecular pathogenesis of B cell malignancy: the role of BCL-6. *Curr Top Microbiol Immunol.* 246:257-263.

Damle, R.N., T. Wasil, F. Fais, F. Ghiotto, A. Valetto, S.L. Allen, A. Buchbinder, D. Budman, K. Dittmar, J. Kolitz, S.M. Lichtman, P. Schulman, V.P. Vinciguerra, K.R. Rai, M. Ferrarini, and N. Chiorazzi. 1999. Ig V gene mutation status and CD38 expression as novel prognostic indicators in chronic lymphocytic leukemia [see comments]. *Blood.* 94:1840-1847.

Dohner, H., S. Stilgenbauer, K. Dohner, M. Bentz, and P. Lichter. 1999. Chromosome aberrations in B-cell chronic lymphocytic leukemia: reassessment based on molecular cytogenetic analysis. *J Mol Med.* 77:266-281.

Erdmann, V.A., M. Szymanski, A. Hochberg, N. de Groot, and J. Barciszewski. 1999. Collection of mRNA-like non-coding RNAs. *Nucleic Acids Res.* 27:192-195.

Fero, M.L., E. Randel, K.E. Gurley, J.M. Roberts, and C.J. Kemp. 1998. The murine gene p27Kip1 is haplo-insufficient for tumour suppression. *Nature.* 396:177-180.

Gaidano, G., P. Ballerini, J.Z. Gong, G. Inghirami, A. Neri, E.W. Newcomb, I.T. Magrath, D.M. Knowles, and R. Dalla-Favera. 1991. p53 mutations in human lymphoid malignancies: association with Burkitt lymphoma and chronic lymphocytic leukemia. *Proc Natl Acad Sci U S A.* 88:5413-5417.

Gaidano, G., E.W. Newcomb, J.Z. Gong, V. Tassi, A. Neri, A. Cortelezzi, R. Calori, L. Baldini, and R. Dalla-Favera. 1994. Analysis of alterations of oncogenes and tumor suppressor genes in chronic lymphocytic leukemia. *Am J Pathol.* 144:1312-1319.

Gupta, V.K., A.P. Schmidt, M.E. Pashia, J.B. Sunwoo, and S.B. Scholnick. 1999. Multiple regions of deletion on chromosome arm 13q in head-and-neck squamous-cell carcinoma. *Int J Cancer.* 84:453-457.

Hamblin, T.J., Z. Davis, A. Gardiner, D.G. Oscier, and F.K. Stevenson. 1999. Unmutated Ig V(H) genes are associated with a more aggressive form of chronic lymphocytic leukemia [see comments]. *Blood.* 94:1848-1854.

Hammond, C., L. Jeffers, B.I. Carr, and D. Simon. 1999. Multiple genetic alterations, 4q28, a new suppressor region, and potential gender differences in human hepatocellular carcinoma. *Hepatology.* 29:1479-1485.

Harada, K., T. Nishizaki, S. Ozaki, H. Kubota, T. Okamura, H. Ito, and K. Sasaki. 1999. Cytogenetic alterations in pituitary adenomas detected by comparative genomic hybridization. *Cancer Genet Cytogenet.* 112:38-41.

Kalachikov, S., A. Migliazza, E. Cayanis, N.S. Fracchiolla, M.F. Bonaldo, L. Lawton, P. Jelenc, X. Ye, X. Qu, M. Chien, R. Hauptschein, G. Gaidano, U. Vitolo, G. Saglio, L. Resegotti, V. Brodjansky, N. Yankovsky, P. Zhang, M.B. Soares, J. Russo, I.S. Edelman, A. Efstratiadis, R. Dalla-Favera, and S.G. Fischer. 1997. Cloning and gene mapping of the chromosome 13q14 region deleted in chronic lymphocytic leukemia. *Genomics.* 42:369-377.

Kapanadze, B., V. Kashuba, A. Baranova, O. Rasool, W. van Everdink, Y. Liu, A. Syomov, M. Corcoran, A. Poltaraus, V. Brodyansky, N. Syomova, A. Kazakov, R. Ibbotson, A. van den Berg, R. Gizatullin, L. Fedorova, G. Sulimova, A. Zelenin, L. Deaven, H. Lehrach, D. Grander, C. Buys, D. Oscier, E.R. Zabarovsky, N. Yankovsky, and et al. 1998. A cosmid and cDNA fine physical map of a human chromosome 13q14 region frequently lost in B-cell chronic lymphocytic leukemia and identification of a new putative tumor suppressor gene, Leu5. *FEBS Lett.* 426:266-270.

Knudson, A.G., Jr. 1971. Mutation and cancer: statistical study of retinoblastoma. *Proc Natl Acad Sci U S A.* 68:820-823.

La Starza, R., I. Wlodarska, A. Aventin, D. Falzetti, B. Crescenzi, M.F. Martelli, H. Van den Berghe, and C. Mecucci. 1998. Molecular delineation of 13q deletion boundaries in 20 patients with myeloid malignancies. *Blood.* 91:231-237.

Latil, A., I. Bieche, S. Pesche, A. Volant, A. Valeri, G. Fournier, O. Cussenot, and R. Lidereau. 1999. Loss of heterozygosity at chromosome arm 13q and RB1 status in human prostate cancer. *Hum Pathol.* 30:809-815.

Li, C., C. Larsson, A. Futreal, J. Lancaster, C. Phelan, U. Aspenblad, B. Sundelin, Y. Liu, P. Ekman, G. Auer, and U.S. Bergerheim. 1998. Identification of two distinct deleted regions on chromosome 13 in prostate cancer. *Oncogene.* 16:481-487.

Liu, Y., M. Corcoran, O. Rasool, G. Ivanova, R. Ibbotson, D. Grander, A. Iyengar, A. Baranova, V. Kashuba, M. Merup, X. Wu, A. Gardiner, R. Mullenbach, A. Poltaraus, A.L. Hultstrom, G. Juliusson, R. Chapman, M. Tiller, F. Cotter, G. Gahrton, N. Yankovsky, E. Zabarovsky, S. Einhorn, and D. Oscier. 1997. Cloning of two candidate tumor suppressor genes within a 10 kb region on chromosome 13q14, frequently deleted in chronic lymphocytic leukemia. *Oncogene.* 15:2463-2473.

Liu, Y., M. Hermanson, D. Grander, M. Merup, X. Wu, M. Heyman, O. Rasool, G. Juliusson, G. Gahrton, R. Detlofsson, and et al. 1995. 13q deletions in lymphoid malignancies. *Blood.* 86:1911-1915.

Lo Coco, F., B.H. Ye, F. Lista, P. Corradini, K. Offit, D.M. Knowles, R.S. Chaganti, and R. Dalla-Favera. 1994. Rearrangements of the BCL6 gene in diffuse large cell non-Hodgkin's lymphoma. *Blood.* 83:1757-1759.

Oscier, D.G., A. Thompsett, D. Zhu, and F.K. Stevenson. 1997. Differential rates of somatic hypermutation in V(H) genes among subsets of chronic lymphocytic leukemia defined by chromosomal abnormalities. *Blood.* 89:4153-4160.

Rosenwald, A., G. Ott, A.K. Krumdiek, M.H. Dreyling, T. Katzenberger, J. Kalla, S. Roth, M.M. Ott, and M.l.-H. HK. 1999. A biological role for deletions in chromosomal band 13q14 in mantle cell and peripheral t-cell lymphomas? *Genes Chromosomes Cancer.* 26:210-214.

Shaughnessy, J., and B. Barlogie. 1999. Chromosome 13 deletion in myeloma. *Curr Top Microbiol Immunol.* 246:199-203.

Siu, L.L., K.F. Wong, J.K. Chan, and Y.L. Kwong. 1999. Comparative genomic hybridization analysis of natural killer cell lymphoma/leukemia. Recognition of consistent patterns of genetic alterations. *Am J Pathol.* 155:1419-1425.

Stankovic, T., P. Weber, G. Stewart, T. Bedenham, J. Murray, P.J. Byrd, P.A. Moss, and A.M. Taylor. 1999. Inactivation of ataxia telangiectasia mutated gene in B-cell chronic lymphocytic leukaemia. *Lancet.* 353:26-29.

Starostik, P., T. Manshouri, S. O'Brien, E. Freireich, H. Kantarjian, M. Haidar, S. Lerner, M. Keating, and M. Albitar. 1998. Deficiency of the ATM protein expression defines an aggressive subgroup of B-cell chronic lymphocytic leukemia. *Cancer Res.* 58:4552-4557.

Stilgenbauer, S., J. Nickolenko, J. Wilhelm, S. Wolf, S. Weitz, K. Dohner, T. Boehm, H. Dohner, and P. Lichter. 1998. Expressed sequences as candidates for a novel tumor suppressor gene at band 13q14 in B-cell chronic lymphocytic leukemia and mantle cell lymphoma. *Oncogene.* 16:1891-1897.

Tamura, K., X. Zhang, Y. Murakami, S. Hirohashi, H.J. Xu, S.X. Hu, W.F. Benedict, and T. Sekiya. 1997. Deletion of three distinct regions on chromosome 13q in human non-small- cell lung cancer. *Int J Cancer.* 74:45-49.

Ueda, T., M. Emi, H. Suzuki, A. Komiya, K. Akakura, T. Ichikawa, M. Watanabe, T. Shiraishi, M. Masai, T. Igarashi, and H. Ito. 1999. Identification of a I-cM region of common deletion on 13q14 associated with human prostate cancer. *Genes Chromosomes Cancer.* 24:183-190.

Yin, Z., M.R. Spitz, R.J. Babaian, S.S. Strom, P. Troncoso, and J. Kagan. 1999. Limiting the location of a putative human prostate cancer tumor suppressor gene at chromosome 13q14.3. *Oncogene*. 18:7576-7583.

Chronic lymphocytic leukemia: A proliferation of B cells at two distinct stages of differentiation.

Damle, R.N.*, Fais, F.*, Ghiotto, F.*, Valetto, A.*, Albesiano, E.*, Wasil, T.*, Batliwalla, F.M.*, Allen, S.L.*, Schulman, P.*, Vinciguerra, V.P.*, Rai, K.R.[#] Gregersen, P.K. *, Ferrarini, M.♦ and Chiorazzi, N.*

*North Shore University Hospital - NYU School of Medicine, Manhasset NY, [#]Long Island Jewish Medical Center, New Hyde Park, NY, ♦ Istituto Nazionale per la Ricerca sul Cancro, Universita di Genova, Genova, Italy.

Introduction

B-type chronic lymphocytic leukemia (B-CLL) is the most commonly occurring leukemia in the western world with as many as 7500 new cases reported each year [1, 2]. B-CLL arises from mature antigen-inexperienced B cells that express surface membrane CD5 in a majority of the cases [3]. B cell neoplasms arise from cells transformed at one of many stages of differentiation. Fig. 1 schematically depicts the phenotypic changes that accompany the process of B cell differentiation triggered by antigen, in either a T-dependent or T-independent manner. The antigen-mediated selection process that occurs in the bone marrow and among recent bone marrow emigrants deletes autoreactive B cells [4, 5]. Therefore, only a fraction of the immature naïve ("transitional") B cells that encounter antigen traverse the germinal center (GC) and eventually convert to long lasting memory cells [reviewed in 6]. However, a characteristic feature of such post-germinal center memory B cells is the presence of Ig V gene mutations. Therefore, V gene mutations serve as an important read-out of a previous GC reaction [7].

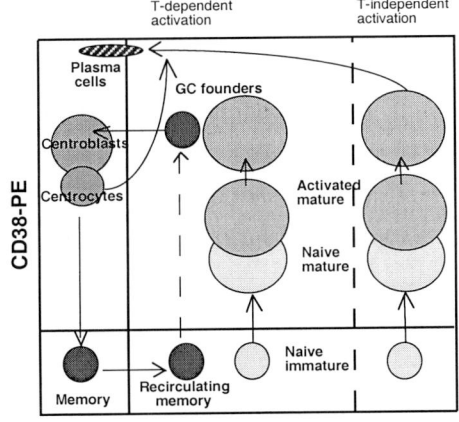

In this regard, studies by our group and others have provided evidence that approximately 50-60% of B-CLL cases arise by the transformation of a B cell that has undergone the GC reaction and has accumulated somatic mutations [8, 9]. These data suggested that CLL is a heterogeneous disease and not homogeneous as it has been viewed.

Fig. 1. Changes in IgD and CD38 expression accompanying T dependent and T independent B cell differentiation.

Rather, the disease derives from two types of B cells: one that expresses Ig V gene mutations and is representative of post-GC memory B cells, and another that does not express V gene mutations. However, the origin of the B-CLL cases with unmutated V genes is not resolved since they could either be unstimulated naïve B cells or antigen-stimulated B cells that have not accumulated V gene mutations.

In an attempt to answer this question, we analyzed a). the potential role of antigen drive by studying the properties of the Ig genes expressed in these cells, b). their state of activation as reflected by the expression of surface membrane markers, c). their replicative history as assessed by analyses of telomere lengths, and d). their telomerase activity. These data suggest that, contrary to earlier beliefs, most B-CLL cases originate from B cells that have encountered and responded to antigen. Possibly, differences in the type of antigen encountered or the point in maturation at which transformation occurred determines the observed differences in Ig V gene mutation characteristics of the two types of B-CLL cases.

Results and Discussion

Characteristics of rearranged Ig V_H genes

Ig V gene sequence analyses revealed that V_H 4-34, 3-07, and 1-69 were the most commonly expressed genes in our B-CLL patients [9]. J_H segment use differed among these three genes in that ~90% of the V_H 3-07 genes associated with a J_H4 segment whereas ~50% of the V_H 1-69 and V_H 4-34 genes associated with a J_H6 segment. D segments were used in percentages very similar to those recently identified in rearranged antibodies [10]. In our cases, 50% of the 1-69 genes were linked with the D3-3 segment, which is consistent with a previous report [11].

The B-CLL cases fell into three groups based on differences in HCDR3 length, amino acid (aa) composition, and charge [9]. Each of these varied in a V_H family-related manner. First, the average HCDR3 length of 4-34-expressing B-CLL cells (17.0 aa) was greater than 1-69-expressing cells (16.33 aa) and 3-07-expressing B-CLL cells (12.56 aa). In most cases, the short HCDR3 segments of the V_H3 group contained a J_H4 segment, whereas the V_H1 group contained a J_H6 or J_H5 segment. Second, 1-69-expressing B-CLL cells frequently contained long stretches of tyrosines coded for by the J_H6 segment. This was in contrast to most 3-07-expressing B-CLL cells that used a J_H4 gene.

Finally, these differences in J_H gene association resulted in a V_H family-related hierarchy in charge: V_H 1-69 (pI: 3.53), 4-34 (pI: 4.37) and 3-07 (pI: 5.89). These V_H gene-related differences in HCDR3 may reflect selection for specific structural motifs that facilitate antigen binding.

Expression of activation markers
Lymphocytes can express a variety of surface markers that are indicative of cell adhesion, activation, apoptosis or other properties. We studied a set of 50 B-CLL patients for their expression of various surface markers. When the B-CLL cases were segregated based on V gene mutation, higher numbers of cells from the unmutated cases expressed the activation markers CD38 and CD69, whereas more cells from mutated cases showed expression of CD71, CD62L and CD23 (Fig. 2) These differences were statistically significant [manuscript in preparation].

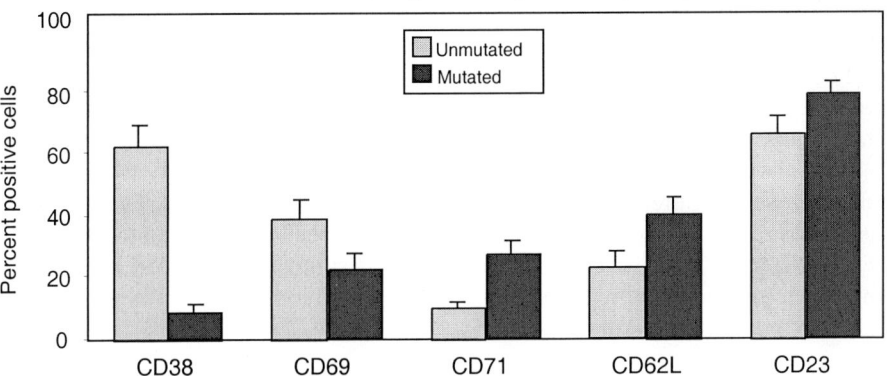

Fig. 2. Expression of activation markers by B-CLL cells.

Although these surface antigens are expressed at different phases post-activation, their expression by the unmutated B-CLL cases suggests that these cases derive from antigen-activated and not antigen-naïve cells.

Since we observed a major difference in the percentages of B-CLL cells expressing CD38, we compared expression of this marker with V gene mutation. A significant inverse correlation existed between the percentages of CD38-expressing B-CLL cells and the percentage of V gene mutation ($r = -0.75$, $p < 0.01$). In addition, the cases with mutated V genes showed <30% B-CLL cells expressing CD38 and those with unmutated V genes showed $\geq$ 30% B-CLL cells expressing CD38 (Fig. 3). Based on clinical features such as treatment histories and survival post-diagnosis, the patients stratified into two distinct groups. Those patients requiring less chemotherapy and with better prognoses displayed mutated IgV genes and < 30% CD38-expressing B-CLL cells and those patients requiring more chemotherapy and with

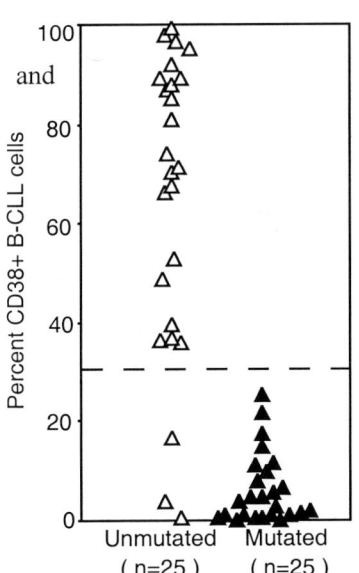

Fig. 3. Scattergrams representing CD38 expression by B-CLL cells from "unmutated" and "mutated" cases.

much worse prognoses displayed unmutated IgV genes and ≥ 30% of $CD38^+$ B-CLL cells [12]. Hamblin et al [13] simultaneously reported that B-CLL patients with unmutated VH genes follow a more aggressive course .

Analyses of replicative history of B-CLL cells
The lengths of the telomeric ends of chromosomes indicate the replicative history of cells. Thus, if B-CLL cells are considered as naïve cells, then they would be expected to have long telomeres with lengths comparable to those of cells that have recently exited the bone marrow. In addition since B-CLL cells do not cycle rapidly but rather survive longer due to an apparent apoptotic defect [14, 15], we anticipated that the unmutated cases would have longer telomeres than the mutated cases.

Therefore, we measured the lengths of the telomeres of the two types of B-CLL cells and compared these with B cells from age-matched healthy donors. As seen in Fig. 4, B cells from B-CLL patients showed significantly shorter telomeres compared to those in healthy controls. This implies a longer replicative history of B-CLL cells compared to normal B cells.

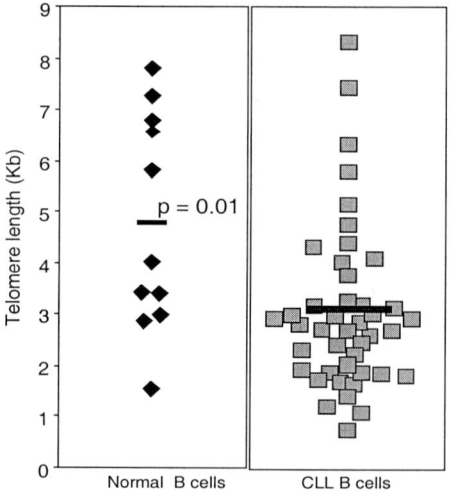

Fig. 4. Telomere lengths of B cells from healthy donors (n = 11) and B-CLL patients (n = 43)

However, when the B-CLL cases were categorized based on V gene mutation or CD38 expression, their telomere lengths differed significantly (p = 0.0045). Surprisingly, the unmutated cases (or those in which ≥ 30% B-CLL cells expressed CD38) had even shorter telomeres than the mutated cases (< 30% $CD38^+$ B-CLL cells), suggesting a longer replicative history for this group (data not shown).

The enzyme telomerase is known to restore telomere loss, by adding hexameric repeats (TTAGGG) to the eroding ends of chromosomes. We are currently studying telomerase activity in cases for which we have analyzed telomere length. Preliminary data indicate that those patients with unmutated V genes (who have an unfavorable prognosis) have elevated telomerase activity (data not shown). This finding is consistent with data reported in CLL and other leukemias [16, 17].

Taken together, these findings imply that B-CLL cells represent a proliferation of B cells at two distinct stages of differentiation. The aspects studied by our group knit well into understanding the origin of the transformed B cell that develops into B-CLL. A plethora of factors and events may decide the fate of the precursor cell that becomes a B-CLL clone. Based on the preceding data, it is plausible that antigen plays a role in the process that precedes or involves leukemogenesis. The type of antigen encountered by the $CD5^+$ (B-1 like) cell, presence or absence of T cell help, etc, may be at the base of the different types of B-CLL cases. However, it is also possible that B-CLL arise from a B-2-like cell that acquires surface membrane CD5 in response to an antigenic signal. The inherent capability of one activated B cell (the B-CLL precursor) to cycle more rapidly than another, before reaching the transforming event, may be an added variable that somehow relates to the aggressiveness of the disease.

The dynamics of the process that lead some of the antigen-encountered cells to enter the GC and accumulate mutations may also affect the final clinical outcome. On the basis of their surface phenotype (expression of IgD, CD38 and CD5 on $CD19^+$ B cells) and occurrence of somatic Ig V gene mutations, the mutated cases probably arise from a post-GC memory B cell that was stimulated by T-dependent exo-antigens.

On the other hand, the unmutated cases also show evidence of antigen drive and selection, based on specific pairing of segments comprising the rearranged V_HDJ_H and characteristic HCDR3 features and expression of activation antigens. These findings suggest that these cases likely derive from an activated pre-GC cell that either was transformed prior to entering a GC or had been exposed to an autoantigen that could not elicit T cell help and initiate the GC reaction. The shorter telomeres of this group could therefore be a result of repeated cell divisions in the pre-transformation phase of the life of the B cell. Telomerase activity which has been reported to associate with aggressive forms of malignancies [18, 19] also appears to be elevated in our cases that belonged to the "unmutated", poor prognosis group. This finding again clearly distinguishes these cases from the clinically, relatively stable "mutated" cases.

As with most malignancies, B-CLL patients are currently categorized on the basis of the existence and severity of symptoms. The studies on V gene mutation status [12, 13] and CD38 expression [12] suggest that these parameters can, independently, supplement current methodologies of classification and can aid in predicting prognoses of the sub-groups of B-CLL.

It is important to note that these two markers (V gene mutation status and CD38 expression) need to be viewed as independent variables that usually overlap. As seen in Fig. 3, in a small subset of patients the V gene mutation data do not inversely correlate with the percentages of $CD38^+$ B-CLL cells. Therefore, the two markers may or may not identify the exact same cases in all instances.

B-CLL may or may not follow an aggressive course and the reasons for such diverse patterns of disease progression are still unclear. The two different origins of the transformed B cell as proposed above may confer different features that indirectly determine the properties of the cell and its ability to progress as a disease. Thus, although B-CLL has been viewed as a homogeneous disease, these and other recent data suggest heterogeneity in the origin of the disease that may lead to the differences in disease progression and outcome observed in different B-CLL patients.

Materials and Methods

Ig V gene sequence analyses
Ig V gene use and sequence and HCDR3 characteristics such as D and J_H gene use and association and HCDR3 length, composition and charge were studied as described in detail earlier [9, 20].

Phenotypic analyses of B-CLL cells
A panel of monoclonal antibodies that identify surface membrane markers such as CD5, CD19, and others indicative of cellular activation and differentiation viz., CD23, CD25, CD38, CD69, CD62L, CD71 etc, were purchased from Becton, Dickinson, USA and employed for three color immunofluorescent detection of cells that co-express CD19, CD5 and one of the other markers mentioned above. The procedure used for staining cells has been described in detail earlier [12]

Analysis of telomere length
Telomere lengths were indirectly quantified by assessing the binding of a telomere specific fluorescent probe $\{(C_3TA_2)_3\text{-FITC}\}$ using a Flow Cytometer (FACS Calibur, Becton Dickinson, USA). Briefly, purified cell populations (eg. $CD19^+$ cells) were incubated with the probe under conditions that enabled it to hybridize with telomeric regions *in situ*, following which propidium iodide was added to facilitate detection of cellular DNA. This method has been described in detail elsewhere [21].

Quantitation of telomerase activity
Telomerase activity in purified cell populations was quantified using the TrapEZE telomerase detection kit purchased from Intergen, NJ, USA. The assay is a two

step detection of telomerase activity. It combines telomerase-mediated extension of template oligomer (step 1), and PCR amplification of extended products (step 2). The assay was performed in accordance with the manufacturer's instructions.

References

1. Rai K, Patel D (1995) Chronic Lymphocytic Leukemia. In: Hoffman R, Benz E, Shattil S, Furie B, Cohen H, Silberstein L (eds) Hematology: Basic Principles and Practice. Churchill Livingstone, New York, pp. 1308-1321
2. Landis SH, Murray T, Bolden S, Wingo PA (1998) Cancer statistics, 1998. CA Cancer J Clin 48:6-29.
3. Caligaris-Cappio F, Gobbi M, Bofill M, Janossy G (1982) Infrequent normal B lymphocytes express features of B-cell chronic lymphocytic leukemia. J Exp Med 155:623.
4. Carsetti R, Kohler G, Lamers MC (1995) Transitional B cells are the target of negative selection in the B cell compartment. J Exp Med 181:2129-2140.
5. Melamed D, Benschop RJ, Cambier JC, Nemazee D (1998) Developmental regulation of B lymphocyte immune tolerance compartmentalizes clonal selection from receptor selection. Cell 92:173-182.
6. Liu YJ, Arpin C (1997) Germinal center development. Immunol Rev 156:111-126.
7. Pascual V, Liu YJ, Magalski A, de Bouteiller O, Banchereau J, Capra JD (1994) Analysis of somatic mutation in five B cell subsets of human tonsil. J Exp Med 180:329-339.
8. Schroeder HW, Jr., Dighiero G (1994) The pathogenesis of chronic lymphocytic leukemia: analysis of the antibody repertoire [see comments]. Immunol Today 15:288-294.
9. Fais F, Ghiotto F, Hashimoto S, Sellars B, Valetto A, Allen SL, Schulman P, Vinciguerra VP, Rai K, Rassenti LZ, Kipps TJ, Dighiero G, Schroeder H, Ferrarini M, Chiorazzi N (1998) Chronic lymphocytic leukemia B cells express restricted sets of mutated and unmutated antigen receptors. J Clin Invest 102:1515-1525.
10. Corbett SJ, Tomlinson IM, Sonnhammer ELL, Buck D, Winter G (1997) Sequence of the human immunoglobulin diversity (D) segment locus: a systematic analysis provides no evidence for the use of DIR segments, inverted D segments, "minor" D segments or D-D recombination. J Mol Biol 270:587-597.
11. Johnson TA, Rassenti LZ, Kipps TJ (1997) Ig VH1 genes expressed in B cell chronic lymphocytic leukemia exhibit distinctive molecular features. J Immunol 158:235-246.
12. Damle RN, Wasil T, Fais F, Ghiotto F, Valetto A, Allen SL, Buchbinder A, Budman D, Dittmar K, Kolitz J, Lichtman SM, Schulman P, Vinciguerra VP, Rai KR, Ferrarini M, Chiorazzi N (1999) Ig V gene mutation status and CD38 expression as novel prognostic indicators in chronic lymphocytic leukemia. Blood 94:1840-1847.
13. Hamblin TJ, Davis Z, Gardiner A, Oscier DG, Stevenson FK (1999) Unmutated Ig VH genes are associated with a more aggressive form of chronic lymphocytic leukemia. Blood 94:1848-1854.
14. Caligaris-Cappio F, Gottardi D, Alfarano A, Stacchini A, Gregoretti MG, Ghia P, Bertero MT, Novarino A, Bergui L (1993) The nature of the B lymphocyte in B-chronic lymphocytic leukemia. Blood Cells 19:601-613.
15. Osorio LM, Jondal M, Aguilar-Santelises M (1998) Regulation of B-CLL apoptosis through membrane receptors and Bcl-2 family proteins. Leuk Lymphoma 30:247-256.
16. Counter CM, Gupta J, Harley CB, Leber B, Bacchetti S (1995) Telomerase activity in normal leukocytes and in hematologic malignancies. Blood 85:2315-2320.
17. Trentin L, Ballon G, Ometto L, Perin A, Basso U, Chieco-Bianchi L, Semenzato G, De Rossi A (1999) Telomerase activity in chronic lymphoproliferative disorders of B-cell lineage. Br J Haematol 106:662-668.
18. Clark GM, Osborne CK, Levitt D, Wu F, Kim NW (1997) Telomerase activity and survival of patients with node-positive breast cancer. J Natl Cancer Inst 89:1874-1881.
19. Wu K, Lund M, Bang K, Thestrup-Pedersen K (1999) Telomerase activity and telomere length in lymphocytes from patients with cutaneous T-cell lymphoma. Cancer 86:1056-1063.

20. Dono M, Hashimoto S, Fais F, Trejo V, Allen SL, Lichtman SM, Schulman P, Vinciguerra VP, Sellars B, Gregersen PK, Ferrarini M, Chiorazzi N (1996) Evidence for progenitors of chronic lymphocytic leukemia B cells that undergo intraclonal differentiation and diversification. Blood 87:1586-1594.
21. Rufer N, Dragonowska W, Thornbury G, Roosnek E, Lansdorp PM (1998) Telomere length dynamics in human lymphocyte subpopulations measured by flow cytometry. Nature Biotechnology 16:743-747.

Expression of Cyclooxygenase-2 and Prostaglandins by B-1 Cells and B-CLL Cells

R.P. Phipps[1,2,3,4], S.J. Pollock[4], K. Kaur[4], J. Kaufman[1], M.A. Borrello[4], B.A. Graf[1], D. Nazarenko[2], L.J. Roberts[5], J.D. Morrow, J. Palis[3], D.J. Ryan[4], and J.M. Bennett[4]

[1]Department of Microbiology and Immunology, University of Rochester School of Medicine and Dentistry, Rochester, USA
[2]Department of Environmental Medicine, University of Rochester School of Medicine and Dentistry, Rochester, USA
[3]Department of Pediatrics, University of Rochester School of Medicine and Dentistry, Rochester, USA
[4]The Cancer Center, University of Rochester School of Medicine and Dentistry, Rochester, USA
[5]Division of Clinical Pharmacology, Departments of Medicine and Pharmacology, Vanderbilt University Medical Center, Nashville, USA

Introduction

B cells exist in two distinct subpopulations, the conventional B-2 subset and the B-1 population. These subsets differ in many ways, including their prevalence during stages of B cell ontogeny, receptor repertoire, functional properties and localization throughout the body. Conventional B cells or B-2 cells arise later in B cell ontogeny and are the major B cell population found in lymphoid tissues and peripheral blood [1]. B-2 cells have been well-characterized for their role in the adaptive immune response. B-1 cells (both $CD5^+$ B-1a and $CD5^-$ B-1b) on the other hand arise very early in ontogeny and exist at a very low frequency in lymphoid organs, but are the predominant B cell population in the peritoneal and pleural cavities [1]. The function of B-1 cells is not yet well understood. To elucidate the role that B-1 cells play in immune responses and in cancer we studied two distinct populations of B-1 cells, the biphenotypic mouse B/Macrophage cell and the malignant human B-CLL cell. Specifically we examined the expression of cyclooxygenase-2 and the production of Prostaglandin E_2 (PGE_2) by these cells.

Cyclooxygenases (COXs), of which there are two isoforms cyclooxygenase-1 (COX-1) and cyclooxygenase-2 (COX-2) are the first enzymes in the biosynthetic pathway leading from arachadonic acid to prostaglandin synthesis. COXs have bifunctional activity whereby they first convert arachidonic acid to PGG through their cyclooxygenase activity and then to PGH through their peroxidase activity. PGH is then converted by isomerases to PGE, PGD and PGF and to thromboxanes by thromboxane synthase [2,3]. COX-1 is constituitively expressed in nearly all tissues and is responsible for producing low levels of prostaglandins necessary for normal physiologic activity. COX-2 on the other hand is an inducible form of the enzyme that is upregulated in response to stimulation with certain cytokines and growth factors. COX-2 expression is greatly increased at sites of inflammation which leads to a resultant increase in the production of prostaglandins such as PGE_2. Increased expression of COX-2 has also been observed in and linked to the genesis of a number of human cancers including colon and pancreatic cancers [4,5].

PGE_2, a product of the COXs, is a small lipid molecule that has immunoregulatory activity [6]. Of great importance is the role that PGE_2 plays in the balance between T_H1 and T_H2 immune responses. PGE_2 has been shown to promote T_H2 responses by inhibiting production of IL-12 a crucial T_H1 cytokine [7]. A great deal of effort has been focused on the development on drugs that specifically block COX-2 activity in order to prevent synthesis of prostaglandins.

Biphenotypic B/Macrophage Cells

B/Macrophage cells are biphenotypic leukocytes that simultaneously express B-1 lymphocyte and macrophage characteristics. The function of these cells is yet to be determined. B/Macrophage cells are generated from culturing purified B-lymphocytes in fibroblast conditioned media. The resulting cells simultaneously express both $CD5^+$ lymphocyte (IgM, IgD, B220, CD5) and macrophage (phagocytosis, F4/80, Mac-1) characteristics (Fig. 1) [8,9]. It has been determined that the factor in fibroblast conditioned media that potentiates this transformation from B cell to B/Macrophage is macrophage-colony stimulating factor (M-CSF)[10]. Importantly, B/Macrophage cells can also be

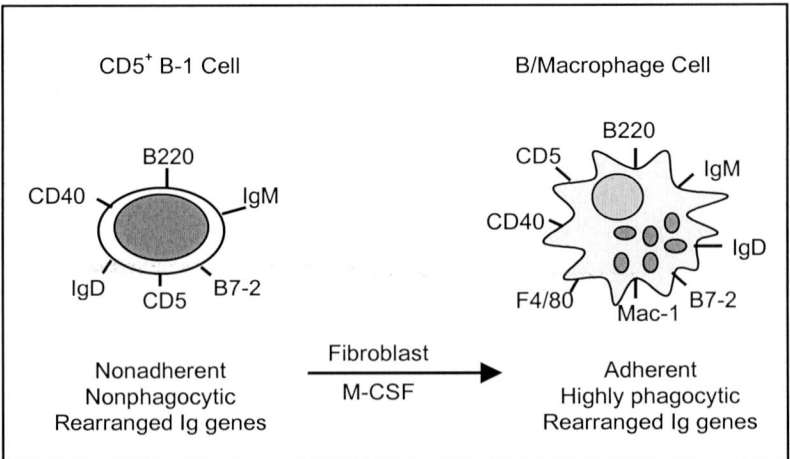

Fig. 1. Normal splenic CD5+ B-1 Cells differentiate to cells simultaneously displaying characteristics of CD5+ B cells and macrophages in the presence of fibroblast conditioned medium that contains M-CSF.

isolated *in vivo*. Approximately 11% of cells isolated from the mouse peritoneal cavity were found to be double positive for the B cell marker B220 and the macrophage marker F4/80 [11].

The biphenotypic nature of the B/Macrophage cell is unique because B-lymphocytes and macrophages are considered to be derived from separate lineages and to have specialized functions. However, it has been reported that certain malignant $CD5^+$ B-lymphocytes are capable of differentiating to macrophage-like cells that express characteristics of both cell types. For example in 1957, Dawe and Potter observed that the $CD5^+$ lymphoblastic lymphoma P388 underwent a morphological transition in vitro toward a macrophage-like phenotype [12]. Differentiation to such biphenotypic cells has been termed 'lineage infidelity' or 'lineage switching' and has been found to be associated with a variety of diseases including AIDS, Sjogren's Syndrome, and human B-Chronic Lymphocytic Leukemia (B-CLL) [13,14]. Twenty-five to 50% of patients with B-CLL bear neoplastic cells displaying CD5 and CD19 simultaneously with monocyte-associated markers CD11b, CD13 and CD15 [14]. Individuals bearing a biphenotypic cancer respond poorly to chemotherapy and have lower remission rates [14]. While all other reports of lineage infidelity have been confined to malignancies and other disease states we believe that B/Macrophage cells are the normal counterpart in this phenomenon.

Since the function of these cells is unknown the potential role of B/Macrophage cells in inflammation and immune responses was explored by evaluating the expression of COX-1 and COX-2 and the production of PGE_2. COX-1 and COX-2 mRNA expression in B/Macrophage cells was evaluated by reverse transcriptase (RT)-PCR [11]. Unlike the precursor B-lymphocytes, B/Macrophage cells were found to constituitively express COX-1 mRNA [11]. In addition B/Macrophage cells also express COX-2 mRNA and treatment with bacterial liposaccharide (LPS) results in increased COX-2 mRNA expression over time [11]. COX-2 mRNA is not expressed in the precursor B-lymphocytes in either untreated or LPS treated samples.

Expression of COX-1 and COX-2 protein by B/Macrophage cells was also examined using both immunocytochemistry and Western blotting techniques [11]. B/Macrophage cells constituitively express COX-1 protein and up-regulate COX-2 in response to LPS [11]. Interestingly, B/Macrophage cells show substantial heterogeneity in their degree of COX-2 expression as not all cells stained with the same intensity.

To confirm functional cyclooxygenase activity, PGE_2 production was evaluated by immunoassay. B/Macrophage cells stimulated with LPS, anti-IgM, anti-IgM plus CD40 ligand (CD40L) and IFN-γ plus CD40L showed a significant increase in PGE_2 production over the untreated sample (Table 1). To determine whether the PGE_2 produced by these cells was synthesized mainly by COX-2, stimulated cultures were treated with indomethacin which blocks COX-1 and COX-2 or with the drug SC58125 which is a COX-2 selective inhibitor [15]. SC58125 treatment inhibited nearly all PGE_2 synthesis, pointing to COX-2 as the major source of PGE_2 in B/Macrophage cells.

Table 1. B/Macrophage cells produce PGE2 that is mainly derived from COX-2[a]

Treatment[b]	PGE$_2$ (pg/ml)		
	No Drug	Indomethacin	SC58125
Untreated	1267 ± 68	1072 ± 70	986 ± 192
LPS	20466 ± 1709 *	1062 ± 255	1009 ± 180
Anti-IgM	13287 ± 1457 *	300 ± 55	736 ± 70
CD40L	2129 ± 171	329 ± 19	329 ± 20
IFN-γ	1048 ± 104	263 ± 64	574 ± 89
Anti-IgM/CD40L	21564 ± 1329 *	356 ± 10	777 ± 81
IFN-γ/CD40L	16983 ± 912 *	206 ± 66	589 ± 114

[a] Adapted from Ref 11
[b] B/Macrophage cells were stimulated in the presence of arachidonic acid (10 μM) for 24 hours with medium only, LPS (10μg/ml), anti-IgM (10μg/ml), CD40L, anti-IgM plus CD40L, IFN-γ (10 μg/ml) and IFN-g plus CD40L. Cultures received no drug, indomethacin (20 μM), or SC58125 (5μM). Supernatents were harvested and assayed for PGE$_2$ in an immunoassay.
*$p<0.005$ compared to untreated samples.

In addition to PGE$_2$, production of other immunoregulatory prostanoids by B/Macrophage cells was evaluated by gas chromatography/mass spectrophotometry [11, 16]. In addition to PGE$_2$, these cells produce appreciable amounts of PGF$_{2\alpha}$ and PGD$_2$ after stimulation with LPS, CD40L and anti-IgM antibody [11]. PGD$_2$ is a particular interest because its metabolite 15-deoxyΔ^{12,14} PGJ$_2$ is the ligand for the nuclear receptor PPARγ, which regulates adipocyte differentiation and glucose homeostasis and has been shown to be a negative regulator of activated macrophages [17]. B/Macrophages are the first example of a B lineage cell capable of producing PGD$_2$.

Demonstration that B/Macrophage cells up-regulate COX-2 and produce PGE$_2$ in response to inflammatory signals clearly provides evidence to support a role for these cells in modulating immune responses. B/Macrophage cells are the first normal nonmalignant cell of B lymphocyte origin that have the ability to up-regulate COX-2 and produce PGE$_2$. Expression of COX-2 by these cells is a characteristic that we believe is unique to the CD5+ B-1a subset of B lymphocytes, defining another difference between B-1 and conventional B-2 lymphocyte subsets.

B/Macrophage cells are likely to promote T$_H$2 immune responses. We have previously shown that unlike conventional macrophages, B/Macrophage cells are inhibited in environments rich in IFN-γ, a known T$_H$1 cytokine.

Differentiation of B/Macrophage cells from B-lymphocytes is blocked by IFN-γ, though the T_H2 cytokine IL-4 does not interfere with this phenomenon [8,9]. PGE_2 produced by B/Macrophage cells may give them the ability to shift the balance toward a T_H2 response and away from T_H1, as it has been shown that PGE_2 inhibits production of IL-12 which is critical for induction of a T_H1 response.

Identification of B/Macrophage Cells in humans will be an important link necessary for understanding biphenotypic hematological malignancies. There is increasing evidence for the involvement of COX-2 in the development of certain cancers. Expression of COX-2 by B-1 cells may be a characteristic that is required for the transformation to a malignant phenotype. Drugs that specifically inhibit COX-2 may therefore be relevant for use as chemopreventive agents.

Chronic Lymphocytic Leukemia and Cyclooxygenase-2

As demonstrated in the preceeding section, mouse B-1 $CD5^+$ cells can differentiate to a biphenotypic cell capable of expressing COX-2. Human B-chronic lymphocytic leukemia (B-CLL) cells are believed to be derived from the B-1 subset, owing to their surface expression of both CD19 and CD5. Surface IgM and/or IgD expression is lower on B-CLL cells than normal B cells, reflective of B-1 cells, and suggests a clonal origin based on almost exclusive presence of either κ or λ light chains. [18]

B-CLL patients may present with symptoms such as unexplained fever, night sweats, loss of weight and recurrent infections, or may be asymptomatic. Typically, they present with at least some symptoms of moderate severity. [19] These symptoms are similar in diseases where COX expression is up-regulated and PGE_2 levels are elevated. These common phenotypic markers and symptoms suggest that the B-1 cells themselves may be responsible for the fevers, etc. by increased production of PGE_2. Supporting this is our findings that some B-CLL patients have elevated PGE_2 levels in their blood (unreported observation).

Our hypothesis was that human B-CLL cells are derived from B-1 cells and are capable of expressing COX-2 and synthesizing PGE_2. We explored whether or not B-CLL cells express COX-1 and COX-2, and whether that expression is regulated following stimulation *in vitro*.

Following receipt of peripheral blood from untreated patients diagnosed with B-CLL, mononuclear cells were isolated on a Ficoll-Paque density gradient. The B cells in this population were isolated using paramagnetic beads coated with anti-CD19 antibodies and yielded cells greater than 98% positive for both CD19 and CD5 by flow cytometric analysis. The cells also expressed only κ or λ light chains. B-CLL cells were then cultured *in vitro* for up to 72 hours with nothing, IFN-γ, CD154 (CD40L), and/or TNFα in an attempt to stimulate increased expression of COX-2. Following stimulation, cells were collected and analysed by immunocytochemistry (ICC), Western blotting and RNAse protection assay (RPA) to assess the presence of COX-2, and whether or not that expression was altered by stimulation. The media from the cultures was collected and analysed by competitive enzyme immunoassay (EIA) to determine the concentration of PGE_2 in each.

B-CLL Cells Express Cox-2 and Produce PGE_2

Our studies revealed that some freshly isolated B-CLL cells contain significant amounts of COX-2 protein as found by both ICC and Western blotting. Following an 18 hour incubation with selected combinations of stimulants (known in other cell types to induce COX-2), we found that the cells can be induced to express higher levels of COX-2 mRNA and COX-2 protein. The level of PGE_2 found in the culture medium of these stimulated cells showed that many of the B-CLL cells produced micromolar amounts of PGE_2. This was particularly interesting as it had not been shown previously that any human B cell was capable of expressing COX-2 and producing significant amounts of PGE_2.

Discussion

Expression of COX-2 by a human B cell is a novel finding, being shown to be true for malignant B-1 cells, but not as yet for B-2 cells. This difference may be due to the more primitive origin of B-1 cells versus their more mature B-2 counterparts. The transformation of other cell types (e.g. colorectal cells) has been linked to the overexpression of COX-2 [20], and may be one of the triggers that predisposes B-1 cells to generate malignant B-CLL.

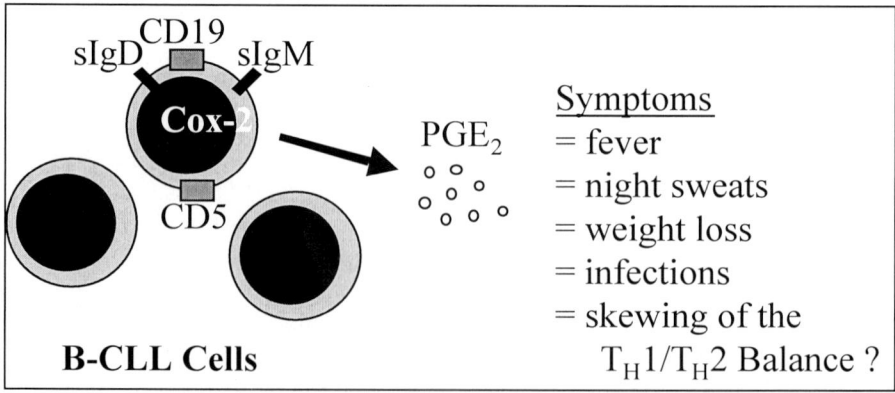

Fig. 2. Production of PGE_2 via COX-2 in B-CLL Cells and possible consequences.

The symptoms associated with high PGE_2 levels in B-CLL patients include fever, night sweats, weight loss and recurrent infections (Fig. 2). It might be possible to relieve some of the symptoms associated with B-CLL by treatment with the recently developed COX-2-specific inhibitors to reduce blood PGE_2 levels. Further, if COX-2 overexpression is the cause of B-CLL transformation, the same treatment may be useful for chemoprevention.

The reason for B cells being capable of expressing COX-2 and producing PGE_2 is also a point of speculation. It may be the primitive nature of B-1 cells that lends to their ability to express COX-2. B-2 cells are more differentiated and possibly commit to a developmental pathway which prevents COX-2 expression. Various leukemic malignancies are transformed from B-1 lineage cells, suggesting that this subset is more undifferentiated and possesses greater pluripotency. The development of B-CLL occurs early in the B lineage, such that COX-2 expression is retained, and the accumulation of the malignant cells allows PGE_2 levels to increase over time.

This research was supported by the following grants: DE11390, CA11198, HL56002, ES07026, T32-DE07061, DK48831, GM42056, CA77839, DK26657, The Rochester Area Pepper Center, The URCC Discovery Fund and The American Cancer Society.

References

1. Hardy RR, Hayakawa K (1994) CD5 B cells, a fetal B cell lineage. Adv in Immunol 55:297-339
2. Smith W, DeWitt D (1996) Prostaglandin endoperoxide H synthases-1 and –2. Adv Immunol 62:167-215
3. Smith W, Garavito R, DeWitt D (1996) Prostaglandin endoperoxide H synthases. J Biol Chem 271:33157-33160
4. Eberhart CE, Coffey A, Radhik A, Giardiello FM, Ferrebach S, DuBois RN (1994) Upregulation of cyclooxygenase 2 gene expression in human colorectal adenomas and adenocarcinomas. Gastroenterology 107:1183-8
5. Tucker ON, Dannenberg AJ, Yang EK, Zhang F, Teng L, Daly JM, Soslow RA, Masferrer JL, Woerner BM, Koki AT, Fahey TJ 3rd (1999) Cyclooxygenase expression is up-regulated in human pancreatic cancer. Cancer Res 59(5):987-90
6. Phipps RP, Stein S, Roper R (1994) A new view of prostaglandin E regulation in the immune response. Immunol Today 12:349-352
7. Van der Pouw Kraan T, Boeije L, Smeenk R, Wijdenes J, Aarden L (1995) Prostaglandin E2 is a potent inhibitor of human interleukin 12 production. J Exp Med 181:775-779
8. Borrello MA, Phipps RP (1995) Fibroblasts support outgrowth of splenocytes simultaneously expressing B lymphocyte and macrophage characteristics. J Immunol 155:4155-4161
9. Borrello MA, Phipps RP (1996) The B/Macrophage cell: an elusive link between CD5+ B lymphocytes and macrophages. Immunol Today 17(10):471-475
10. Borrello MA, Phipps RP (1999) Fibroblast-secreted macrophage colony stimulating factor is responsible for generation of biphenotypic B/Macrophage cells from a subset of mouse B lymphocytes. J Immunol 163:3605-3611

11. Graf BA, Nazarenko D, Borrello MA, Roberts JL, Morrow JD, Palis J, Phipps RP (1999) Biphenotypic B/Macrophage cells express COX-1 and up-regulate COX-2 expression and prostaglandin E2 production in response to proinflammatory signal. Eur J Immunol In Press
12. Dawe CJ, Potter M (1957) Morphologic and Biologic progression of a lymphoid neoplasm of the mouse in vivo and in vitro. Am J Pathol 33:603
13. Sohn CC, Sheibani K, Windberg CD, Rappaport H (1985) Monocytoid B lymphocytes: their relation to the patterns of acquired immunodefiency syndrome (AIDS) and AIDS-related lymphoadenopathy. Hum Path 16:979-984
14. Liu YC, Cleveland RP, Madelaire C, Hines JD, (1995) Discordant immune phenotype of chronic B-cell lymphoproliferative disorders in simultaneous specimens from bone marrow and peripheral sites. Arch Pathol Lab Med 119:53-58
15. Siebert K, Zhang Y, Leahy K, Hauser S, Masferrer J, Perkins W, Lee L, Isakson P (1994) Pharmacological and biochemical demonstration of the role of cyclooxygenase 2 in inflammation and pain. Proc Natl Acad Sci 91:12013-17
16. DuBois RN, Awad J, Morrow J, Roberts LJ, Bishop PR (1994) Regulation of eicosinoid production and mitogenesis in rat intestinal epithelial cells by transforming growth factor alpha and phorbol ester. J Clin Invest 93:493-498
17. Ricote M, Li A, Wilson T, Kelly C, Glass C (1998) The peroxisome proliferator-activated receptor-γ is a negative regulator of macrophage activation. Nature 391:79-86
18. Cheson BD (1992) (ed) Chronic Lymphocytic Leukemia: Scientific Advances and Clinical Developments, National Cancer Institute, Marcel Dekker, New York. (Basic and Clinical Oncology Series, vol 1).
19. Kanti RT, Dilip VP (1980) Chemotherapy of Chronic Lymphocytic Leukemia and Hairy Cell Leukemia. In: Dorr RT, VonHoff DD (eds) Cancer Chemotherapy Handbook. Appleton and Lange, Stamford, CT, pp 1399-1407.
20. Taketo MM (1998) Cyclooxygenase-2 inhibitors in tumorigenesis. J Natl Cancer Inst 90:1609-1620.

Diffuse Large-Cell and "True" Histiocytic Lymphomas of Mice

C.-F. Qi,[1] M. Hori,[1] L. Taddesse-Heath,[1] N. A. Jenkins,[2] N. G. Copeland,[2] H. Shen,[2] T. A. Torrey,[1] J. W. Hartley,[1] S. K. Chattopadhyay,[1] T. N. Fredrickson,[1] and H. C. Morse III[1]

[1] Laboratory of Immunopathology, National Institute of Allergy and Infectious Diseases, National Institutes of Health, Bethesda, MD; and [2] Mammalian Genetics Laboratory, National Cancer Institute, NIH, Frederick, MD

Burkitt-Like Diffuse Large-Cell Lymphoma

The relations of mouse to human lymphomas have been unclear due, in some measure, to differences in terminology for lymphoma types with seemingly similar features. This circumstance is exemplified by the use of the term lymphoblastic lymphoma (LL) as a diagnostic category for mouse lymphoma that covers histologically indistinguishable cases of differing cellular origin (T cell vs. B cell) and stage of differentiation (sIg$^-$ pre-B cell vs. sIg$^+$ B cell). In the proposed WHO nomenclature for human hematopoietic neoplasms, the term LL is reserved for cases originating in precursor B cells or precursor T cells with sIg$^+$ tumors being excluded [1].

Recently, we found that a series of NFS.V$^+$ mouse lymphoma cases diagnosed as B cell LL expressed BCL6 protein at high levels and that about 15% of cases had changes in the genomic organization of *Bcl6* (Qi et al., submitted for publication). More recent studies (Chattopadhyay et al., unpublished observations) demonstrated that the frequency of altered *Bcl6* genomic structure among NFS.N$^+$ LL is lower than this earlier figure (0/60 new cases) and that none were seen among marginal zone lymphomas (MZL) (0/110) or a more limited number of small lymphacytic lymphomas (SLL) (0/13). In addition, the WEHI 231 cell line was shown to have a t(5;16) chromosomal translocation affecting *Bcl6* with the translocation breakpoint located in the first intron of the gene. The earlier findings that similar translocations involving *BCL6* are found in nearly half the cases of human diffuse large-cell B cell lymphomas (DLCL) [2,3] suggested that sIg$^+$ LL might be a subset of mouse DLCL.

NFS.V$^+$ mice comprise strains of NFS/N mice with low tumor incidence congenic for ecotropic virus induction loci from AKR and C58 mice. The mice develop a high incidence of B cell-lineage lymphomas of various histologic types that clearly relate to diagnostic categories for human lymphomas [4]. The types include splenic MZL [5], SLL, follicular (FL), and DLCL with centroblastic (CB) or immunoblastic (IB) cytologic features. LL comprised approximately 25% of cases [6].

To determine whether the suggested identification of sIg$^+$ LL as a subset of DLCL could be generalized beyond cases in NFS.V$^+$, we reevaluated the histopathology of lymphomas developing in AKXD recombinant inbred strains. DNA prepared from these tumors was previously studied for the organization of immunoglobulin (Ig) heavy and light chain genes, T cell receptor genes, and alterations in the structure of a series of proto-oncogenes or preferred proviral integration sites [7,8]. The results of this analysis (Table 1) yielded a series of interesting findings. First, the predominant tumor types—MZL and non-pre-B LL (here designated Burkitt-like; see below)—occurred at very similar frequencies in both tumor sets. Second, DLCL(IB) and DLCL(CB) were not observed among the AKXD lymphomas but comprised 20% of the NFS.V$^+$ lymphomas. Finally, lymphomas comprising histiocytes and IB or CB, here designated DLCL(HS), were detected at a frequency of 20% among the AKXD neoplasms but represented less than 1% of the NFS.V$^+$ cases.

Table 1. Subsets of AKXD RI and NFS.V$^+$ B lymphomas[1]

Diagnosis	Percent of B lineage tumors	
	AKXD	NFS.V$^+$
Pre-B LL	8	ND[2]
SLL	1	10
MZL	35	42
FL	8	6
DLCL		
CB	0	14
IB	0	6
BL	28	23
HS	20	<1

[1] Numbers derive from studies of 366 AKXD and 677 NFS.V$^+$ lymphomas and leukemias of all histologic types.
[2] ND, not determined.

A series of the AKXD B cell lymphomas was examined by Southern hybridization for the organization of *Bcl6*. Similar to the NFS.V$^+$ series, about 15% of cases exhibited changes in the genomic structure of this locus. These changes appear not to be due to proviral insertional mutagenesis. Analyses of genomic DNA sequences flanking a large series of ecotropic virus somatic integration sites [9] in AKXD B cell lymphomas showed that none were included within a 9-kb region in the *Bcl6* locus, including the major translocation complex area between exons 1 and 2. Efforts are in progress to clone the rearranged genes to determine whether translocations are responsible. These results support the contention that the sIg$^+$ LL of both NFS.V$^+$ and AKXD RI mice belong to the diagnostic category of DLCL. The features that distinguish non-pre-B LL from most other B lineage tumors include: apparent origins from the lymph nodes rather than the spleen; cell size intermediate between that of SLL and DLCL(CB) or Burkitt lymphoma; and occasional starry sky appearance. These features are somewhat similar to those of human lymphomas variably diagnosed in the past as Burkitt-like (BL) or atypical Burkitt but now felt to fall under the heading of DLCL. For this reason, we have chosen to designate them as DLCL(BL).

Although a substantial proportion of human DLCL exhibits translocations affecting *BCL6*, essentially all have mutations within the first intron of the gene [10]. Most are point mutations with features similar to those seen in Ig gene variable regions, leading to the suggestion of a common mutational machinery acting on quite distinct genomic regions. Indeed, mutations of this type were subsequently shown to occur in normal B cells in the germinal center [11], the site of Ig hypermutation. Mutations of the *BCL6* first intron have also been described in differing proportions of human FL, BL, and MALT lymphomas as well as DLCL but not in B-CLL or mantle cell lymphoma. To determine whether the first intron of mouse *Bcl6* is mutated in mouse lymphomas, we cloned and sequenced a 9-kb fragment from the NFS/N *Bcl6* locus including 3 kb upstream of the start site. PCR primers were used to amplify portions of the first intron, comparable to the mutational hotspots identified for the human locus, from mouse lymphomas. These fragments were then cloned and sequenced—no mutations were found. Previous studies of mice had indicated that the germinal center B cells from normal Peyer's patch also lacked mutation in the first intron, even though the Ig variable regions were mutated [12]. This suggests that different mutational mechanisms may act on Ig and Bcl6 sequences in humans but not in mice or that the mouse *Bcl6* sequences are not accessible to the mutational machinery. Studies designed to discriminate between these possibilities are in progress.

Histiocyte-Associated DLCL

Histiocytic sarcoma, a term used in the veterinary literature to denote lymphomas of macrophage origin, was diagnosed in both NFS.V$^+$ and AKXD mice. The tumors were characterized by richness of cells with cytologic features of macrophages admixed with lesser numbers of lymphoid cells, usually of centroblastic or immunoblastic morphology. When these diagnoses were aligned with the molecular characteristics determined for the lymphomas, essentially all were found to have clonal rearrangements of both heavy and light chain genes and thus diagnoses as B cell lymphoma. Unlike the majority of other B cell lymphoma types, the non-germline bands were often faint, and there was rarely substantial loss of the germline band, consistent with the low representation of lymphoid cells in the samples. Included among human DLCL are those referred to as histiocyte-rich [13]. These characteristically comprise 90% to 95% histiocytes and the remainder B lymphoma. The frequency of histiocyte-like cells is much lower in the mouse cases, indicating that they are not the precise mouse counterpart of the human histiocyte-rich category of DLCL. We have provisionally designated the mouse tumors as histiocyte-associated DLCL—DLCL(HS).

Although the clonal nature of the B cell component in DLCL(HS) is clear, it is not known whether the "histiocytic" contribution is a reactive or transformed population of true macrophage origin. It is possible that this component could be clonally related to the B cell population, because *i)* normal peritoneal B1 cells and B cells cultured in M-CSF have many features of macrophages [14,15] and *ii)* macrophages can derive from some cultured pre-B cell lymphomas [16].

"True" Histiocytic Lymphoma/Sarcoma

DLCL(HS) contrast with a smaller number of cases in NFS.V$^+$ and AKXD mice that were conspicuous for the uniformity, streaming, and palisading of histiocytes and a virtual absence of lymphocytes. Prototype tumors of this type have been observed in mice predisposed to the development of myeloid neoplasms [16], were malignant as demonstrated by serial transplantation in scid mice, and lacked Ig gene rearrangements. Markers of clonality have not been identified. It would appear that these cases are the mouse counterpart of human "true" histiocytic lymphoma/sarcoma, a rapidly lethal disease very resistant to treatment [17].

Conclusions

B cell lymphomas of mice are considerably more heterogeneous than previously appreciated. Many of the newly identified types have many similarities to human lymphomas, with parallels including histologic and cytologic appearance, immunophenotype, and molecular abnormalities. A major task for the future will be to determine whether these models can be exploited to develop novel insights into the pathogenesis of similar human lymphomas.

References

1. Jaffe ES, Harris NL, Diebold J, Müller-Hermelink H-K. (1999) World Health Organization classification of neoplastic diseases of the hematopoietic and lymphoid tissues. Am J Clin Path 111:S8-S12
2. Ye BH, Rao PH, Chaganti RSK, Dalla-Favera R (1993) Cloning of *bcl-6*, the locus involved in chromosome translocations affecting band 3q27 in B-cell lymphoma. Cancer Res 53:2732-2735
3. Miki T, Kawamata N, Hirosawa S, Aoki N. (1994) Gene involved in the 3q27 translocation associated with B-cell lymphoma, *BCL5*, encodes a Krüppel-like zinc-finger protein. Blood 1:26-32
4. Taddesse-Heath L, Morse HC III (1999) Lymphoma in genetically

engineered mice. In: Ward JM, Mahler J, Maronpot RR, Sundberg JP, Frederickson R (eds) Pathology of genetically-engineered mice. Iowa State University Press, Ames, IA

5. Fredrickson TN, Lennert K, Chattopadhyay SK, Morse HC III, Hartley JW (1999) Splenic marginal zone lymphomas of mice. Am J Pathol 154:805-812
6. Hartley JW, Chattopadhyay SK, Lander MR, Taddesse-Heath L, Nagashfar Z, Morse HC III, Fredrickson TN (1999) Accelerated appearance of multiple B cell lymphoma types in NFS/N mice congenic for ecotropic murine leukemia viruses. Lab Invest (in press)
7. Mucenski ML, Taylor BA, Jenkins NA, Copeland NG (1986) AKXD recombinant inbred strains: models for studying the molecular genetic basis of murine lymphomas. J Virol 6:4236-4243
8. Gilbert DJ, Neumann PE, Taylor BA, Jenkins NA, Copeland NG (1993) Susceptibility of AKXD recombinant inbred mouse strains to lymphomas. J Virol 67:2083-1090
9. Li J, Shen H, Largaespada DA, Nakamura T, Shaughnessy JD, Jenkins NA, Copeland NG (1999) Large-scale cloning of leukemia disease genes via inverse PCR. Nat Genet 23:348-353
10. Migliazza A, Martinotti S, Chen W, Fusco C, Ye BH, Knowles DM, Offit K, Chaganti RSK, Dalla-Fevera R (1995) Frequent somatic hypermutation of the 5' noncoding region of the *BCL6* gene in B-cell lymphoma. Proc Natl Acad Sci USA 92:12520-12524
11. Shen HM, Peters A, Baron B, Zhu X, Storb U (1998) Mutation of *BCL-6* gene in normal B cells by the process of somatic hyprmutation of Ig genes. Science 280:1750-1752
12. Neuberger Ms, Ehrenstein MR, Klix N, Jolly CJ, Yélamos J, Rada C, Milstein C (1998) Monitoring and interpreting the intrinsic features of somatic hypermutation. Immunol Rev 162:107-116
13. Delabie J, Vandenberghe E, Kennes C, Verhoef G, Foschini MP, Stul M, Cassiman JJ, De Wolf-Peeters C (1992) Histiocyte-rich B-cell lymphoma. A distinct clinicopathlogic entity possibly related to lymphocyte predominant Hodgkin's disease paragranuloma subtype. Am J Surg Pathol 16:37-48
14. Fredrickson TN, Harris AW (1999) Atlas of mouse hematopathology. Harwood Academic Publishers, Sydney.
15. Borrello MA, Phipps RP (1999) Fibroblast-secreted macrophage colony-stimulating factor is responsible for generation of biphenotypic B/macrophage cells from a subset of mouse B lymphocytes. J Immunol 163:3605-3611
16. Davidson WF, Pierce JH, Rudikoff S, Morse HC III (1988) Relationship between B cell and myeloid differentiation. J Exp Med 168:389-407
17. Ralfkiaer E, Delsol G, O'Connor NTJ, Brandtzaeg P, Brousset P, Vejlsgaard GL, Mason DY (1990) Malignant lymphomas of true histiocytic origin. A clinical, histological, immunophenotypic and genotypic study. J Pathol 160:9-17

Opinions on the Nature of B-1 Cells and Their Relationship to B Cell Neoplasia

M. Potter[1] and F. Melchers[2]

[1]Laboratory of Genetics, National Cancer Institute, National Institutes of Health, Bethesda, MD USA; [2]Basel Institute for Immunology, Basel, Switzerland

B-1 Cells: Historical Background

In the late 1970's and early 1980's the classical T cell antigen Leu-1 in humans or Lyt-1 in the mouse was found on B-CLL cells and several lymphomas in the mouse (see [1,2,3] for references). From this beginning two salient facts began to emerge. First, the Leu-1 (CD5) marker was consistently demonstrable on 95% of B-CLL cells but also in 3-5% of normal B lymphocytes in the spleen and tonsils [3] and, second, the mouse homologue Ly-1 or Lyt-1 (CD5) was discovered on a subpopulation of splenic B cells in normal mice [2]. In 1986 a second $CD5^+$ B cell population was reported in mice located in the peritoneal cavity (PerC) [4]. Lee Herzenberg's laboratory at Stanford then developed a series of elegant studies on $CD5^+$ B cells in the mouse that demonstrated these cells possessed special biological properties (see below). $CD5^+$ B cells have been found in other mammalian species, but they appear to have different characteristics in each. As yet a common pattern has not emerged.

In 1991 at a now famous New York Academy of Sciences meeting held in Palm Beach Gardens, FL on "CD5 B Cells in Development and Disease" the participants adopted the B-1/B-2 nomenclature [5] to designate the two functionally distinct subpopulations of B cells. In this division it was proposed that $CD5^-$ B-2 cells (synonymous with conventional B cells) are generated continuously in the adult bone marrow. The adult bone marrow hematopoietic stem cells (HSC) appeared to have only minimal effectiveness in regenerating the peritoneal cavity $CD5^+$ B cell population after hematopoietic destruction by total body x-irradiation. In contrast, the fetal liver HSCs generate $CD5^+$ B-1 cells that are the predominant B cell of the fetus and neonate. Significantly, several weeks after birth the liver ceases to provide these cells. Nonetheless, B-1 cells persist in adults through self renewal of V(D)J rearranged B cells in the peripheral tissues. In the adult mouse B-1 cells are a minor population that is predominantly located in the PerC and spleen. A second smaller but functionally similar subset of B cells in the mouse was found that expressed IgM^{hi}, IgD^{lo} that did not express CD5 [6,7,8], and because these cells were also self renewing, they were designated B-1b; however, adult bone marrow stem cells can regenerate B-1b cells much more efficiently than B-1a cells [6]. Honjo and his colleagues have evidence that B-1 cells also can be found in the lamina propria of the gut [9,10,11].

The question that has evoked much discussion about $CD5^+$ B cells concerns their development and differentiation. Two major hypotheses are: first, that these cells arise as special cellular products of fetal/neonatal B lymphopoiesis [these are the 'lineage' [12] and related 'developmental switch' [13] hypotheses] and, second, the induction hypothesis is based on evidence that $CD5^+$ expression can be induced in mature $CD5^-$ B-2 cells [14].

The Lee Herzenberg school contends that the $CD5^+$ B cells originate from HSC in the fetal liver of the mouse as a separate lineage [15], see *Herzenberg*] or by a special developmental pattern [see *Hayakawa* and [13,16]]. The Geoffrey Haughton school proposes that the $CD5^+$ phenotype is induced (see *Berland* and *Wortis*). A debate on this subject between Lee Herzenberg and Geoffrey Haughton with comments by Henry Wortis and Richard Hardy was published in *Immunology Today* in 1993 [15]. The argument opposing the separate lineage hypothesis was epitomized by Haughton *et al.* who titled their paper "*B- 1 Cells Are Made Not Born*" [15] and proposed B-1 cells arise by a "separate differentiation pathway" rather than a lineage. The experimental basis for this hypothesis came from experiments by Henry Wortis and his associates who demonstrated that $CD5^-$ B-2 cells in the mouse could be induced to express CD5 by anti Ig and IL-6 [14].

The discussion of this topic continues in this book. *Clarke* (see *Tatu* and *Clarke*) extends the Haughton hypothesis as an "induced differentiation" hypothesis [17,18] which results from several developmental steps that select (peripheral?) "B-1" cells from a common uncommitted precursor of both B-1 and B-2 cells, the B-0 cell. *Berland* and *Wortis* (this volume) postulate that B cells with specificity for a specific autoantigen upon binding to that autoantigen are "driven" into the B-1a phenotype. Both of these models suggest that the differentiation steps take place in mature B cells possibly in peripheral lymphoid tissues. In contrast, *Hayakawa* and *Hardy* explain the origin of "B- 1a" cells as a developmental process coupled to cellular selection events occurring during the pro B to pre B steps. These steps are determined in B cells by the specificity of the pre B cell receptor interaction with antigen. This is followed by a cellular selection process into B-1a ($CD5^+$) cell types (see *Hayakawa, Hardy*).

$CD5^+$ B Cells in the Mouse

As a basis for discussion the characteristics of the $CD5^+$ B cells in mice are listed and described (Table 1).

Table 1. Morphological and functional characteristics of CD5$^+$ B cells in the mouse

1.	Cell surface markers: IgM^{hi}, IgD^{lo}, B220^{lo}, CD5$^+$, CD43$^+$, CD11b, (MHC-1) [19].
2.	B-1 cells are relatively large lymphocytes that exhibit an activated phenotype. STAT-3 is constitutively phosphorylated in B-1 cells [20].
3.	CD5$^+$ B cells are located in peritoneal and pleural cavities (PerC) ranging from 5 to 70% depending upon inbred strain background and spleen where there are 5% of the B-220 fraction [4]; generated in the fetal liver and omentum [21] are the predominant B cells in fetuses and neonates.
4.	The CD5$^+$ PerC B cells are the apparent progenitors of 50% of the extensive lamina propria plasma cells that secrete IgA (plasma cells do not express CD5) [22,23].
5.	PerC CD5$^+$ B cells are self renewing from V(D)J differentiated cells [6,24,25,26].
6.	The immunoglobulins (IgM, IgG) produced by many of the CD5$^+$ PerC B cells bind autogenous antigens, *e.g.*, phosphatidyl choline, ss and ds DNA [27], Thy-1 [28]; and some bacterial antigens: phosphoryl choline [29]; several polysaccharides [30] and in addition the Igs and the corresponding BCRs frequently bind more than one unrelated antigens (polyreactivity) usually with low affinity [31].
7.	CD5$^+$ B cells are produced in very low numbers in Btk-/- (Xid) [32] and CD19-/- mice.
8.	CD5$^-$ B cells can be induced by specific agents to express CD5 [14].

The relatively high number of CD5$^+$ B cells in the PerC of the mouse [4,25] provided an accessible source of these cells and led to studies on regeneration following depletion by total body x-irradiation. It was discovered that when mice were lethally irradiated and protected post-irradiation by the injection of adult bone marrow cells, restoration of the CD5$^+$ B PerC population did not occur [25]. When, however, the irradiated mice were given PerC B cells after lethal irradiation, the PerC CD5$^+$ B cell population was restored [25,6]. This finding was the critical evidence that established CD5$^+$ B cells as a separate self renewing population. An important feature of the self renewing B cells was that they were V(D)J differentiated, *i.e.*, had productively rearranged their IgL and H chain genes.

Splenic Marginal Zone B Cells (MZ B Cells)
It is appropriate here to describe a functionally related B cell population that apparently do not express CD5 the marginal zone B cells as this subpopulation received considerable attention and discussion (see *Martin* and *Kearney*; *Dammers et al.*) These cells, like the CD5$^+$ B cells, express IgM^{hi},IgD^{lo}, but differ by having the C3d complement receptor CD21. The marginal zone of the spleen is in the white pulp, lies outside of the follicular perimeter and contains both T cells and mature B cells. This B cell population apparently does not actively recirculate. MZ B cells are positioned so that antigens from invaders into the blood stream can rapidly contact and activate the cells to produce Ig. Thus, the marginal zone B cells are an early response element. These cells play a prominent role in response to Type II thymus independent antigens. A second important function can take place in the spleen. This is a positive selection process whereby a "random repertoire" is converted to a more specific one (see *Martin* and *Kearney*). The transitional steps in mature B cell

maturation may take place in relation to the establishment of the marginal zones [33] and could be sites where B-1 cells are made (see *Tatu* and *Clarke*).

B-1 Cells in the Mouse

While $CD5^+$ B cells are commonly called B-1 cells by some authors in studies of other species, this may be an inadequate definition in the mouse where more information is available. We shall define B-1 cells in the mouse in a broader context to indicate that group of B cells whose BCRs bind to antigens of autogenous origin with low binding affinity or avidity. Many of these are also polyreactive and can bind foreign antigens as well (see section below on polyreactivity). Cells with these specificities can develop along different pathways (see Fig. 1 and footnote). The original B-1 subdivision defined by Herzenberg and Kantor [5] contained B-1a and B-1b components that potentially originate from different precursors in the fetal liver and adult bone marrow. Lam and Rajewsky [34] have also presented evidence that B-1 cell development is determined by the antigen binding specificities of the BCR and the density of the BCRs on the cell surface (see discussion of induction of CD5 expression for other possibilities).

Properties of Immunoglobulins Produced by B-1 Cells (Natural and Polyreactive Antibodies)

The immunoglobulins produced by B-1 cells in humans and mice have been described as 'natural antibodies' (NAbs)[4,31,35,36,37]. There is probably no simple definition of NAbs which are generally regarded as antibodies that appear in serum without obvious evidence of immunization, or are antibodies that are produced under conditions where the organism is protected from exposure to exogenous (xenoantigens), *e.g.*, the fetus or animals raised in germ free or antigen free conditions. Many Nabs are composed of germ line sequences, *i.e*, have a paucity of both N additions and somatic hypermutations. NAbs are also known by their antigen binding properties, such as low affinity for target antigens, polyreactivity and binding to antigens of autogenous origin. Early characterization of antibodies produced by B-1 cells revealed binding activities for phosphatidyl choline, Thy-1, ss and ds DNA [27] and Ig, the latter two well known potential autoantigens.

A characteristic of many 'natural antibodies' and antibodies produced by $CD5^+$ cells in humans and mice is polyreactivity [35,36,38]. Polyreactive (also referred to as multispecific, degenerate, heteroantibodies,``sticky" or heterophilic

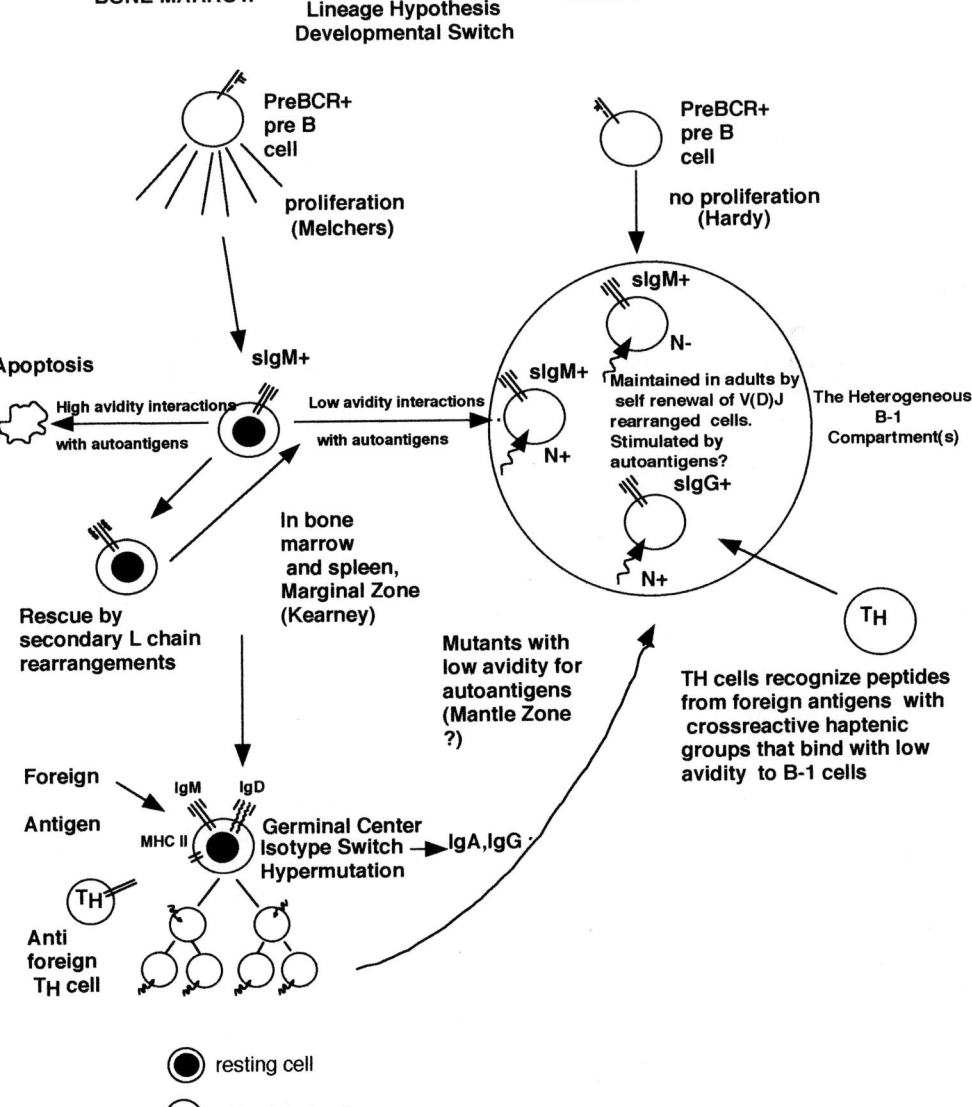

Fig. 1. How the B1 compartment could be filled from two major B cell sources and how it could be maintained. Hematopoietic stem cells give rise to B lineage cells in fetal liver during embryonic development without N-regions at their VDJ joints (N⁻) and in bone marrow during adult life with N-regions (N⁺). However they use the pre B cell receptor, the precursors of the B lineage finally yields cells in which the H chain and L chain genes have been productively rearranged and an IgM molecule has been deposited on the surface [for a review see [39]]. These immature B cells are exposed to autoantigens at the sites where they are generated. Since many of the newly generated cells die by apoptosis, the collection of autoantigens presented at these primary sites include not only membrane-bound but also intracellular and nuclear antigens. High affinity interactions between IgM and autoantigens induce apoptosis of the corresponding specific immature B cell. Some of these high affinity autoreactive immature B cells may be rescued by L chain editing. Low avidity interactions, on the other hand, could activate, and thereby positively select B cells from death into the B1 cell compartment. Continuous interactions with the same autoantigens could keep them alive, activated and self-renewing even after they have moved into the B cell-specific sites, *e.g.*, into the peritoneal cavity. Positive selection of immature B cells into the B1 compartments can continue after these immature cells migrate to the spleen and as they enter the spleen at the marginal zone.

Immature B cells with no affinity for autoantigens present in the primary lymphoid organ, the circulation or the spleen, migrate into the B cell rich follicular regions where they differentiate to mature B cells which remain resting until stimulated by foreign antigens. They are expected to be conventional B2 cells.

Foreign antigen is first taken up by dendritic cells which process and present the peptides of the foreign antigen in complexes with MHC. Peptide-MHC class II complexes activate anti-foreign helper T cells (T_H). Foreign antigen is also taken up via IgM and IgG in antigen-specific, high affinity interactions by mature B cells in follicles, processed and the peptides of the foreign antigen presented in MHC class II molecules on B cells to the activated T_H cells. This puts in motion a germinal center reaction in the webs of follicular dentritic cells. B cells are induced to proliferate, to switch their expression of Ig H chain genes, to hypermutate their V_H- and V_L-regions, to secrete Igs and to develop into memory cells. Hypermutation will not only generate mutant B cells with higher affinity for the foreign antigen but, on occasion, also mutant B cells producing low affinity, autoantigen reactive Ig, also of IgG and IgA classes. They may accumulate in the mantle zone of the germinal center. Continuous exposure to autoantigens, again, may positively select them from death, keep them activated and self-renewing, even after they have moved to other sites such as the peritoneum.

The B1 cell compartment is kept at this low level of activation through low affinity interactions of the sIgs of its cells until foreign antigen enters. If it shows mimicry with an autoantigen in a haptenic group recognized by an Ig on a B cell, this B1 B cell will take up the mimicry antigen (for example, hsp 60 of mouse with hsp 65 of bacteria, or groEL - see *Potter*), process it, but now present foreign peptides of the bacterial protein on MHC class II, hence activating T_H cells to that foreign peptide/MHC class II complex. If the infection is chronic, so will be the B1 response, thus becoming a potential site for malignant transformation of the B1 cells. Antibiotics may be able to cure this chronic B1 cell stimulation.

Pospisil and Mage [40] hypothesize that survival signals may be transmitted to B cells via V_H framework region interactions with superantigens such as endogenous CD5 and exogenous microbial products.

antibodies [41,42,43,44]) bind to two or more evolutionarily and structurally unrelated antigens. Formal proof of polyreactivity historically depended upon the availability of monoclonal antibodies that were generated with the hybridoma technology. The binding affinities of these interactions are low ($< 10^{-7}$M) (see *Notkins*). Such reactions are thought to differ from cross-reacting epitopes shared by two genetically related but distinct proteins. Antigenic targets (epitopes) for polyreactive antibodies may include structures on self as well as foreign antigens. Polyreactivity also characterizes many fetal/neonatal immunoglobulins [45,46]. B cells producing polyreactive antibodies can be found in the spleen, circulation, PerC and Peyer's patches [47,48]. Many of these polyreactive antibodies are autoreactive with IgG (RF) and DNA.

Several functions have been proposed for polyreactive antibodies. First, polyreactive antibodies are produced constitutively and, thus, can be a potential first line of humoral immune defense. Second, polyreactive antibodies can bind to cellular debris from dying cells and aid phagocytes in their removal. Third, Notkins and his co-workers have described an additional function of B cells that produce polyreactive antibodies, mainly to act as antigen presenting cells [49]. They have labeled polyreactive B cells with fluorescent probes and have shown that these cells lack a critical co-stimulatory receptor B-7. They suggest that the T-B interaction involving B cells with polyreactive Ig receptors would suppress T cell proliferation and activation and induce T cell tolerance.

Polyreactive antibodies utilize different Ig V_L and V_H gene families [46]; however, the H chain appears to be the more important determining component. Further, H-CDR3 has been directly implicated by gene swapping and site directed mutagenesis experiments [50,51,52,53,54] as the critical region in the Ig molecule that determines the polyreactive phenotype. Ichigashi and Casali [54] constructed recombinants of polyreactive Rheumatoid Factor (RF) with a monoreactive anti-insulin antibody and demonstrated the polyreactivity phenotype resided in the H-CDR3 of the polyreactive partner. Martin *et al.* [51] generated a series of mutations in H-CDR3 of another polyreactive RF (SMI) and showed very conclusively that changing Asp at position 100 to Asn, or Arg at position 100 for Cys, abolished the polyreactive phenotype. Schroeder *et al.* found that the fetal repertoire is highly enriched for polyreactive antibody producing B cells in both humans and mice [52] and proposed that factors limiting the size of H-CDR3 generated antibodies with 'flat' contours that would be more likely to form non-specific reactions. B cells expressing Ig's in germ line configuration and lacking are more likely to have short H-CDR3 regions.

Factors that Induce the Expression of CD5 in B Cells

The expression of the CD5 transmembrane protein can be induced in some CD5⁻ B-2 cells in mice [14] and humans [55,56]. This was first demonstrated in humans by treating peripheral blood mononuclear cells or tonsillar B cells *in vitro* with phorbol esters (PMA) [57,58,59]. CD5 could be demonstrated in cultures of CD5⁻ cells within 60 hours. CD5 expression was not associated with proliferation. Zupo *et al.* [60] found that there were two subpopulations of CD5⁻ B cells in human tonsils. A minor population could be induced to express CD5 by PMA, while the major population was non-inducible. Induction was not due to expansion of pre-existing CD5$^+$ B cells.

Ying-zi *et al.* [14] isolated IgM$^+$, B220$^+$, CD23$^+$, IgD^{hi} high density B cells from the spleens of mice, stimulated them *in vitro* with anti-Ig (monoclonal rat anti-IgM) and found that CD5 was induced and expressed on the surface of these cells. The anti-Ig induced cells, however, continued to express IgD and CD23 which is not the typical phenotype of a B-1a cell. However, when IL-6 was added to the cultures, these cells lost IgD and CD23 and resembled the B-1a phenotype. LPS failed to induce CD5. B cells stimulated with thymus dependent inductive signals did not express CD5 [14].

New understanding of the physiological effects of CD5 receptor signaling in B cells is currently developing from new findings and promises to provide new insights into the B-1/B-2 subdivision. Biochemical data support the cellular findings by showing that signaling in CD5$^+$ B cells (B-1) differs from that in B-2 cells. First, in sharp contrast to B-2 cells, anti-Ig does not activate these cells into cell cycle. *Rothstein et al.* (this volume) show that CD5$^+$ B cells have constitutively (phosphorylated) activated nuclear STAT-3. *Berland* and *Wortis* (this volume) have found the enhancer NFATc is elevated in CD5$^+$ B cells.

Function of CD5 in Mouse Peritoneal B Cells

A relevant question concerns the physiological role of CD5 in B cells. Most is known about this question in the murine PerC B cell population. Whether this information is applicable to other CD5$^+$ B cells is not at all clear. Further, there is no universal agreement on how CD5 works, but there is increasing evidence that CD5 associates with both the TCR in T cells and BCR in B cells and has a regulatory function involving signaling from these receptors. In murine Per C B cells Bikah *et al.* [61] noted that anti sIgM treatment of B-2 cells induced a proliferative response while failing to do so in CD5$^+$ B-1 cells. They then compared the responses of PerC Mac1$^+$ B cells (the counterpart of B-1a cells) isolated from CD5$^{-/-}$ (knockout) mice and found these cells proliferated when treated with anti-sIgM. Further, when CD5$^+$ B cells were first treated with anti-CD5 and then stimulated with anti sIgM, there was a proliferative response. These findings led to the compelling proposal that CD5 negatively regulated proliferative signals emanating from BCR ligation in PerC B cells. It follows that the function of CD5 was to prevent the proliferative [62] expansion of B cells PerC that produce autoantibodies [63].

Two of the critical functions of PerC CD5$^+$ cells are production of about 50% of the serum IgM [27,64] and to act as precursors for approximately 40-50% of the vast lamina propria plasma cell population [22,65]. Very few, if any, PerC CD5$^+$ B cells can be found to synthesize DNA even though the turnover rate was found to be

1.3% per day [66]. This suggests that CD5 may play a role in suppressing proliferation while maintaining the cells in an activated state presumably to produce and secrete antibody. A fraction of the PerC CD5$^+$ cells must leave the PerC and enter the circulation, lose CD5 and continue along the differentiation pathway. The experiments by Honjo and his colleagues indicate that antigens from the microbial flora activate B-1 cells that migrate to the lamina propria to become IgA secreting plasma cells [9,10]. Honjo *et al.* present evidence that the PerC B-1 cells migrate to the LP and even possibly return to the peritoneal space. The evidence for the later possibility is incomplete.

CD5$^+$ and Self Renewing V(D)J Rearranged B Cells in Other Species

CD5$^+$ B cells in some other mammals have been called B-1 cells. However, CD5$^+$ B lymphocytes have different biological characteristics than those found in the mouse. These differences relate primarily to the proportions and distributions of these cells in the different species, but the large numbers of CD5$^+$ B cells in other species suggest there may be other important differences.

CD5
In avians CD5 is expressed at low levels in all adult B-lymphocytes [67]. In the rabbit, which has been in the past one of the favored species for generating high affinity specific antibodies, most if not all of the circulating (PBL) and lymph node (LN) B lymphocytes are CD5$^+$ [68]. In adult humans and sheep the proportions and numbers of CD5$^+$ B cells are considerably higher than in the mouse and, further, have been found in different tissues, notably the lymph nodes and peripheral blood (PBL). In humans 5-30% of splenic and PBLs are CD5$^+$ [69]; these cells are also found in the tonsils [60] and PerC [70]. CD5$^+$ B cells are concentrated in the mantle zones on the outer rims of lymphoid follicles [3]. Many CD5$^+$ B cells in humans express high IgD [71]. In sheep 31-65% of PBLs are CD5$^+$ [72]; however, in contrast to B-1 cells in the mouse these B cells proliferate after stimulation with anti IgM and have a high rate of apoptosis.[73]. The levels of CD5$^+$ cells differ in individual sheep. CD5$^+$ (B-1) cells have not been found in rats [74]. These various findings indicate CD5 expression varies markedly in different species. It is too premature to conclude that CD5$^+$ B cells from the different species represent a homologous functional subset of cells. This will rest on careful analyses in each species that include functional studies.

Self Renewal
The self renewal or replenishment of V(D)J rearranged B cells has interesting precedents in other species where B cell development, Ig gene rearrangement and

BCR diversification take place in a separate B cell lymphopoietic tissue in fetal and early life. The B cells derived from these tissues leave the site and migrate into the periphery where they proliferate through self renewal and function throughout adult life. In several cases, *i.e.*, the Bursa and the Ileal Peyer's Patches (IPP) [60,75], this tissue source involutes. The most famous example is the Bursa of Fabricius in avians [76,77]. Pre-bursal B stem cells with rearranged Ig genes enter the bursa, becoming bursal stem cells that proliferate, rearrange their Ig genes and then diversify by a gene conversion mechanism. The bursa involutes early in life and ceases to supply new B cells. Avian adults essentially utilize these post-bursal stem cells as a source of antibody forming cells (see Table 2 for references). Related self renewing mammalian B cells are also derived from B cells generated in the IPP in sheep [75,60] which also involutes at 3 months of age. The rabbit appendix (see *Mage*) appears similar to the avian bursa being a site where cells with rearranged Ig genes proliferate and undergo V gene diversification by gene conversion. This occurs in the first few months after birth and requires gut flora. Although the rabbit appendix does not involute, it is of key importance in young animals for generation and expansion of B lymphocyte populations with diverse repertoires [77,78]. The PerC $CD5^+$ cells in the mouse resemble the peripheralized B cells in avians and rabbits where diversification takes place in the bursa or appendix and the progeny migrate to the periphery and self replenish. B-1 cells are the predominant B cell populations in fetuses and during the neonatal period in humans and mice. These B cells are generated chiefly in the fetal liver and the omentum. The fetal liver ceases to be a B lymphopoietic organ in the adult. The self renewal mechanism has not been clearly defined in the mouse and is chiefly based on reconstitution experiments. The phenomenon of chronic allotype suppression after neonatal exposure to anti-allotype suppression in rabbits and in the B-1 subset of mice is consistent with self renewal of the affected cell populations [79,80].

Functions of B-1 Cells and the Antibodies They Produce in Immune Defense

In this volume several examples of B-1 cell participation in the defense against infection or inflammation were described. *Baumgarth et al.* showed that IgM produced by B-1 cells is required for protection against influenza virus infections in mice. *Rajan* and *Paciorkowski* have found that B-1 cells assist in the recovery from experimental filarial infections with *Brugia malayi*. *Askanese* and *Tsuji* have shown that B-1 cell produced IgM participates in the recruitment of T cells in contact sensitivity. Recently, in the literature *Ochsenbein et al.* [81] have shown that NAbs play an important role in the defense against viral infections by trapping viral antigen in lymph nodes, thus permitting the immune system to develop immune responses.

Two forms of inflammation that develop from tissue injury have implicated NAbs, IgM produced by CD5$^+$ B-1 cells in the mouse. In an ischemia- reperfusion model system it has been found that complement plays an active role in inducing tissue injury. The injury process is thought to induce the presentation of neo-antigens which are bound by circulating IgM, thus triggering complement activation. The effective IgM has been hypothesized to be a NAb, and experimental evidence supports the possibility that the antibodies in question are produced by B-1 CD5$^+$ cells (see *Reid*). A second inflammation model system is the formation of atherosclerotic plaques in mice. In this system *Silverman* has shown that cells producing T-15 anti-phosphoryl choline antibodies are present in the atherosclerotic plaques. The T-15 idiotype is a much studied NAb in mice, and has been shown by Masmoudi *et al.* to be produced by CD5$^+$ B-1 cells [29].

CD5$^+$ B Cells and B Cell Neoplasia

The differentiation of the BCR generates different functional classes of antigen binding cells. Those whose BCR receptors which have weak binding activity for self antigens, which are often polyreactive, can upon arrival in the periphery have a higher chance of finding a self antigen. Important relevant autoantigens are those that are associated with cell death and the liberation of complexes of the cytoskeleton, of nucleic acids with attached proteins and fragments of membranes. B cells at sites of inflammation may be repeatedly stimulated to remain viable or even proliferate. This type of stimulation may have the important cellular consequence in prolonging the life of autoreactive clonal cells because of available antigen. Inflammation could be a factor that increases the availability of autoantigens. Extension of B cell clonal life may permit more opportunities for stochastic mutational processes to generate oncogenic mutations but also provide the necessary cell divisions for establishing these mutations in the genome.

B-CLL from the very beginning of the CD5 story has been the paradigm of a B cell neoplasm with consistent 95% association with CD5 expression. The B-CLL cell has presented many problems for investigators interested in establishing a relationship between the characteristic 'tumor' cells and one of the components of the B cell family. The B-CLL cell has the phenotype of a mature B lymphocyte and has been found to resemble the B-lymphocytes of the mantle zone of secondary lymphoid follicles [3,82], *i.e.*, the rim of cells that surround germinal centers. These B-CLL cells accumulate 'relentlessly' in the circulation, and although they are out of cycle *in vivo* and *in vitro*, they can be activated to secrete Ig *in vitro* by phorbol esters (PMA) and driven into cell cycle by other cooperating agents [83]. The search for the basic molecular genetic defects in the B-CLL cell continues (see *Migliazza* this volume for a full discussion of the current status of this problem). By various means, including induction of maturation to an Ig secreting cell status and through hybridomas, it has been possible to recover and characterize the Igs produced by B-CLL cells [84,85,86]. The antigen binding activities include: IgG, ssDNA, ds DNA, histones, cardiolipin, and cytoskeletal proteins, actin, myosin and tubulin. In addition many of the antibodies are poly (multi) reactive. Thus, the predominant type of antigen B-CLL cell Igs bind are autoantigens. Clinically, B-CLL patients suffer from autoimmune related phenomena, including various cytopenias and hemolytic anemias. The connection between the autoantibodies produced in B-CLL cells and the associated autoimmune manifestations is not a direct one, and the autoimmune pathology may be secondary (see [85]). As the disease progresses the patients become increasingly hypogammaglobulemic [82,84]. A similar set of antigen binding activities has been described for B cell follicular lymphomas [87]. *Damle et al.* (this book) have shown that B-CLLs can be divided into two groups based on the Igs they produce. One type produces Igs that do not have somatic hypermutations (the more malignant group), and the other that has extensive somatic mutations. This suggests that at least some of the precursors of these cells are antigen driven to enter germinal centers.

In a related process in autoimmune mice [NZB, (NZB x NZW)F1], $CD5^+$ B cells expand in number [88] progressively beginning around 3 to 4 months of age. Specific clones of cells can be detected early. Later in adult life the expanding $CD5^+$ cells begin to infiltrate lymphoid tissues. In some of the mice intraclonal progression to a large cell lymphoma (Richter's Syndrome) stage can occur [89]. Some of the clonal populations of PerC $CD5^+$ B lymphocytes appear around 3 months of age [88]. These cells do not appear to secrete Ig, however, hybridomas generated from these cells have been studied, and these are associated with the formation of NAbs with auto and polyreactivity [31].

Other evidence supporting the relationship of NAb producing cells to B cell neoplasia is indirect and comes from evidence that many of the immunoglobulins recovered from human B cell tumors: B-CLL [86,84,90], follicular lymphoma [87] and various monoclonal gammopathies [91] are autoreactive. B-CLL, follicular lymphoma and monoclonal gammopathies are autoreactive, polyreactive and bear a close resemblance to known NAbs (e.g., RFs, ss and dsDNA) [91]. There is some early suggestive evidence that pristane induced plasmacytomas secrete polyreactive antibodies (see *Potter*). Thus, the subset of B cells that produces auto- and polyreactive antibodies may be more prone to develop oncogenic mutations and progressing to B cell neoplasia as B cell lymphomas, leukemias and plasma cell tumors.

A particularly puzzling feature of B cell related tumors in humans are indolent phases of tumor evolution that are so prominent in Follicular Lymphoma, B-CLL and the MGUS phase of myeloma. The transition from a cell type that may have clonal longevity because it is continuously stimulated by autoantigens to one with low grade neoplastic properties similar to those in indolent B cell tumors are probably difficult to characterize at present. Aneuploidy has been described as an early phenotypic change in the evolution of expanded clones in the NZB mouse [89]. There are data showing that some of the consistent oncogenic mutations, *e.g.*, the chromosomal translocation t(14;18) associated with FL in humans [92] and t(12;15) in mice, can be detected in normal mice [93] and are associated with these indolent early stages of B cell tumor formation.

Concluding Remarks

The antibody generating and diversifying tissues in mammals and other vertebrates produce large numbers of B cells each with its own specificity for antigens. These cells are subject to selection mechanisms that can eliminate (negative selection) or encourage survival of the cells that are ready to enter the peripheral B cell pool (positive selection). Classification of cells in the peripheral B cell pool into subsets has not yet been successful. B cell surface markers that delineate subsets across species lines have not been found. CD5 is a potential candidate marker and has had some usefulness in the mouse and humans but shows considerable variation in levels of the expression and tissue distribution in other species. Too little is known at present about how CD5 actually works in these other settings to use this marker as a universal discriminant for a B cell subset.

The most fundamental marker in B cells is the antigen binding receptor (BCR), but no single classification based on specificities is acceptable as yet. There are some suggestive divisions of B cells based on the specificities of antigen binding sites. Those cells with BCRs capable of recognizing autoantigens with high affinity are eliminated (negatively selected) as soon as they are made by the presence of autoantigen. Those cells whose binding sites are capable of interacting with antigens of autogenous origin with low enough avidity/affinity to escape negative selection comprise another arbitrary group. This group includes many B cells whose BCRs are polyreactive. A third group of B cells consists of those that possess the potential to bind to antigens not present in self (foreign antigens).

For many years the physiological importance of low avidity/affinity, autoreactive, polyreactive antibodies produced by the first group of B cells (above) was ignored, but more recently there has been growing interest in these cells and the antibodies they produce. As we learn more about their structure and how they bind antigens *in vivo,* we may be approaching an exciting new era in antibody research. These antibodies may be a component of the innate immune system that forms a first line of defense and also participates in the elimination of cell debris resulting from cell death in tissues. New regulatory functions of these antibodies in inflammation, parasitic infections, antibody responses to viral infections and contact sensitivity were discussed in this book. One of the unusual features of these cells is their continuous stimulation by autoantigens. Growing evidence suggests that the cells that produce these antibodies may be more prone to neoplastic development because they are chronically stimulated.

Acknowledgments

We wish to thank Rose Mage, NIAID, Abner Notkins, NIDR, and Siegfried Janz, LG, NCI, for reading the manuscript and making many helpful suggestions.

References

1. Kipps TJ (1989) The CD5 B cell. Adv Immunol 47:117-185
2. Manohar V, Brown E, Leiserson WM, Chused TM (1982) Expression of Lyt-1 by a subset of B lymphocytes. J Immunol 129:532-538
3. Caligaris-Cappio F, Gobbi M, Bofill M, Janossy G (1982) Infrequent normal B lymphocytes express features of B-chronic lymphocytic leukemia. J Exp Med 155:623-628
4. Hayakawa K, Hardy RR, Herzenberg LA (1986) Peritoneal Ly-1 B cells: genetic control, autoantibody production, increased lambda light chain expression. Eur J Immunol 16:450-456
5. Herzenberg LA, Kantor AB (1992) Layered evolution in the immune system. A model for the ontogeny and development of multiple lymphocyte lineages. Ann N Y Acad Sci 651:1-9
6. Kantor AB, Stall AM, Adams S, Watanabe K, Herzenberg LA (1995) De novo development and self-replenishment of B cells. Int Immunol 7:55-68.
7. Kantor AB (1996) V-gene usage and N-region insertions in B-1a, B-1b and conventional B cells. Semin Immunol 8:29-35
8. Stall AM, Adams S, Herzenberg LA, Kantor AB (1992) Characteristics and development of the murine B-1b (Ly-1 B sister) cell population. Ann N Y Acad Sci 651:33-43
9. Murakami M, Nakajma K, Yamazaki K, Muraguchi T, Serikawa T, Honjo T (1997) Effects of breeding environments on generation and activation of autoreactive B-1 cells in anti-red blood cell autoantibody transgenic mice. J Exp Med 185:791-794
10. Murakami M, Tsubata T, Shinkura R, Nisitani S, Okamoto M, Yoshioka H, Usui T, Miyawaki S, Honjo T (1994) Oral administration of lipopolysaccharides activates B-1 cells in the peritoneal cavity and lamina propria of the gut and induces autoimmune symptoms in an autoantibody transgenic mouse. J Exp Med 180:111-121
11. Murakami M, Tsubata T, Okamoto M, Shimizu A, Kumagai S, Imura H, Honjo T (1992) Antigen-induced apoptotic death of Ly-1 B cells responsible for autoimmune disease in transgenic mice [see comments]. Nature 357:77-80
12. Herzenberg LA, Stall AM, Lalor PA, Sidman C, Moore WA, Parks DR (1986) The Ly-1 B cell lineage. Immunol Rev 93:81-102
13. Hardy RR, Hayakawa K (1991) A developmental switch in B lymphopoiesis. Proc Natl Acad Sci USA 88:11550-11554
14. Ying-zi C, Rabin E, Wortis HH (1991) Treatment of murine CD5- B cells with anti-Ig, but not LPS, induces surface CD5: two B-cell activation pathways. Int Immunol 3:467-476
15. Herzenberg LA, Kantor AB (1993) B-cell lineages exist in the mouse. Immunol Today 14:79-83
16. Hardy RR, Li YS, Hayakawa K (1996) Distinctive developmental origins and specificities of the CD5+ B-cell subset. Semin Immunol 8:37-44
17. Clarke SH, Arnold LW (1998) B-1 cell development: evidence for an uncommitted immunoglobulin (Ig)M+ B cell precursor in B-1 cell differentiation. J Exp Med 187:1325-1334

18. Tatu C, Ye J, Arnold LW, Clarke SH (1999) Selection at multiple checkpoints focuses V(H)12 B cell differentiation toward a single B-1 cell specificity. J Exp Med 190:903-914
19. Stall AM, Wells SM, Lam KP (1996) B-1 cells: unique origins and functions. Semin Immunol 8:45-59
20. Karras JG, Wang Z, Huo L, Howard RG, Frank DA, Rothstein TL (1997) Signal transducer and activator of transcription-3 (STAT3) is constitutively activated in normal, self-renewing B-1 cells but only inducibly expressed in conventional B lymphocytes [see comments]. J Exp Med 185:1035-1042
21. Solvason N, Chen X, Shu F, Kearney JF (1992) The fetal omentum in mice and humans. A site enriched for precursors of CD5 B cells early in development. Ann NY Acad Sci 651:10-20
22. Kroese FGM, Cebra JJ, van der Cammen MJF, Kantor AB, Bos NA (1995) Contribution of B-1 cells to intestinal IgA production in the mouse. Methods: Compan Meth Enzymol 8:37-43
23. Kroese FGM, Butcher EC, Stall AM, Lalor PA, Adams S, Herzenberg LA (1988) Many of the IgA producing plasma cells in murine gut are derived from self-replenishing precursors in the peritoneal cavity. Int Immunol 1:75-84
24. Kantor AB, Herzenberg LA (1993) Origin of murine B cell lineages. Annu Rev Immunol 11:501-538
25. Hayakawa K, Hardy RR, Stall AM, Herzenberg LA (1986) Immunoglobulin-bearing B cells reconstitute and maintain the murine Ly-1 B cell lineage. Eur J Immunol 16:1313-1316
26. Hayakawa K, Hardy RR, Herzenberg LA (1985) Progenitors for Ly-1 B cells are distinct from progenitors for other B cells. J Exp Med 161:1554-1568.
27. Hayakawa K, Hardy RR, Honda M, Herzenberg LA, Steinberg AD (1984) Ly-1 B cells: functionally distinct lymphocytes that secrete IgM autoantibodies. Proc Natl Acad Sci U S A 81:2494-2498
28. Hayakawa K, Asano M, Shinton SA, Gui M, Allman D, Stewart CL, Silver J, Hardy RR (1999) Positive selection of natural autoreactive B cells. Science 285:113-116
29. Masmoudi H, Mota-Santos T, Huetz F, Coutinho A, Cazenave P-A (1990) All T15 Id-positive antibodies (but not the majority of VHT15+ antibodies) are produced by peritoneal CD5+ B lymphocytes. Int Immunol 2:515-520
30. Lehmann H-P, Lehle G (1991) A myeloma M 104E private idiotope is represented predominantly among anti-dextran, peritoneal cavity Ly-1+ precursor B cells. Eur J Immunol 21:1201-1205
31. Kaushik A, Lim A, Poncet P, Ge XR, Dighiero G (1988) Comparative analysis of natural antibody specificities among hybridomas originating from spleen and peritoneal cavity of adult NZB and BALB/c mice. Scand J Immunol 27:461-471
32. Hayakawa K, Hardy RR, Parks DR, Herzenberg LA (1983) The "Ly-1 B" cell subpopulation in normal immunodefective, and autoimmune mice. J Exp Med 157:202-218.
33. Loder F, Mutschler B, Ray RJ, Paige CJ, Sideras P, Torres R, Lamers MC, Carsetti R (1999) B cell development in the spleen takes place in discrete steps and is determined by the quality of B cell receptor-derived signals. J Exp Med 190:75-89
34. Lam KP, Rajewsky K (1999) B cell antigen receptor specificity and surface density together determine B-1 versus B-2 cell development. J Exp Med 190:471-477
35. Ternynck T, Avrameas S (1986) Murine natural monoclonal autoantibodies: a study of their polyspecificities and their affinities. Immunol Rev 94:99-112
36. Casali P, Schettino EW (1996) Structure and function of natural antibodies. Curr Top Microbiol Immunol 210:167-179
37. Bhat NM, Kantor AB, Bieber MM, Stall AM, Herzenberg LA, Teng NN (1992) The ontogeny and functional characteristics of human B-1 (CD5+ B) cells. Int Immunol 4:243-252
38. Chen ZJ, Wheeler J, Notkins AL (1995) Antigen-binding B cells and polyreactive antibodies. Eur J Immunol 25:579-586

39. Melchers F, Rolink T (1998) B lymphocyte development and biology. In: Paul WE (ed) Fundamental Immunology. Philadelphia, Lippincott-Raven, Philadelphia, pp 183-224
40. Pospisil R, Mage RG (1998) CD5 and other superantigens as 'ticklers' of the B-cell receptor. Immunol Today 19:106-108
41. Sperling R, Francus T, Siskind GW (1983) Degeneracy of antibody specificity. J Immunol 131:882-885
42. Nahm MH, Hoffmann JW (1990) Heteroantibody: phantom of the immunoassay [editorial; comment]. Clin Chem 36:829
43. Morris RJ (1995) Antigen-antibody interactions: how affinity and kinetics affect assay design and selection procedures. In: Ritter MA, Ladyman HM (eds) Monoclonal antibodies.. University Press, Cambridge, pp 34-59
44. Nigg EA, Walter G, Singer SJ (1982) On the nature of crossreactions observed with antibodies directed to defined epitopes. Proc Natl Acad Sci U S A 79:5939-5943
45. Chen ZJ, Wheeler CJ, Shi W, Wu AJ, Yarboro CH, Gallagher M, Notkins AL (1998) Polyreactive antigen-binding B cells are the predominant cell type in the newborn B cell repertoire. Eur J Immunol 28:989-994
46. Bhat NM, Bieber MM, Teng NN (1997) Heavy chain variable gene usage by human B-1 lymphocytes and polyreactive autoantibodies. Hum Antibodies 8:146-150
47. Shimoda M, Inoue Y, Azuma N, Kanno C (1999) Natural polyreactive immunoglobulin A antibodies produced in mouse Peyer's patches. Immunology 97:9-17
48. Quan CP, Berneman A, Pires R, Avrameas S, Bouvet JP (1997) Natural polyreactive secretory immunoglobulin A autoantibodies as a possible barrier to infection in humans. Infect Immun 65:3997-4004
49. Chen ZJ, Shimizu F, Wheeler J, Notkins AL (1996) Polyreactive antigen-binding B cells in the peripheral circulation are IgD+ and B7-. Eur J Immunol 26:2916-2923
50. Crouzier R, Martin T, Pasquali JL (1995) Heavy chain variable region, light chain variable region, and heavy chain CDR3 influences on the mono- and polyreactivity and on the affinity of human monoclonal rheumatoid factors. J Immunol 154:4526-4535
51. Martin T, Crouzier R, Weber JC, Kipps TJ, Pasquali JL (1994) Structure-function studies on a polyreactive (natural) autoantibody. Polyreactivity is dependent on somatically generated sequences in the third complementarity-determining region of the antibody heavy chain. J Immunol 152:5988-5996
52. Schroeder HWJ, Ippolito GC, Shiokawa S (1998) Regulation of the antibody repertoire through control of HCDR3 diversity. Vaccine 16:1383-1390
53. Adib-Conquy M, Gilbert M, Avrameas S (1998) Effect of amino acid substitutions in the heavy chain CDR3 of an autoantibody on its reactivity. Int Immunol 10:341-346
54. Ichiyoshi Y, Casali P (1994) Analysis of the structural correlates for antibody polyreactivity by multiple reassortments of chimeric human immunoglobulin heavy and light chain V segments. J Exp Med 180:885-895
55. Freedman AS, Freeman G, Whitman J, Segil J, Daley J, Levine H, Nadler LM (1989) Expression and regulation of CD5 on in vitro activated human B cells. Eur J Immunol 19:849-855
56. Paavonen T, Quartey-Papafio R, Delves PJ, Mackenzie L, Lund T, Youinou P, Lydyard PM (1990) CD5 mRNA expression and auto-antibody production in early human B cells immortalized by EBV. Scand J Immunol 31:269-274
57. Miller RA, Gralow J (1984) The induction of Leu-1 antigen expression in human malignant and normal B cells by phorbol myristic acetate (PMA). J Immunol 133:3408-3414
58. Freedman AS, Freeman G, Whitman J, Segil J, Daley J, Nadler LM (1989) Studies of in vitro activated CD5+ B cells. Blood 73:202-208
59. Freedman AS, Freeman G, Whitman J, Segil J, Daley J, Levine H, Nadler LM (1989) Expression and regulation of CD5 on in vitro activated human B cells. Eur J Immunol 19:849-855

60. Zupo S, Dono M, Massara R, Taborelli G, Chiorazzi N, Ferrarini M (1994) Expression of CD5 and CD38 by human CD5- B cells: requirement for special stimuli. Eur J Immunol 24:1426-1433
61. Bikah G, Carey J, Ciallella JR, Tarakhovsky A, Bondada S (1996) CD5-mediated negative regulation of antigen receptor-induced growth signals in B-1 B cells. Science 274:1906-1909
62. Tarakhovsky A, Muller W, Rajewsky K (1994) Lymphocyte populations and immune responses in CD5-deficient mice. Eur J Immunol 24:1678-1684
63. Sen G, Bikah G, Venkataraman C, Bondada S (1999) Negative regulation of antigen receptor-mediated signaling by constitutive association of CD5 with the SHP-1 protein tyrosine phosphatase in B-1 B cells. Eur J Immunol 29:3319-3328
64. Forster I, Rajewsky K (1987) Expansion and functional activity of Ly-1+ B cells upon transfer of peritoneal cells into allotype-congenic, newborn mice. Eur J Immunol 17:521-528
65. Kroese FG (1989) Many of the IgA producing plasma cells in murine gut are derived from self-replenishing precursors in the peritoneal cavity. Int Immunol 1:75-84.
66. Deenen GJ, Kroese FGM (1993) Kinetics of peritoneal B-1a cells (CD5 B cells) in young adult mice. Eur J Immunol 23:12-16
67. Koskinen R, Gobel TW, Tregaskes CA, Young JR, Vainio O (1998) The structure of avian CD5 implies a conserved function. J Immunol 160:4943-4950
68. Raman C, Knight KL (1992) CD5+ B cells predominate in peripheral tissues of rabbit. J Immunol 149:3858-3864
69. Youinou P, Jamin C, Lydyard PM (1999) CD5 expression in human B-cell populations. Immunol Today 20:312-316
70. Nisitani S (1997) Preferential localization of human CD5+ B cells in the peritoneal cavity. Scand J Immunol 46:541-545.
71. Gadol N, Ault KA (1986) Phenotypic and functional characterization of human Leu1 (CD5) B cells. Immunol Rev 93:23-34
72. Birkebak TA, Palmer GH, Davis WC, McElwain TF (1994) Quantitative characterization of the CD5 bearing lymphocyte population in the peripheral blood of normal sheep. Vet Immunol Immunopathol 41:181-186
73. Chevallier N, Berthelemy M, Laine V, Le Rhun D, Femenia F, Polack B, Naessens J, Levy D, Schwartz-Cornil I (1998) B-1-like cells exist in sheep. Characterization of their phenotype and behaviour. Immunology 95:178-184
74. Vermeer LA, de Boer NK, Bucci C, Bos NA, Kroese FG, Alberti S (1994) MRC OX19 recognizes the rat CD5 surface glycoprotein, but does not provide evidence for a population of CD5bright B cells. Eur J Immunol 24:585-592
75. Reynaud CA, Mackay CR, Muller RG, Weill JC (1991) Somatic generation of diversity in a mammalian primary lymphoid organ: the sheep ileal Peyer's patches. Cell 64:995-1005
76. Weill JC, Reynaud CA (1987) The chicken B cell compartment. Science 238:1094-1098
77. Reynaud CA, Weill JC (1996) Postrearrangement diversification processes in gut-associated lymphoid tissues. Curr Top Microbiol Immunol 212:7-15
78. Vajdy M, Sethupathi P, Knight KL (1998) Dependence of antibody somatic diversification on gut-associated lymphoid tissue in rabbits. J Immunol 160:2725-2729
79. Mage RG, Sehgal D, Schiaffella E, Anderson AO (1999) Gene-conversion in rabbit B-cell ontogeny and during immune responses in splenic germinal centers. Vet Immunol Immunopathol 72:7-15
80. Pospisil R, Mage RG (1998) Rabbit appendix: a site of development and selection of the B cell repertoire. Curr Top Microbiol Immunol 229:59-70
81. Ochsenbein AF, Fehr T, Lutz C, Suter M, Brombacher F, Hengartner H, Zinkernagel RM (1999) Control of early viral and bacterial distribution and disease by natural antibodies. Science 286:2156-2159
82. Caligaris-Cappio F, Hamblin TJ (1999) B-cell chronic lymphocytic leukemia: a bird of a different feather. J Clin Oncol 17:399-408

83. Carlsson M, Matsson P, Rosen A, Sundstrom C, Totterman TH, Nilsson K (1988) Phorbol ester and B cell-stimulatory factor synergize to induce B-chronic lymphocytic leukemia cells to simultaneous immunoglobulin secretion and DNA synthesis. Leukemia 2:734-744
84. Kipps TJ, Carson DA (1993) Autoantibodies in chronic lymphocytic leukemia and related systemic autoimmune diseases. Blood 81:2475-2487
85. Jahn S, Schwab J, Hansen A, Heider H, Schroeder C, Lukowsky A, Achtman M, Matthes H, Kiessig ST, Volk HD (1991) Human hybridomas derived from CD5+ B lymphocytes of patients with chronic lymphocytic leukemia (B-CLL) produce multi-specific natural IgM (kappa) antibodies. Clin Exp Immunol 83:413-417
86. Broker BM, Klajman A, Youinou P, Jouquan J, Worman CP, Murphy J, Mackenzie L, Quartey-Papafio R, Blaschek M, Collins P (1988) Chronic lymphocytic leukemic (CLL) cells secrete multispecific autoantibodies. J Autoimmun 1:469-481
87. Dighiero G, Hart S, Lim A, Borche L, Levy R, Miller RA (1991) Autoantibody activity of immunoglobulins isolated from B-cell follicular lymphomas. Blood 78:581-585
88. Stall AM (1988) Ly-1 B-cell clones similar to human chronic lymphocytic leukemias routinely develop in older normal mice and young autoimmune (New Zealand Black-related) animals. Proc Natl Acad Sci U S A 85:7312-7316.
89. Phillips JA, Mehta K, Fernandez C, Raveche ES (1992) The NZB mouse as a model for chronic lymphocytic leukemia. Cancer Res 52:437-443
90. Borche L, Lim A, Binet JL, Dighiero G (1990) Evidence that chronic lymphocytic leukemia B lymphocytes are frequently committed to production of natural autoantibodies. Blood 76:562-569
91. Grunebaum E, Buskila D, Shoenfeld Y (1993) Serum human immunoglobulins (monoclonal gammopathies) as natural autoantibodies. In: Shoenfeld Y (ed) Natural autoantibodies: their physiological role and regulatory significance. CRC Press, Boca Raton, pp81-108
92. Limpens J, Stad R, Vos C, de Vlaam C, de Jong D, van Ommen GJ, Schuuring E, Kluin PM (1995) Lymphoma-associated translocation t(14;18) in blood B cells of normal individuals. Blood 85:2528-2536
93. Muller JR, Jones GM, Janz S, Potter M (1997) Migration of cells with immunoglobulin/c-myc recombinations in lymphoid tissues of mice. Blood 89:291-296

Subject Index

Alicia rabbit   89
anergic B cells   51
antibodies, natural   44ff, 59ff, 310
-, in *Xenopus*   234
-, polyreactive   41ff, 265ff, 310
-, shark   234
atherosclerosis   189ff
autoimmunity   201
autoreactivity   51

B cells, B/macrophages   294
-, development   108
-, homeostasis   67
-, superantigen   251
-, transportation factors   137
B-1 cells   6
-, B-0   49, 77ff
-, B-1a   6, 49ff
-, B-1b   6
-, definition   15
-, development   49ff
-, origins   3, 49, 311
-, positive selection   17, 39, 57
-, progenitor   3
-, progenitors   3
-, repertoire   7, 16
-, splenic   124
-, vs B-2   3, 6, 15, 163ff
bcl-2   31ff
bcl-xL   102
B-CLL   87, 92ff, 275ff, 285ff, 293
Brugia malayi (filarial parasite)   179 ff
btk   31, 49ff

CD5   4, 17, 87, 137, 141ff, 307ff
CD5[+] B cells   309
CD5 enhancer   133

CD19   50
CD21   57
CD30   57
CD40   154
CD43   26, 81
chemokines and receptors   121ff, 225ff
c-myc   151
complement   57ff
contacta sensitivity   171ff
cyclins   125, 151
cyclooxygenase   293ff

fetal liver   25

GALT   87, 221ff
GroEL   268

Hsp65   268

IFN-gamma   176
immunoglobulins (Ig), IgA   211ff, 221ff
-, IgM   167, 171
-, sharks   235
-, $V_H 11/V_K 9$ 16, 25
-, $V_H 12$ 7ff, 77ff
-, $V_H 81X$ 26, 99
-, $V_K 21C$ 17
-, *Xenopus*   234
influenza virus   163ff
innate immunity   164
intestinal microflora   211
ischemia-reperfusion injury   58

lamina propria plasma cells   212ff
lineage hypothesis   62
lymphomas, CH12   135
-, CH31   151

-, CH33   151

Mac-1   124, 223
mantle zone B cells   92
marginal zone B cells (MZ)
   99, 107ff, 111
memory B cells   110
mice, inbred strains, BAB   4
-, BALB/cAn   7, 201, 266
-, BALB/cJ   201
-, C.B-17   7ff, 77
-, CBA/ca   31
-, CBA/N (Xid)   31, 174
-, SCID   4
mice, knockout (-/-), ApoE   190
-, CD5   19, 145
-, Cr2   61
-, JH   174
-, LDLR   190
-, Rag-1   27, 28
-, Rag2   39, 59, 223
-, µMT   42, 174
mice, mutant (immunodeficient),
   aly/aly   231ff
-, CBA/N (Xid)   31ff, 39ff
mice, transgenic (Tg), 3H9µ   27, 145
-, 81X   100
-, bcl-xL   99
-, B1-8   62
-, CBA/N-bcl-2   31ff
-, κ8L   40
-, M603H   40
-, $V_H11/V_\kappa9$ 16, 25
-, V_H12 (6-1)   7ff, 77ff, 79ff
-, $V_H12/V_\kappa1A$ double   79
-, $V_H3609\mu$ 19
myeloma proteins   295ff

natural antibodies   49ff, 59ff
NF-AT   51, 131ff

peritoneal B cells   124, 134, 102
phosphoryl choline (PC)   39ff
phosphatidyl choline (PtC)   77ff
plasma cells   212
polyreactive antigen binding
   B (PAB) cells   241ff, 310
polysaccharides   107ff
positive selection   17, 39, 57, 68, 97
pre-B cells   68
pre-B receptor   25
pristane   201ff

Rag-2   27, 39, 44
receptor editing   77, 83
receptors, B7-1 (CD80)   107
-, B7-2 (CD26)   107
-, BCR   70, 131, 147
-, CD21/CD19/TAPA   57
-, CD21/CD35   57ff

self renewal   87ff
SHP-1   143ff
STAT3   121
SpA (protein A from *S. aureus*)   251
surrogate L chains, VpreB   25
-, λ5   25

T cells   171
T-15   192ff, 257ff
TCR   70
TdT   6,26
Thy-1/CD90   17

Xid (see also CBA/N)   31ff, 39ff
XLA   31ff

Current Topics in Microbiology and Immunology

Volumes published since 1989 (and still available)

Vol. 211: **Wolff, Linda; Perkins, Archibald S. (Eds.):** Molecular Aspects of Myeloid Stem Cell Development. 1996. 98 figs. XIV, 298 pp. ISBN 3-540-60414-6

Vol. 212: **Vainio, Olli; Imhof, Beat A. (Eds.):** Immunology and Developmental Biology of the Chicken. 1996. 43 figs. IX, 281 pp. ISBN 3-540-60585-1

Vol. 213/I: **Günthert, Ursula; Birchmeier, Walter (Eds.):** Attempts to Understand Metastasis Formation I. 1996. 35 figs. XV, 293 pp. ISBN 3-540-60680-7

Vol. 213/II: **Günthert, Ursula; Birchmeier, Walter (Eds.):** Attempts to Understand Metastasis Formation II. 1996. 33 figs. XV, 288 pp. ISBN 3-540-60681-5

Vol. 213/III: **Günthert, Ursula; Schlag, Peter M.; Birchmeier, Walter (Eds.):** Attempts to Understand Metastasis Formation III. 1996. 14 figs. XV, 262 pp. ISBN 3-540-60682-3

Vol. 214: **Kräusslich, Hans-Georg (Ed.):** Morphogenesis and Maturation of Retroviruses. 1996. 34 figs. XI, 344 pp. ISBN 3-540-60928-8

Vol. 215: **Shinnick, Thomas M. (Ed.):** Tuberculosis. 1996. 46 figs. XI, 307 pp. ISBN 3-540-60985-7

Vol. 216: **Rietschel, Ernst Th.; Wagner, Hermann (Eds.):** Pathology of Septic Shock. 1996. 34 figs. X, 321 pp. ISBN 3-540-61026-X

Vol. 217: **Jessberger, Rolf; Lieber, Michael R. (Eds.):** Molecular Analysis of DNA Rearrangements in the Immune System. 1996. 43 figs. IX, 224 pp. ISBN 3-540-61037-5

Vol. 218: **Berns, Kenneth I.; Giraud, Catherine (Eds.):** Adeno-Associated Virus (AAV) Vectors in Gene Therapy. 1996. 38 figs. IX,173 pp. ISBN 3-540-61076-6

Vol. 219: **Gross, Uwe (Ed.):** Toxoplasma gondii. 1996. 31 figs. XI, 274 pp. ISBN 3-540-61300-5

Vol. 220: **Rauscher, Frank J. III; Vogt, Peter K. (Eds.):** Chromosomal Translocations and Oncogenic Transcription Factors. 1997. 28 figs. XI, 166 pp. ISBN 3-540-61402-8

Vol. 221: **Kastan, Michael B. (Ed.):** Genetic Instability and Tumorigenesis. 1997. 12 figs.VII, 180 pp. ISBN 3-540-61518-0

Vol. 222: **Olding, Lars B. (Ed.):** Reproductive Immunology. 1997. 17 figs. XII, 219 pp. ISBN 3-540-61888-0

Vol. 223: **Tracy, S.; Chapman, N. M.; Mahy, B. W. J. (Eds.):** The Coxsackie B Viruses. 1997. 37 figs. VIII, 336 pp. ISBN 3-540-62390-6

Vol. 224: **Potter, Michael; Melchers, Fritz (Eds.):** C-Myc in B-Cell Neoplasia. 1997. 94 figs. XII, 291 pp. ISBN 3-540-62892-4

Vol. 225: **Vogt, Peter K.; Mahan, Michael J. (Eds.):** Bacterial Infection: Close Encounters at the Host Pathogen Interface. 1998. 15 figs. IX, 169 pp. ISBN 3-540-63260-3

Vol. 226: **Koprowski, Hilary; Weiner, David B. (Eds.):** DNA Vaccination/Genetic Vaccination. 1998. 31 figs. XVIII, 198 pp. ISBN 3-540-63392-8

Vol. 227: **Vogt, Peter K.; Reed, Steven I. (Eds.):** Cyclin Dependent Kinase (CDK) Inhibitors. 1998. 15 figs. XII, 169 pp. ISBN 3-540-63429-0

Vol. 228: **Pawson, Anthony I. (Ed.):** Protein Modules in Signal Transduction. 1998. 42 figs. IX, 368 pp. ISBN 3-540-63396-0

Vol. 229: **Kelsoe, Garnett; Flajnik, Martin (Eds.):** Somatic Diversification of Immune Responses. 1998. 38 figs. IX, 221 pp. ISBN 3-540-63608-0

Vol. 230: **Kärre, Klas; Colonna, Marco (Eds.):** Specificity, Function, and Development of NK Cells. 1998. 22 figs. IX, 248 pp. ISBN 3-540-63941-1

Vol. 231: **Holzmann, Bernhard; Wagner, Hermann (Eds.):** Leukocyte Integrins in the Immune System and Malignant Disease. 1998. 40 figs. XIII, 189 pp.
ISBN 3-540-63609-9

Vol. 232: **Whitton, J. Lindsay (Ed.):** Antigen Presentation. 1998. 11 figs. IX, 244 pp. ISBN 3-540-63813-X

Vol. 233/I: **Tyler, Kenneth L.; Oldstone, Michael B. A. (Eds.):** Reoviruses I. 1998. 29 figs. XVIII, 223 pp.
ISBN 3-540-63946-2

Vol. 233/II: **Tyler, Kenneth L.; Oldstone, Michael B. A. (Eds.):** Reoviruses II. 1998. 45 figs. XVI, 187 pp. ISBN 3-540-63947-0

Vol. 234: **Frankel, Arthur E. (Ed.):** Clinical Applications of Immunotoxins. 1999. 16 figs. IX, 122 pp. ISBN 3-540-64097-5

Vol. 235: **Klenk, Hans-Dieter (Ed.):** Marburg and Ebola Viruses. 1999. 34 figs. XI, 225 pp. ISBN 3-540-64729-5

Vol. 236: **Kraehenbuhl, Jean-Pierre; Neutra, Marian R. (Eds.):** Defense of Mucosal Surfaces: Pathogenesis, Immunity and Vaccines. 1999. 30 figs. IX, 296 pp.
ISBN 3-540-64730-9

Vol. 237: **Claesson-Welsh, Lena (Ed.):** Vascular Growth Factors and Angiogenesis. 1999. 36 figs. X, 189 pp.
ISBN 3-540-64731-7

Vol. 238: **Coffman, Robert L.; Romagnani, Sergio (Eds.):** Redirection of Th1 and Th2 Responses. 1999. 6 figs. IX, 148 pp.
ISBN 3-540-65048-2

Vol. 239: **Vogt, Peter K.; Jackson, Andrew O. (Eds.):** Satellites and Defective Viral RNAs. 1999. 39 figs. XVI, 179 pp.
ISBN 3-540-65049-0

Vol. 240: **Hammond, John; McGarvey, Peter; Yusibov, Vidadi (Eds.):** Plant Biotechnology. 1999. 12 figs. XII, 196 pp.
ISBN 3-540-65104-7

Vol. 241: **Westblom, Tore U.; Czinn, Steven J.; Nedrud, John G. (Eds.):** Gastroduodenal Disease and Helicobacter pylori. 1999. 35 figs. XI, 313 pp.
ISBN 3-540-65084-9

Vol. 242: **Hagedorn, Curt H.; Rice, Charles M. (Eds.):** The Hepatitis C Viruses. 2000. 47 figs. IX, 379 pp. ISBN 3-540-65358-9

Vol. 243: **Famulok, Michael; Winnacker, Ernst-L.; Wong, Chi-Huey (Eds.):** Combinatorial Chemistry in Biology. 1999. 48 figs. IX, 189 pp. ISBN 3-540-65704-5

Vol. 244: **Daëron, Marc; Vivier, Eric (Eds.):** Immunoreceptor Tyrosine-Based Inhibition Motifs. 1999. 20 figs. VIII, 179 pp. ISBN 3-540-65789-4

Vol. 245/I: **Justement, Louis B.; Siminovitch, Katherine A. (Eds.):** Signal Transduction and the Coordination of B Lymphocyte Development and Function I. 2000. 22 figs. XVI, 274 pp. ISBN 3-540-66002-X

Vol. 245/II: **Justement, Louis B.; Siminovitch, Katherine A. (Eds.):** Signal Transduction on the Coordination of B Lymphocyte Development and Function II. 2000. 13 figs. XV, 172 pp. ISBN 3-540-66003-8

Vol. 246: **Melchers, Fritz; Potter, Michael (Eds.):** Mechanisms of B Cell Neoplasia 1998. 1999. 111 figs. XXIX, 415 pp.
ISBN 3-540-65759-2

Vol. 247: **Wagner, Hermann (Ed.):** Immunobiology of Bacterial CpG-DNA. 2000. 34 figs. IX, 246 pp.
ISBN 3-540-66400-9

Vol. 248: **du Pasquier, Louis; Litman, Gary W. (Eds.):** Origin and Evolution of the Vertebrate Immune System. 2000. 81 figs. IX, 324 pp. ISBN 3-540-66414-9

Vol. 249: **Jones, Peter A.; Vogt, Peter K. (Eds.):** DNA Methylation and Cancer. 2000. 16 figs. IX, 169 pp.
ISBN 3-540-66608-7

Vol. 250: **Aktories, Klaus; Wilkins, Tracy, D. (Eds.):** Clostridium difficile. 2000. 20 figs. IX, 143 pp. ISBN 3-540-67291-5

Vol. 251: **Melchers, Fritz (Ed.):** Lymphoid Organogenesis. 2000. 62 figs. XII, 215 pp.
ISBN 3-540-67569-8

Printing: Saladruck, Berlin
Binding: H. Stürtz AG, Würzburg